URBAN FLOOD MANAGEMENT

BATH SPA UNIVERSITY LIBRARY

URBAN FLOOD MANAGEMENT

CHRIS ZEVENBERGEN

UNESCO-IHE/Delft University of Technology, Delft, The Netherlands
Dura Vermeer Group, Hoofddorp, The Netherlands

ADRIAN CASHMAN

CERMES – University of the West Indies, Barbados

NIKI EVELPIDOU

University of Athens, Greece

ERIK PASCHE

Hamburg University of Technology, Germany

STEPHEN GARVIN

BRE Scotland, Glasgow, UK

RICHARD ASHLEY

Pennine Water Group, Sheffield, UK

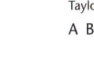

CRC Press
Taylor & Francis Group
Boca Raton London New York Leiden

CRC Press is an imprint of the
Taylor & Francis Group, an **informa** business

A BALKEMA BOOK

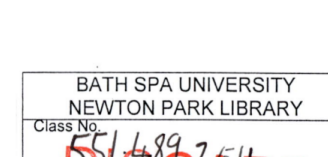

ESF provides the COST Office through an EC contract

COST is supported by the EU RTD Framework programme

CRC Press/Balkema is an imprint of the Taylor & Francis Group, an informa business

© 2011 Taylor & Francis Group, London, UK

Cover image: Dusty Allen Smith: River Drive, Davenport, IA June 20, 2008

Typeset by Vikatan Publishing Solutions (P) Ltd, Chennai, India
Printed and bound in Poland by Poligrafia Janusz Nowak, Poznán

All rights reserved. No part of this publication or the information contained herein may be reproduced, stored in a retrieval system, or transmitted in any form or by any means, electronic, mechanical, by photocopying, recording or otherwise, without prior permission in writing from the publisher. Innovations reported here may not be used without the approval of the authors.

Although all care is taken to ensure integrity and the quality of this publication and the information herein, no responsibility is assumed by the publishers nor the author for any damage to the property or persons as a result of operation or use of this publication and/or the information contained herein.

Although greatest care was taken to obtain permission from the copyright holders for the reproduction of figures included in this publication, there are instances where we have been unable to trace or contact the copyright holder. If this is the case, please contact the publisher.

Published by: CRC Press/Balkema
P.O. Box 447, 2300 AK Leiden, The Netherlands
e-mail: Pub.NL@taylorandfrancis.com
www.crcpress.com – www.taylorandfrancis.co.uk – www.balkema.nl

British Library Cataloguing in Publication Data
A catalogue record for this book is available from the British Library

Library of Congress Cataloging-in-Publication Data
Urban flood management / Chris Zevenbergen ... [et al.].
p. cm.
Includes bibliographical references.
ISBN 978-0-415-55944-7 (hard cover : alk. paper)
1. Urban runoff--Management. 2. Storm sewers. 3. Flood control. I. Zevenbergen, C. (Chris)

TD657.U75 2010
363.34'936--dc22

2010031755

ISBN: 978-0-415-55944-7 (Pbk)

Table of contents

Preface ix
List of contributors xi
Illustration credits xv

Introduction to urban flood management 1
What is this book all about? 3

I SETTING UP THE FRAMEWORK 7

1 Setting the stage for Integrated Urban Flood Management 9
1.1 Why are cities special cases? 9
1.2 The city as a living organism 15
1.3 Vulnerability of urban areas: a rough guide 18
1.4 Types of uncertainty 20
1.5 Adoption of a system approach 26

2 Urban floods 37
2.1 The influence of climate and other factors 37
2.2 Types of flooding 42
2.3 Pitfalls in using the historical record (or 'stationarity is dead') 54

II DRIVERS FOR CHANGE 61

3 Urbanisation 63
3.1 Principles of land-use planning 63
3.2 Urban typologies: from central square to edge city 67
3.3 Growing and shrinking: density issues induced by globalisation 71
3.4 Megacities in the delta 73

4 Climate change: key uncertainties and robust findings 79
4.1 A review of the past 79
4.2 Signs of change 85
4.3 Expected consequences 86

III URBAN FLOOD RISK 95

5 The hydrology of cities 97
5.1 The hydrological cycle 97
5.2 Land use and runoff 103
5.3 Modelling surface runoff 107
5.4 Modelling pluvial flooding 113
5.5 Modelling coastal flooding 118

6 Urban flood-risk assessment 123
6.1 Introduction to the theory of risk 123
6.2 Quantifying flood probability 126

6.3	Tangible and intangible damages	138
6.4	Loss of life estimation in flood-risk assessment	141
6.5	Cross-scale factors and indirect damages	142
6.6	Flood-risk mapping	146

IV RESPONSES | 155

7	**Responding to flood risk**	**157**
7.1	Responses	157
7.2	Performance standards and expectations	159
7.3	Resilience, vulnerability, robustness and sustainability	164
7.4	Precautionary and adaptive responses	167
7.5	Confronting flood management with land-use planning: lessons learnt	173
7.6	Building types, infrastructure and public open space	177

8	**Urban drainage systems**	**183**
8.1	A historical perspective	183
8.2	Major and minor flows	184
8.3	SUDS/LIDS	186
8.4	Practices in water sensitive urban design	198

9	**Flood proofing the urban fabric**	**205**
9.1	Managing flooding through site design: basic principles	205
9.2	Managing flooding through detailed design (individual properties/buildings)	208
9.3	Flood resilient repair and retrofitting	215
9.4	Urban flood defences and barriers	219

10	**Enhancing coping and recovery capacity**	**227**
10.1	Flood forecasting, warning and response	227
10.2	Emergency planning, management and evacuation	235
10.3	Compensation and flood insurance	240

V TOWARDS FLOOD RESILIENT CITIES | 253

11	**Managing for resiliency**	**255**
11.1	Asset management, some basic principles	255
11.2	Assessing resilience in flood-risk management	259
11.3	Transitioning from entrapment to resilience approaches	261

12	**Capacity building and governance**	**269**
12.1	Risk perception, acceptance and communication	269
12.2	Adaptive capacity	273
12.3	Characteristics of effective learning initiatives	274

13	**Shelter for all**	**281**
13.1	What does the future hold?	281
13.2	Challenges and opportunities	282
13.3	Turning ideas into action	285
13.4	Success stories: seizing windows of opportunity	290

References	297
Abbreviations and acronyms	309
Glossary	311
Subject index	319

Preface

Floods have a significant impact on cities. Since the beginning of civilisation, people have had to cope with them and, if possible, to adapt to them. As a consequence, the various human civilisations have observed the effect of such natural phenomena and have tried to understand the underlying causes. At first they provided mythological or religious explanations; floods were seen as an act of the gods. With the Enlightenment, floods were conceived as due to ignorance or neglect on the part of man; resulting in the development of more rational approaches to manage the occurrence of flooding. Since the latter half of the twentieth century flood-risk management has evolved as a discipline in its own right. As it is currently practised this involves a very broad range of expertise, corresponding to the different scientific domains including physics, biology, geographical science and of course civil and other branches of engineering. It increasingly draws in contributions from the social sciences, planning and politics.

The UN has estimated that there will be a global increase from 2.9 billion urban residents in the 1990s to 5.0 billion by 2030. More than 50% of the world's population now live in urban areas. Most of the growth in urban populations will occur in developing countries in Asia: in small- and medium-sized cities and in mega-cities. The urban population of industrialised countries is now expected to grow by 11% over the next thirty years. An unintended side effect of this concentration of population is that the number exposed to floods is growing as well. Especially, with the unprecedented growth of Asian mega-cities, this will contribute substantially to a global increase of flood risk. While trend changes caused by climate change pose major challenges for cities, especially in developing countries, the increasing frequency of extreme events could induce major demographic and economic disruptions. Exchanging and sharing experiences and good practices must therefore be a main priority. Moreover, it has become increasingly clear that current curricula in planning and engineering schools are outdated and have not kept pace with the changing demands of the cities of today. Educational programmes should embrace innovative planning ideas, including the ability to engage in participatory planning, negotiation and communication.

This book attempts to cover the basic components needed to understand the urban flooding system including the urban fabric, green spaces, coasts and surface waters such as rivers and canals, groundwater, etc. We believe, however, that a holistic and systematic conceptualisation of urban flood risk must expand its scope beyond the traditional components of flood-risk management: which is predominantly focused on assessing, preventing and controlling the extent of damage and probability of occurrence. Therefore, a brief analysis of a spectrum of components of the urban system and of the interactions between them is also part of the scope this textbook. We have chosen here to provide an overview of the processes that rule the behaviour of cities and citizens. Our goal is to provide the reader with enough information on the interactions between the different elements of the flooding system and on the dominant feedbacks so that you can analyse the urban flooding system in an integrated way. By this means, the reader should be able to understand the dominant causes of urban flooding and to develop appropriate responses, at least conceptually. The reader should, however, recognise that this is a rapidly developing field. Knowledge is advancing about the causes and impacts of floods and the way in which the drivers are changing due to climate and other causes, as well as how best to address this. For this reason, as a reader, you should see this book as a first step on 'the active learning' pathway that you need to commit to, as by the time you reach the end, the information presented will already be superseded. Therefore, you will need to continue on the pathway by seeking out the latest knowledge about the causes, impacts and best way to deal with flood risk into the future.

Each chapter has three parts:

- **Theory and background information:** This constitutes the main body of each chapter. It provides the reader with the background material and, where appropriate, an introduction to the theory.
- **Learning objectives and key questions:** Besides theoretical and practical background to each subject, each chapter sets out what the main learning objectives are and poses a number of questions and points for the reader to consider. Our hope is that they will stimulate thinking and that the reader will attempt to formulate their own solutions to these problems, and then referring to their solutions re-evaluate their understanding of the material.
- **Case studies, anecdotes:** There is a large collection of additional information in the form of case studies. These illustrate through practical examples some of the real world difficulties associated with the chapter's subject matter.

The idea of writing a textbook emerged in 2008 in COST Action $C22^1$, a European research platform on Urban Flood Management, established in 2005. This Action has brought together more than 40 European scientists involved in water management, hydrology, urban planning and design, civil engineering, social sciences and construction engineering. The main objective of the Action is to exchange experiences and promote the diffusion of best practices in Urban Flood Management. We were thus fortunate to start with a network already immersed in the subject that had stood the test of time. We would like to acknowledge our debt to our colleague members of this Action. Many of them have read carefully the various sections of this textbook and suggested modifications that improved its quality significantly. We are also thankful to the several colleagues outside this Action who have either provided concrete input to the book or valuable feedback. We would like to thank Ellen Brandenburg and Victoria Muñoz for their unconditional support to collecting all the text contributions and illustrations. We are indebted to the COST organisation, the Interreg IIIB project MARE and the Dutch Living with Water Programme for providing financial support. Finally, we also want to acknowledge the organisations, publishers and scientists who allow us to reproduce their work without any charge.

The Authors,
September 2010

¹ COST stands for European Co-operation in the field of Scientific and Technical Research, an intergovernmental framework for the co-ordination of nationally-funded research at a European level.

List of contributors

Barker, Robert (9.1)
BACA Architects, London, United Kingdom

Banasik, Kazimierz (8.2)
Faculty of Engineering and Environmental Science, Warsaw University of Life Sciences – SGGW, Warsaw, Poland

Bjerkholt, Jarle (8.2)
University of Life Sciences and Technology, University of Life Sciences Norway, Ås, Norway

Björnsen, Gerhard (11.1)
Björnsen Consulting Engineers, Koblenz, Germany

Blanksby, John (2.2, 13.1, 13.2)
University of Sheffield, Department of Civil and Structural Engineering, Sheffield, UK

Brilly, Mitja (2.1, 2.2)
Faculty of Civil Engineering and Geodesy, University of Ljubljana, Ljubljana, Slovenia

Bruijn de, Karin (7.3, 7.4)
Deltares, Delft, The Netherlands

Crichton David (10.3)
Fellow of the Chartered Insurance Institute (CII), Inchture, Scotland

Diez, Javier (4.1)
Technical University Madrid, Madrid, Spain

Djordjevic, Slobodan (5.2, 5.3, 5.4, 5.5)
University of Exeter, Exeter, United Kingdom

Esteban, Dolores (4.1)
Technical University Madrid, Madrid, Spain

Hooimeijer, Fransje (3.3)
Department of Architecture and Planning, Delft University of Technology, Delft, The Netherlands

Gersonius, Berry (1.3, 1.4, 1.5, 7.3, 7.4, 11.1, 11.2)
UNESCO-IHE, Delft, The Netherlands

Graaf de, Rutger (1.3)
Technical University of Delft/DeltaSync, Delft, The Netherlands

Herath, Srikantha (4.4, 6.1)
United Nations University, Jingumae, Japan

Herk van, Sebastiaan (12.3)
Bax & Willems Consulting Venturing sl, Barcelona, Spain

Heppeler, Jörn (10)
Regierungspräsidium Stuttgart, Stuttgart, Germany

Jacimovic, Nenad (2.2)
River Engineering, Faculty of Civil Engineering, University of Belgrade, Belgrade, Serbia

János, Major (6.6)
Szent István University Ybl Miklós Faculty of Architecture and Civil Engineering, Budapest, Hungary

Jonkman, Bas (6.4)
Department of Hydraulic Engineering, Faculty of Civil Engineering and Geosciences, Delft University of Technology, Delft, The Netherlands

Kamphuis, Bill (9.4)
Queens University, Kingston, Ontario, Canada

Kelly, David (9.1, 9.3)
Building Research Establishment (BRE), Glasgow, Scotland

Kok, Matthijs (10.2)
Department of Hydraulic Engineering, Delft University of Technology, Delft, The Netherlands

Lawson, Nigel (2.2, 4.3)
School of Environment and Development – University of Manchester, Manchester, UK

Lindholm, Oddvar Georg (8.2)
Department of Mathematical Sciences and Technology, University of Life Sciences Norway, Ås, Norway

Lopez Gutierrez, Jose Santos (4.1)
Technical University Madrid, Madrid, Spain

Manojlovic, Natasa (6.3, 9.1, 9.3)
Institute of River & Coastal Engineering, Hamburg Technical University, Hamburg, Germany

Mrekva, László (4.2)
Eötvös József College Institute of Hydraulic and Water Management, Baja, Hungary

Negro, Vicente (4.1)
Technical University Madrid, Madrid, Spain

Newman, Richard (12.2)
University of Sheffield, Sheffield, United Kingdom

Ottow, Bouke (10.2)
Deltares, Delft, The Netherlands

Pathirana, Assela (2.3)
UNESCO-IHE, Delft, The Netherlands

Paz, Rosa M. (4.1)
Complutense University of Madrid, Madrid, Spain

Puyan, Najib (11.1)
Municipality Maarssen, Maarssen, The Netherlands

Rijke, Jeroen (5.1, 12.1)
Flood Resilience Group, UNESCO-IHE, Delft, The Netherlands

Schertzer, Daniel (5.3)
Centre d'Enseignement et de Recherche Eau Ville Environment (Cereve/ENPC), Paris, France

Stalenberg, Bianca (9.4)
Section of Hydraulic Engineering, Faculty of Civil Engineering and Geosciences, Delft University of Technology, Delft, The Netherlands

Stanic, Milos (2.2)
Faculty of Civil Engineering, University of Belgrade, Belgrade, Serbia

Stickles, Lee (13)
Topographical Shifts, San Francisco, USA

Tippett, Joanne (12.2)
Planning and Landscape, School of Environment and Development, University of Manchester, UK

Telekes, Gábor (4.2)
Szent Istvan University Ybl Miklo Faculty of Architecture and Civil Engineering, Budapest, Hungary

Terpstra, Teun (10.2)
University of Twente Faculty of Behavioural Sciences, Enschede, The Netherlands

Tourbier, Joachim (8.2, 8.3, 8.4)
Department of Architecture, Institute of Landscape Architecture Technische Universität Dresden, Dresden, Germany

Tucci, Carlos (1.1)
Institute of Hydraulic Research – Federal University of Rio Grande do Sul, Porto Alegre-RS, Brazil

Veen van der, Bonne (10.2)
Deltares, Delft, The Netherlands

Veerbeek, William (3.1, 3.2, 6.3)
Flood Resilience Group, UNESCO-IHE, Delft, The Netherlands

Ven van de, Frans (5.1)
Civil Engineering, Delft University of Technology, Delft, The Netherlands

Veldhuis ten, Marie-Claire (6.1, 6.2)
Department of Water Management, Delft University of Technology, Delft, The Netherlands

Werner Micha (10.1)
UNESCO-IHE Institute for Water Education, Delft, The Netherlands

Illustration credits

The joint permissions and contributions of the authors and copyright holders of figures included in this work are greatly acknowledged.

Bura Sant en Co (Edwin Santhagens, 2008)	Fig. 0-1 (2008)
Gil Cohen, The National Guard Image Gallery:	Fig. 0-3 (1927)
Gustave Doré:	Fig. 0-2 (1866)
Tony Hartawan/WpN (2007):	Opening page Part I
Adapted from Stirling:	Figs. 1-11, 1-12 (2003)
Assela Pathirana:	Figs. 2-16, 2-17, 2-18 (2009)
Berry Gersonius:	Fig. 1-18 (2009)
Carl Folke:	Fig. 1-15 (2006)
Chris Zevenbergen:	Fig. 1-5 (2007), Figs. 1-2, 1-3, 1-7, 1-8, 1-17, 1-19 (2008)
Defra:	Fig. 1-14 (2004)
Foresight:	Fig. 1-13 (2004)
Marcin Stępień:	Fig. 2-15 (2008)
Mitja Brilly:	Fig. 2-4 (2007)
Munich Re:	Fig. 1-1 (2008)
N. Jacimovic:	Fig. 2-14 (2009)
Nigel Lawson:	Figs. 2-11, 2-12, 2-13 (2009)
Nilfanion:	Fig. 2-3 (2005)
P.A. Piazolli:	Fig. 2-1 (2009)
Rijkswaterstaat, The Netherlands:	Fig. 1-6 (2004)
Rutger de Graaf:	Figs. 1-9, 1-10 (2008)
Stephan Sheppard:	Fig. 1-4 (2007)
Trokilinochchi:	Fig. 2-2 (2008)
Rob Lengkeek (1995):	Opening page Part II
Center for Landuse Research and Education:	Figs. 3-3, 3-5 (2009)
Chummu Hugh:	Fig. 4-2 (2010)
IPCC *Synthesis Report*:	Figs. 4-4, 4-5, 4-6 (2007)
IPCC *Third Assessment Report*:	Fig. 4-1 (2001)
Johan van de Pol:	Fig. 3-6 (2010)
Milly et al.:	Fig. 4-10
Paul Regouin:	Fig. 3-4 (2002)
Stadsarchief Amsterdam:	Fig. 3-7 (2001)
UKCIP:	Figs. 4-7, 4-8 (2009)
Victoria Bejarano (Dura Vermeer):	Fig. 4-9 (2010)
Victoria Munoz:	Fig. 4-3
William Veerbeek (based on Nicholls et al., 2007):	Fig. 3-8 (2009)

Illustration credits

Chris Zevenbergen (2009): Opening page Part III
Adapted from Kaplan: Fig. 6-2 (1997)
Bas Jonkman: Figs. 6-17, 6-18 (2007)
Chris Zevenbergen: Fig. 5-4 (2007)
Daniel Schertzer: Figs. 6-4, 6-5, 6-6, 6-7, 6-9, 6-10, 6-11, 6-12
FISRWG, USA: Fig. 5-2 (1998)
Han Vrijling: Fig. 6-3 (2001)
Jan Rijke Fig. 5-1
Kazimierz Banasik: Fig. 5-6 (2002), Figs. 5-7, 5-8 (2005)
Macor et al.: Fig 6-8 (2007)
Marie-Claire ten Veldhuis: Fig. 5-12 (2008), Fig. 6-1
National Hurricane Center (NHC): Fig. 5-15 (1988)
NISAC: Fig. 6-19 (2003)
Slobodan Djordjevic: Figs. 5-10, 5-11, 5-14
UCAR: Fig. 5-9 (2006)

Dura Vermeer, The Netherlands (2007): Opening page Part IV
Adepted from Evans et al.: Fig. 7-1 (2004)
Atelier Dreiseitl: Fig. 8-12 (2008)
Baca Architects: Figs. 9-1, 9-2, 9-3 (2009)
Bas Kolen: Fig. 10-7 (2009)
Berry Gersonius: Figs. 7-2, 7-3 (2009)
Bianca Stalenberg: Figs. 9-14, 9-15 (2006)
Bill Kamphuis: Figs. 9-12, 9-13 (2009)
Brand, S. How Buildings Learn.
New York: Viking: Fig. 9-10 (1994)
Chris Zevenbergen: Figs. 7-7, 8-1 (2004), Fig. 9-7, (2005), Figs. 9-4, 9-5, 9-6 (2009), Fig. 8-10
David Crichton: Fig. 10-9 (1999)
DETR: Fig. 7-4 (2000)
EA (UK): Fig.10-5 (2009)
Erik Pasche: Fig. 9-9 (2004), Fig. 8-11 (2009)
FEMA: Fig. 9-8 (1993)
Ian Goodall: Fig. 7-5 (2009)
Joachim Tourbier: Figs. 8-2, 8-3, 8-4, 8-5 (1983), Figs. 8-6, 8-7, 8-8, 8-9
Kim Carsell: Fig. 10-3 (2004)
Levee Safety Program, U.S. Army Corps
of Engineers: Fig. 9-11
Micha Werner: Fig. 10-4 (2009)
Michael Wilson: Fig. 10-8 (2007)
Richard Ashley: Fig. 7-6 (2009)
SEPA (UK): Fig. 10-6 (2009)
The National Infrastructure Protection Plan
(NIPP): Fig. 7-8
UN/ISDR: Fig. 10-1 (2001)

DuraVermeer (2004) Opening page Part V
Baca Architects: Fig. 13-4 (2008)
Bill Kamphuis: Fig. 12-2 (1997)
Chris Zevenbergen: Fig. 12-5 (2009)
Ellen Brandenburg: Fig 12-3 (2005)
Geels and Kemp: Fig. 11-3 (2000)
Jeroen Rijke: Fig. 11-2 (2009), Fig. 13-7 (2008)
Ketso (UK): Fig. 12-6 (2009)
Lee Stickles: Fig. 13-1 (2009)
Municipality of Dordrecht: Fig. 13-3 (2008)
Richard Ashley: Fig. 11-4, Fig.12-4 (2007),
 Fig. 13-2 (2009)
Ruud Raaijmakers: Fig. 12-1 (2008)
Yorkshire Council: Fig. 13-5 (2010)

Introduction to urban flood management

Floods are part of nature. They may happen almost on any place on earth when there is too much water at one place and at one time. Although floods are associated with negative impacts they are not always bad, the nutrients that they bring to floodplains create some of the most fertile places on the planet. The seasonal floods replenish those nutrients and productivity. They have enabled the great civilisations of the past to arise out of such bounty. But trying to harness and shape such natural forces to serve our own ends human beings has done so with unintended consequences. Although most floods may be predicted, they can cause massive damage and destruction of property, as most urban communities are located near water sources such as coasts and rivers. Along with windstorms, floods are the most common and widespread of all natural disasters. About a third of all loss events and a third of economic losses are due to the effects of floods. In terms of the number of victims, floods have been responsible for more than half of the deaths caused by natural catastrophes in the past decades. Worryingly, flood events appear to be increasing in frequency and perhaps intensity world-wide.

Flood effects may range in scale from the local, affecting a neighbourhood or community, to very large, affecting an entire region or river basin and even multiple states or nations. However, not all floods are alike (see Chapter 2). They may be due to quite different causes and exhibit different behaviour. While some floods develop slowly, sometimes over a period of days, flash floods develop quickly, sometimes in just a few minutes and without any visible signs of rain. They may occur on a regular basis or be very rare. Usually, a distinction is made between floods that are infrequent (with return periods below 20 years), rare floods (with return periods between 20 and 100 years) and very rare floods (return periods exceeding 100 years). Each type of flood requires its own type of responses (see Chapter 7).

Floods may affect almost all aspects of our lives. This is particularly the case for cities. Cities are the most vulnerable because of the concentration in these areas of people and their possessions, and economic activities are potentially exposed to floods (see Chapter 3). The scale of devastation that may be visited on urban populations and economies caused by extreme weather events underscore their particular vulnerability. A large and growing proportion of the impacts are in urban areas in low- and middle-income countries. In these cases, residents are not aware of the dangers associated with floods. Cities are affected by floods in one of three major ways, (see Chapter 6):

- population; safety, health, livelihoods;
- economic sectors: change in productive capacity affected by change in resource productivity or market demand;
- physical infrastructure, buildings, urban services (e.g. water supply), specific industries.

In the coming decades, global warming (see Chapter 4) is projected to further intensify the hydrological cycle, with impacts that will probably be more severe than those so far observed. Climate change is projected to lead to an increase in the frequency and intensity of floods in large parts of Europe. In particular, flash and urban floods, triggered by local intense precipitation events, are likely to be more frequent. Flood hazards will also probably increase during wetter and warmer winters, with more frequent rain and less frequent snow. Even in regions where the mean river flow will drop significantly, as in the Iberian Peninsula, the projected increase in precipitation intensity and variability may actually cause more floods. Quantitative projections of changes in precipitation and river flows at the river basin scale remain, however, highly uncertain, due to the limitations of climate models, as well as scaling issues between climate and hydrological models. While the general understanding of the potential hazards of flooding has improved considerably in the past few decades, the ability to forecast events in the light of more variable and changing hydrological conditions will become even more challenging.

Most of the impacts of flooding are adverse and are generally projected to worsen, certainly over the next few decades. There is therefore a need for all countries, developing and developed, to adapt to climate and socio-economic change. In the face of inherent uncertainties of future changes, adaptation procedures need to be designed that can be altered or that are robust to change. Adaptation aims at increasing the resilience of natural and human systems to current and future impacts of climate change (see Chapter 7). Whilst EU countries have experience in flood prevention, Asian countries bring experience on managing flood disasters and of developing coping and recovery capacity.

In particular, the Asian region experiences frequent flood disasters of high magnitude. And although the number of deaths caused by flooding has decreased, the number of affected populations and economic losses has increased significantly. These trends pose a significant development challenge for Asia. Poverty is a major contributor to people's vulnerability to flooding and frequent flood impact leads to increase in poverty and hence vulnerability gradually sapping the ability to cope and recover. Moreover, the

TEXTBOX 0-1
What is a flood?

A flood is an excess of water on land that is normally not flooded. A flood is a condition that occurs when water overflows the artificial or natural boundaries of a stream, river, or other body of water onto adjacent areas which rely for their functioning on being kept free from an excess of water on the land. Floods are associated with heavy rainfalls or thawing snow conditions which may lead to an unusual or rapid accumulation or runoff of surface waters from any source.

In the US, the term flooding is used whenever water rises above its normal boundaries and spills onto normally dry land. Most countries consider this as a separate type of flooding called inundation.

last 20 years has also seen massive migration of people to urban centres in search of employment and better access to services. This has led to an increase in urban flood risk as people settle in high-density urban slums on city fringes, on or next to flood protection embankments and along riverbanks (see Chapter 2 and 3). In other words, many migrants settle in marginal areas at greater risk of flooding because they have few other economic alternatives. The challenge for the Asian region is to develop strategies, policies, and implement activities, that aim to secure the social, political and economic livelihoods for people living with risk.

Urban Flood Management is not about preventing any flooding or even minimising flood losses. The absolute prevention of flooding is an impossible task. Rather Urban Flood Management is about maximising and maintaining the performance of the city as a whole (see Chapter 13). Urban Flood Management is also about looking for opportunities and solutions that add to the welfare of a society in such a way that the sum of social and economic benefits outweighs the potential costs.

WHAT IS THIS BOOK ALL ABOUT?

This textbook integrates the expertise from disciplines such as hydrology, sociology, architecture, urban design, construction and water resources engineering. The subject

TEXTBOX 0-2

Multifunctional options

Urban Flood Management is also challenged to foster *multifunctional* options. We may for example increase water retention capacity in a city to mitigate local flooding and at the same time enhance the water quality of surface waters, or increase green areas for public space and amenity. The Roof Park in Rotterdam is probably the largest green roof in Europe. On top of 85,000 m^2 of shops, housing and schools there will be a park linking neighbourhoods that are currently separated by railway lines. The row facing the river Maas will be built as a flood defence system. The plan was developed in close consultation with the local residents.

Soure: Architect. Edwin Santhagens, 2008.

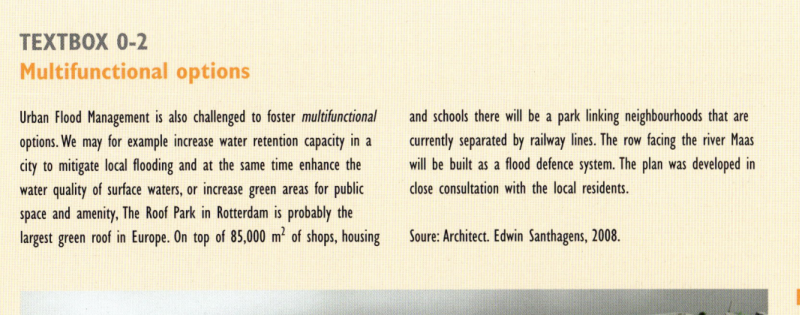

FIGURE 0-1
Roof Park in Rotterdam, The Netherlands

Source: Bura Sant en Co (Edwin Santhagens), 2008.

TEXTBOX 0-3
The great flood

The Flood, the Old Testament story of Noah and his ark, is merely one of many told down through the ages. It is the better known telling of part of the Sumerian legend of the Epic of Gilgamesh, which goes back to at least 2000 BC. In both Western and Eastern civilisations, people have recounted legends of floods engulfing the entire earth. Only in Africa is the flood story rarely told. Scholars have argued for centuries about the origins of these stories, their commonality, and whether such cataclysms ever occurred. Nonetheless, although the extent of these floods is uncertain, their violent nature is often vividly portrayed. The stories, of course, are told in terms of how floods affect people's lives. They reflect public perception which sees widespread human and material losses in the flood's wake.

FIGURE 0-2
Gustave Doré's illustrations of the story of Noah's Ark: Doomed men and beasts try desperately to save their children as God floods the world

Source: Le Déluge, Gustave Doré, 1866.

FIGURE 0-3

The great Mississippi River flood of 1927 was one of the worst natural disasters in American history. It inundated 27,000 square miles, killing as many as 1,000 people and displacing 700,000 more. At a time when the entire budget of the federal government was barely $3 billion, the flood caused an estimated $1 billion in damage

Source: The Great Flood of 1927, Gil Cohen, 1927.

is approached from a broadly international perspective and various anecdotes and case studies, expert advice and literature recommendations are included to support the information provided. Developed by a team of specialists, this volume is intended to inform, sensitise and hopefully educate students studying hydrology, geography, civil and environmental engineering and management at the university level in urban flood management. Moreover, professionals will find this book useful as a reference.

1

Setting up the framework

We live in a world that is becoming increasingly urbanised; towns and cities provide shelter, create employment and business opportunities and provide social opportunities. The coming together of all the diverse elements that make up urban living is what makes towns and cities special; together they add value beyond the mere sum of their constituent parts. Yet this very aggregation creates its own vulnerabilities. If we are to understand the range of impacts floods can have on the fabric of urban living then we need to have some understanding of how that urban fabric functions and what makes it special. We also need to have some understanding of the nature of the flood events to which towns and cities might be subjected, for we cannot understand one without the other. In Part 1, the attributes of cities and floods are discussed in order to establish the framework within which urban flood management will be explored through the subsequent chapters. Cities can be thought of as living systems that have the ability to self-organise and like any organism they have their own special characteristics and vulnerabilities and it is for this reason that systems based approaches need to be adopted. We also look at the nature of flooding that can affect urban areas as well as the emerging challenges that climate change poses. What makes climate change problematic for flooding is the way in which it potentially makes obsolete many of the hydrodynamic assumptions on which historically flood modelling and management has been based—or does it? These issues are explored in Part 1.

People walk with their belongings through the flooded streets of Jakarta, Indonesia on Sunday, Feb. 4, 2007. At least 25 people have been killed and almost 340,000 others made homeless, officials said.

Source: Tony Hartawan/WpN, 2007

Setting the stage for Integrated Urban Flood Management

Learning outcomes

In this chapter, you will learn about the following key concepts:

- How rising urban population will be increasingly at risk from flooding in the future.
- Urban systems function as complex adaptive systems having different scales of operation.
- Historical hydrological data is not a good guide to future urban hydrology and how to accommodate climate change.
- Vulnerability is determined by the interaction of threshold, coping, recovery and adaptive capacities.
- Understand the difference between risk and uncertainty.

At the end of the chapter there are a number of Key Questions that you are invited to consider.

1.1 WHY ARE CITIES SPECIAL CASES?

1.1.1 Urban population

The world has crossed the point at which 50 per cent of humanity lives in urban areas. By around 2020, more than half of the population of developing countries will be urban. Urbanisation opens up countless opportunities for economic development and welfare. During last century urban areas have outperformed rural areas on almost every dimension of economic development, whether the rate of innovation, speed of demographic transition, levels of education, health, life expectancy, infant mortality, or access to clean water and sanitation. However, many urban areas in the world are not functioning well, especially where cities are growing not because they are themselves economically dynamic, but because their rural hinterlands are in such distress. With the high densities of urban populations and assets and more than 80 per cent of the cities in river basins, close to the coast or both, flooding poses a profound challenge.

The projected climate-induced changes will aggravate the impact of already existing stresses, such as land use, demographic and socio-economic changes. For example, urbanisation is now being accelerated by the rapid globalisation and expansion of local economies, especially in Asia. The UN estimated a global increase from 2.9 billion urban residents in the 1990s to 5.0 billion by 2030. Most of its growth will occur in developing countries in Asia: mainly in small- and medium-sized cities and also in mega-cities. The urban population of industrialised countries is expected to grow by 11 per cent in the next 30 years. An unintended side effect of this concentration of population is that the number of people exposed to floods will also grow, both because of sheer numbers and through their settlement of vulnerable areas at risk from flooding.

1.1.2 Floods on the rise

It would appear from the available records that since the 1960s there has been a rise in the number of reported flood disaster events and since the early 1990s this increase has been even more significant (see Figure 1-1). Certainly, there has been a noticeable increase in the economic losses associated with such events, even in less-developed countries. Much of the increase in numbers has been attributed to more accurate recording of events and better communications between regions, especially previously isolated ones. On the other hand, the number of deaths attributable to floods has fallen over the last 30 years, mainly a consequence of improved warning systems and better preparedness. However, economically, floods are the biggest cause of loss from natural events. One of the trends that has been noted in connection with floods is population growth, which has led more people to live in potentially hazardous areas so there are a greater number at risk concomitantly with lower coping capacity (see Section 1.3). The significant increase recorded since the 1990s may be an early sign of climate change, though it might also be influenced by population trends, urbanisation and other land-use changes. Separating human factors out from a possible climate change signal, though, presents a problem. Unless factors such as the overcrowding of urban spaces that are often in flood prone and vulnerable areas are addressed, then it is likely that the numbers affected and the economic losses associated with flood events will grow, irrespective of whether or not the number of flood events increases. But at the same time human intervention in the landscape coupled with potential changes in climate might contribute to increasing numbers of flooding incidents.

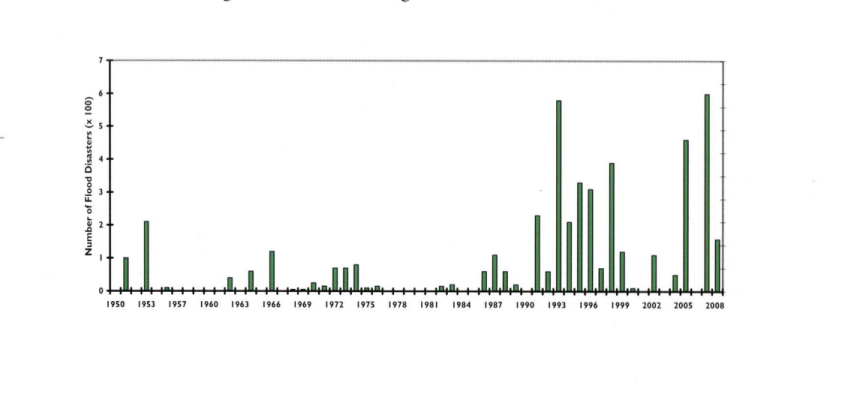

FIGURE 1-1
Floods on the rise: number of flood disasters worldwide

Source: Munich Re, 2008.

1.1.3 Urbanisation

In 1900, 13 per cent of the global population was urban; in 2007, it had increased to 49.4 per cent, occupying only 2.8 per cent of global territory. In 2050, it is forecasted to be 69.6 per cent of the world's population. The world is becoming increasingly urban as result of economic development and jobs distribution. In developed countries, the population is stabilised and urban population is already large, but in developing countries, the population is still growing. In 2050, the world population will be about 9 billion and most of its growth will be in the cities.

Urban development accelerated in the second half of the twentieth century due to high population growth rates and migration to cities, with most of the increase concentrated in densely populated areas. Countries such as Brazil moved from 55 per cent urban population in 1970 to 83 per cent today, occupying only 0.25 of the country's area. In India and China, the urban population is 29.2 per cent and 42.2 per cent, respectively, and this trend of increasing urbanisation is set to continue.

Urban growth in developing countries has taken place in an unsustainable way, with a consequent degradation of quality of life and the environment. This process is more significant in Latin America where 77 per cent of the population is urban. There are currently 388 cities in the world with a population of over 1 million; 44 of these cities are in Latin America. Some 16 megacities (over 10 million inhabitants) grew up in the late twentieth century, representing 4 per cent of the world's population, and at least four of these cities are in Latin America (see Table 1-1), representing over 10 per cent of that region's population.

Urban growth in developing countries has been significant. In developed countries, the population has stopped growing and is tending to decrease, as the birth rate is less than 2.1 children per marriage, thus keeping the population stable. The present

FIGURE 1-2
Urbanisation Tokyo

Source: Chris Zevenbergen, 2008.

TABLE 1-1

Largest cities in the world and in Latin America (UN, 2009)

Largest in the world		**Largest cities in Latin America**	
City	**Population millions**	**City**	**Population millions**
Tokyo	26.44	Mexico City	17.8
Mexico City	18.07	São Paulo	16.3
São Paulo	17.96	Buenos Aires	12.02
Bombay	16.09	Rio de Janeiro	10.65
Los Angeles	13.21	Lima	7.44
Calcutta	13.06	Bogotá	6.77
Shanghai	12.89	Santiago de Chile	5.47
Dakar	12.52	Belo Horizonte	4.22
Delhi	12.44	Porto Alegre	3.76

population is recovering or being maintained only through controlled migration. In developing countries, growth is even higher and the United Nations projection is that the population will not stabilise until 2150. Urbanisation is a worldwide process, with differences between continents. In Latin America, urbanisation has been high as the

FIGURE 1-3
Slum area Ahmedabad in flood plain, India

Source: Chris Zevenbergen, 2008.

rural population has moved to the cities. It was estimated that around 2010 there will be 60 cities with more than 5 million inhabitants, and most of them will be located in developing countries.

It is therefore possible that the problems faced by South American countries and Mexico may be reproduced on other continents as the trend towards urbanisation continues. Urbanisation increases the competition for the same natural resources (air, land and water) in a small space for human needs on living, production and amenities. The environment formed by natural space and by the population (socio-economic and urban) is a living and dynamic being that generates a set of interconnected effects, which if not controlled, can lead the city to chaos.

Flooding is often a major problem in informal settlements such as in the slum areas of Ahmedabad, India. Many settlements are located along rivers in low-lying flood plains and the houses are built of weak and inadequate (e.g. low water resistance) building materials.

During the past century, urban development created standards of urban concentration. In the large cities, there was a process of urban decentralisation towards the periphery, leaving the centres of the cities unpopulated and in decline. Difficulties with transportation routes, increased traffic and deterioration in transportation have led to changes of attitude in this process.

TEXTBOX 1-1
Urban expansion and increasing flood vulnerability

Sheppard and his team are one of the first to examine the dynamics and underlying processes of global urban expansion (Sheppard, 2007). They presented highly relevant information on built-up areas of cities and their changes over time. This work revealed that the spatial distribution of the urban population in nearly all of the 90 cities surveyed is by and large not the result of conscientious planning. One of the key messages is that cities in all regions must plan early and much more carefully to accommodate and diffuse the impact of over-concentrations of people and economic activities in order to avoid large-scale catastrophes that will otherwise ensue.

This lack of careful planning, or even uncontrolled urbanisation, will exacerbate the trend of increasing flood vulnerability of cities due to a combination of the following factors:

1. New 'green field' development in previously non-urban use areas, leading to encroachment and expansion onto flood-prone areas, such as flood plains and lowlands;
2. Redevelopment of built-areas ('brown fields') and through 'infill' of the remaining open spaces in already built-up areas, leading to an overall density increase and subsequent increase of surface sealing and disruption of natural drainage channels;
3. Urban areas once developed will rarely disappear and tend to favour the status quo, even after major flood disasters. The tabula rasa opportunity for correcting old errors and adopting new approaches to reduce vulnerability is seldom being exploited.

FIGURE 1-4
Urban growth of the city of Hyderabad (1990 and 2000) showing rapid urban expansion through 'infill' of the remaining open spaces in already built-up areas and through 'outspill' in areas previously in non-urban use

Source: Sheppard, 2007.

In cities in developing countries, part of the population lives in irregular or informal areas usually called slums. The growth in slums has been significant, and their increasing density is cause for concern, since the greatest rate of population growth occurs particularly among the low-income population. Slums are overcrowded dwellings of poor quality for the low-income population who occupy unregulated areas without property rights. Therefore, there is the *formal and the informal city*. Urban management usually reaches the former. This population is lacking most services such as a water supply, sanitation, drainage and solid waste disposal, thus human waste enters the environment, spreading diseases.

Cities are like living systems, constantly self-organizing in many ways in response to both internal interactions and the influence of external factors (see Section 1.2). Cities are changing faster than we can understand the diverse forces that are triggering these changes. Relative to geological timescales, urban patterns of change that we see today have occurred extremely fast. New developments can take place, ignorant of underlying risk, as their existence has been too short for them to experience extreme events. Urban planning, on the other hand, is static (see Chapter 7). To a large extent, we live in yesterday's cities: many urban attributes such as roads and buildings are legacies of past policy and decision-making. Tomorrow's cities will be shaped by decisions we take today that transform the legacies from the past. An approach based on complex adaptive systems and resilience would allow planners and decision-makers to learn and adapt to the inevitable failures of urban management actions (see Chapter 13).

Why are cities special cases? Cities are particularly vulnerable to flooding because of:

- direct impacts on the citizens. The scale of population at risk. A large and growing proportion of the population concentrates in urban areas. Urban centres; safety, health, livelihoods and human factors are the most important and often the most neglected areas of discussions about flood;
- direct impacts to physical infrastructure; buildings, urban services (e.g. roads, railway lines, embankments, public water supply and sanitation) and specific industries;
- indirect impacts to economic sectors: change in productive capacity effected by the change in resource productivity or market demand;

TEXTBOX 1-2
River flooding in Estrela (RS), Brazil

The city of Estrela, State of Rio Grande do Sul in Brazil has population of 28,300. It is located on the banks of the river Taquari, in a watershed of 25,000 km^2. The flood level in extreme cases may vary 18 m in a single day.

Part of the area close to the city and the banks of the river are unsettled on account of that risk, but areas where the risk is less frequent are settled. In 1979, when the urban master plan was being prepared, it was realised that there was a need to prepare zoning of areas liable to flooding for inclusion in that plan. A technical study was conducted that took account of the probability of flooding in the city and proposed limits for urban settlement in the city.

The flood zoning propose restriction to construction such as: (a) limit of flooding regulation area: 26.00 m; (b) between 24 and 26 m: area that may be built on, subject to piles above a water level of 26 m and other constructions recommendations; (c) area below 24 m: permanent reserve (established in 1981).

To avoid the invasion by unregulated settlements and recovery of the area liable to flooding already settled, a Municipal Law No 1970 of 1983 was approved. The municipality stipulated that the area liable to flooding could be changed by approval and an increase on construction area above its limits in developed valuable downtown areas of the city. In principle, it uses an economic principle to give value to flood-risk areas. The municipal law states as follows (PME, 1983):

'*Article 1* – The municipality is authorized to permit the construction of buildings for commercial, residential or mixed use above the urbanization rates permitted by law in the master plan, provided that:

Paragraph 1: an area of land in the same zone and with an area equivalent to 4/10 of the built-up area exceeding that permitted in that place is set aside for public use as a green area or for institutional use.

Paragraph 2: Where the area of land located in the permanent conservation or landscape conservation areas is set aside for public use, its value shall be equivalent to the built-up area exceeding that permitted in that place, and that area must be in the same zone or, if unavailable, in the nearest zone adjoining the place.'

- relatively high dependency on centralised infrastructure systems and utility services;
- low safety margins due to improved modelling capabilities for prediction and optimisation;
- economic sectors: change in productive capacity effected by the change in resource productivity or market demand.

1.2 THE CITY AS A LIVING ORGANISM

Cities are evolving systems. We have little understanding of their behaviour. We know that they are highly dynamic, in that they face changing environments and inputs, and adapt to these changes. Over a longer historical period, cities have always successfully adapted to changing environmental conditions and thus have been extremely resilient. From 1100 to 1800, only 42 cities worldwide were deserted after their construction.

It is an interesting and useful exercise to think of a city as a system. An urban system, like many other systems such as living organisms, is structured hierarchically and is a typical example of what is referred to as a Complex Adaptive System. At a high level, cities are made up of various components that take inputs and produce outputs. At a low level, it is composed of interacting parts or subsystems like buildings, a transport network and a supporting business environment for organisations to interact. Each of these has an important influence on one another and because they are interacting, the whole is more than the sum of its components.

System and environment are generally separated by a boundary. A system approach allows the consideration of interacting factors at multiple scales, both temporal and spatial. The system approach identifies multiple linkage and exchange pathways. The system approach provides a framework for better understanding the linkages between factors. Typically, Complex Adaptive Systems display emergence. Emergence refers to high-level patterns that emerge from low-level rules and events that might not have been anticipated. The system approach is often used to analyse the flooding system (see Section 7.1) and to develop strategies for dealing with flood risk (see Section 1.5).

1.2.1 Urban dynamics

All attributes of the urban fabric require periodic upgrades, major refurbishment and/or complete renewal. Roads need resurfacing every 5–10 years, dams and water supply infrastructures need major refurbishment every 20–30 years. Buildings have lifetimes ranging from 30 to 300 years, but exterior surfaces (skin) now change every 20 years, and electrical wiring, plumbing and heating systems (services) are replaced every 7 to 15 years. Although cities have always adapted to changing environmental conditions through autonomous adaptation, the dynamics of climate change may warrant adapting the building stock to better cope with increasing flood risk, through planned retrofitting and/or re-designing its structure during its lifetime. A hurdle to moving to planned adaptation seems to be the paradox in present day planning practices: a time frame of 20 years or even less is considered long term, whereas implicitly it is assumed that buildings last forever and 'site or urban location is eternal'. Planning decisions are thus typically based on yesterday's specifications that assume that the world is static. Renewal schemes of buildings may provide an opportunity to exploit substitutions of built components and structures for planned sustainability and climate change adaptation.

FIGURE 1-5
Adapting to climate change: substitution of built components and structures

Source: Chris Zevenbergen, 2007.

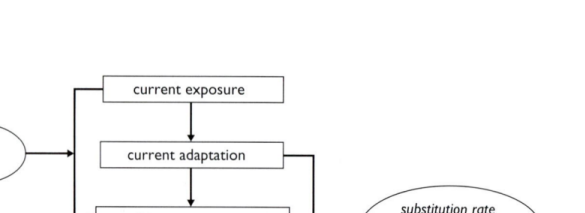

In other words: adaptation should become an element in urban renewal schemes and life cycle assessments, which in turn calls for planning ahead for up to 100 years. The capacity to adapt to changing conditions depends on their substitution rates as illustrated in Figure 1-5.

1.2.2 Multi-scale dynamics and emergent patterns in urban spatial systems

The city system is made up of various components that take inputs and produce outputs. At a lower spatial level, it is composed of interacting parts or subsystems such as buildings, a transport network and a supporting business environment for organisations to interact. At a higher level, it is part of a supra system, 'the catchment'. In principle, at each spatial level, there are three types of measures to reduce the overall system's flood vulnerability, based on the type of possible responses of a system to floods. These are:

- reducing exposure;
- reducing the system's sensitivity;
- mitigating the impacts (recovery).

Flood exposure is directly related to the physical mechanism underlying the flood propagation through the catchment system. The propagation of a flood wave to lower spatial levels is buffered by thresholds that are set at each level. Consequently, flood exposure at a certain spatial level is dependent on the interventions taken at a higher level. In other words, managing flood exposure involves a feedback process that operates top-down, Figure 1-7. In traditional flood-risk management policies, flood exposure is generally modified through governmental interventions and restricted to measures taken at the catchment level only. Reducing the system's sensitivity will reduce the direct impacts, the indirect impacts, or both. If the urban system is provided with a sufficient amount of redundant attributes, it can switch from one attribute to another. Interventions to reduce sensitivity at a certain level may also enhance the system's redundancy because of the so-called 'ripple effect'. Consequently, these interventions will reduce indirect impacts and therefore increase the robustness of the system at a higher spatial level. For example, designing a building that can be made flood-proof would be beneficial for the house owner. A number of such buildings would enhance a city's robustness against floods. In conclusion: urban flood resilience involves multiple spatial levels and resilient approaches are based upon an understanding of their interactions and taking advantage of the interventions at these different levels.

TEXTBOX 1-3
Changing flood risk in The Netherlands

The Netherlands is protected from storm surges and river floods by a reinforcement of the primary flood defence system consisting of coastal dunes, dikes and storm-surge barriers (the Delta Works). These were implemented in response to the dramatic flooding disaster in 1953 and were carried out under the so-called Delta Act. Since then, the focus has been on structural flood defences. The current flood safety standards date back to this law and include standards for maximum exceedance probability of the design of storm surge levels that dikes must sustain. A statistical approach to the storm surge levels (based on historical data) was chosen and an extrapolated storm surge level and a cost-benefit analysis formed the basis for the optimal dike height. In recent decades, the development of a new approach made it possible to assess the flood risks, taking into account the multiple failure mechanisms of a dike section and the length affected. Recently a new model has been developed in which an increase of the potential damage by economic growth could be incorporated. In this model, the expected yearly loss by flooding is taken as the central variable, rather than the exceedance probability. These considerations have opened the discussion about a fundamental reassessment of the acceptability of the flood risks in The Netherlands, in which the (increasing) economic impacts have to be taken into account. Simultaneously, scientific support was growing, which revealed that the hydraulic baseline conditions, such as storm wave properties and maximum river discharges, may be different and more severe than recently thought, and that climate change and sea-level rise may aggravate this situation.

The wake-up call came in 1993 and 1995 when the rivers Meuse and Rhine almost flooded. A few years later, Katrina fed the debate in The Netherlands on the limitations of the ability to control extreme events by technical means, raising the possible challenges posed by the increased value of investments, population growth and climate change. Parallel to these paradigm changes, attention focused on the development of a flood warning system and land-use planning. Development control issues gradually received more attention in the late 1990s. In that same period, a nationally Integrated Water Management Strategy (WB21) (Room for Water, damage reduction through planning and zoning) resulted in a new policy that advocated the transition from the traditional focus on probabilities towards a more integrated approach. The latter, however, is perceived as a modification of traditional practices, rather than a drastic change of direction. The huge levels of investments and dependence on the existing defence systems in the most densely, low lying Randstad (conurbation area in Western Holland) is so high by now that options for change are limited. The barrier to change is also caused by the strong interconnectivity between water institutions, management structures, routines and infrastructural entrapment. Although much emphasis is placed on the institutional integration of spatial planning and water management, the actual implementation in the Netherlands of the resulting integrated plans seems to be hampered.

FIGURE 1-6
Dutch delta works (Maeslantkering)

Source: Rijkswaterstaat, The Netherlands, 2004.

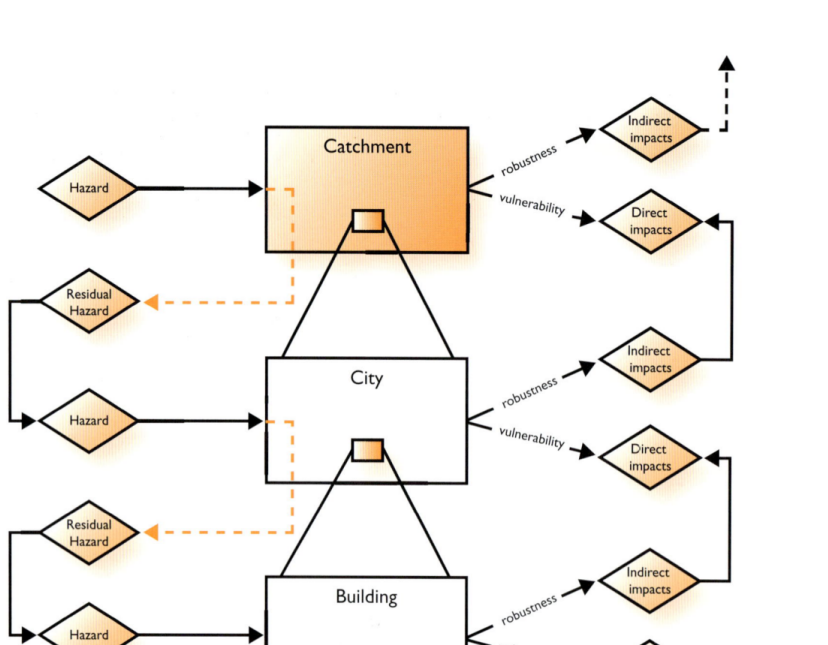

FIGURE 1-7
Travelling across spatial levels

This Figure depicts the propagation of a flood wave through the catchment system in case of a failure or overtopping of the primary flood-protection system.

Source: Chris Zevenbergen, 2008.

1.3 VULNERABILITY OF URBAN AREAS: A ROUGH GUIDE

Urbanisation, land subsidence and sea level rise will increase the vulnerability of the urbanised low-lying areas all over the world. Vulnerability is often defined as the sensitivity of a system to exposure to shocks, stresses and disturbances, or the degree to which a system is susceptible to adverse effects, or the degree to which a system or unit is likely to experience harm from perturbations or stress. The system under consideration can range in size from individuals through communities to regions and even higher. The vulnerability concept is widely used in studies on risks and natural hazards and often also includes social and political dimensions.

A literature review of vulnerability studies shows that vulnerability can be considered as a combination of threshold capacity, coping capacity, recovery capacity and adaptive capacity. Table 1-2 illustrates the four capacities.

Vulnerability components are highly connected. Consequently, increasing one vulnerability component often decreases one or more of the other components resulting in higher, rather than reduced vulnerability. The connection between vulnerability components is illustrated in Figure 1-9, which presents a conceptual damage return period graph. As a result of dike construction or reservoir construction, environmental variations with low return periods will cause no damage. This is the threshold domain in Figure 1-9. Even if thresholds have been built, there will be some occasions when the threshold will be exceeded. Then, coping with hazard impacts

TABLE 1-2
Four components of vulnerability

Threshold capacity	Threshold capacity is the ability of a society to build up a threshold against variation in order to prevent damage. In flood-risk management, examples are building river dikes and increasing flow capacity to set a threshold against high river flow (see Section 11.4). The objective of building threshold capacity is prevention of damage. The time horizon lies in the past; past disaster experiences of society are the guiding principle to determine the height of the threshold. In the Netherlands, for example, constructed dikes had the same height as the highest experienced flood. The ability of a society to build, operate and maintain threshold capacity is determined by its social, institutional, technical and economic attributes, arrangements and responsibilities.
Coping capacity	Coping capacity is the capacity of society to reduce or absorb damage in case of a disturbance that exceeds the damage threshold. The coping capacity of society is determined by the presence of effective emergency and evacuation plans, the availability of damage reducing measures, a communication plan to create risk awareness among residents, and a clear organisational structure and responsibility for disaster management. The objective of developing coping capacity is the reduction of damage. The time orientation is instantaneous because, in case of emergencies, only 'here and now' is important. The uncertainty is low because the magnitude of the hazard is clear at the time society has to deal with it. Also, for coping capacity, the ability of a society to build, operate and maintain is determined by its social, institutional, technical and economic abilities. There is a large range of coping capacity options (see Section 8.10). In many industrialised countries, threshold-exceeding events for water management do not occur frequently. As a result, it is not clear who is responsible for damage reduction in case of emergencies. Multiple stakeholders such as fire-fighters, water boards, municipalities and other government agencies are involved.
Recovery capacity	Recovery capacity is the third component and refers to the capacity of a society to recover to the same or an equivalent state as before the emergency. It is the capacity of a flooded area to reconstruct buildings, infrastructure and dikes. The objective of developing and increasing recovery capacity is to quickly and effectively respond after a disaster. The time horizon is instantaneous, right after the disaster, but will change gradually towards a focus on the future. Although economical damage estimates may be difficult, the uncertainty of the hazard magnitude will be relatively low compared to possible future hazards because the effects will still be noticeable. The economic capacity of the country to finance the reconstruction determines the recovery success to a large extent. However, institutional ability and technical knowledge are also important. A society that is able to recover from the impacts of hazards will be less vulnerable to these hazards. Recovery time may range from weeks to decades, depending on the spatial scale and disaster magnitude. Recovering from the Katrina hurricane in New Orleans will take years (see Chapters 10 and 11).
Adaptive capacity	Adaptive capacity is the capacity of a nation, a community living in a river basin, or even the world to cope with, and adjust to uncertain future developments and catastrophic, not frequently occurring disturbances such as extreme floods. Therefore, the time orientation lies in the future. Although a system may be functioning well at present, human and environmental developments, both from inside or outside the considered system, can put a system under strain and threaten its future functioning. Examples are climate change, population growth and urbanisation. The problem of adapting to uncertain future developments may be illustrated by an example of land use. Although future risks from river or sea floods are unknown, land-use decisions that determine future vulnerability are taken in the present time. The objective of developing adaptive capacity is to be able to accommodate the potential future developments and impacts by constructing a robust living and working environment. The uncertainty of the nature and magnitude of future hazards and impacts is high. However, solutions will have to be developed for long time horizons and financial and spatial reserves will have to be made, to allow for adaptations. There are many options available for society to increase its adaptive capacity, varying from technical options to insurance policies and communication strategies (see Chapter 7 and 12). The number of organisations involved in the adaptive capacity determinants is also large. Consequently, there is no clear picture about who is responsible for increasing adaptive capacity. Moreover, there is societal disagreement about the developments, the problems, and the solutions that are relevant for adaptive capacity.

and recovering from them, is necessary. This is the coping and recovery domain in which damage reduction is the prime goal. Finally, there are very unlikely events with very high return periods where the expected damage is that extreme that recovery is neither feasible nor possible. These are the types of occasions we want to prevent by adapting. Therefore, this is the adaptive domain. Figure 1-9 illustrates that by increasing only the threshold domain, for instance by building higher and stronger dikes, the coping and recovery domain becomes smaller. However, these domains are important; by coping and recovering, people become aware of risks. An approach that only focuses on increasing threshold capacity results in a system that is increasingly vulnerable to rarely occurring disasters. Disasters that cause damage will occur less frequently, but the ones that do occur will cause more damage. Consequently, for a complete vulnerability reducing strategy, attention should be paid to all components and domains of vulnerability.

FIGURE 1-8
Venice showing elevated cat walk to keep feet dry

Source: Chris Zevenbergen, 2008.

FIGURE 1-9
The four components and three domains of the vulnerability framework illustrated by a damage return period graph. The three domains are interrelated; changes in one domain affect the other domains, resulting in an overall change in vulnerability

Source: Rutger de Graaf, 2008.

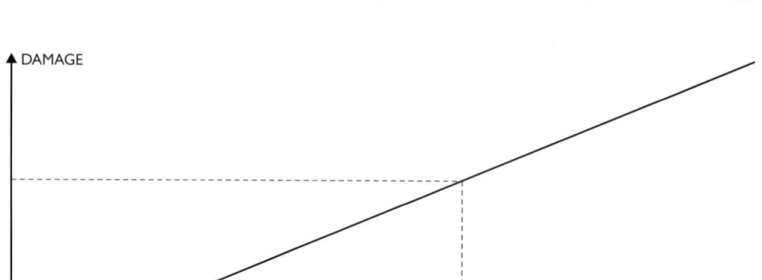

1.4 TYPES OF UNCERTAINTY

1.4.1 What is uncertainty?

One crucial theme of flood-risk management is the issue of how to deal with uncertainty. Our incomplete understanding of complex natural systems, and the uncertainties associated with human behaviour, organisations and social systems make it extremely difficult to predict future vulnerability of cities to flooding. Traditionally we have sought to include uncertainty in the risks. Flood protection measures were based on the accumulated knowledge of past weather events and climate change was perceived as

TEXTBOX 1-4

Capacity response measures in Japan to reduce flood vulnerability

Japan is a country that is frequently exposed to all kinds of natural hazards including flooding and droughts. Consequently, coping, recovery and adaptive capacity are well developed in Japan to adapt to changes in the physical conditions. It can be concluded that in Japan all four capacities of the framework are used to reduce vulnerability of urban lowland areas. For urban pluvial flood control, in addition to improving sewer capacity, risk communication, stakeholder involvement, emergency plans, wet proofing of buildings and elevated houses and infrastructure are applied. Other coping and recovery capacity increasing capacity measures that reduce the effects of flooding are stormwater infiltration and retention and securing access to flooded areas by elevated roads and emergency shiplock.

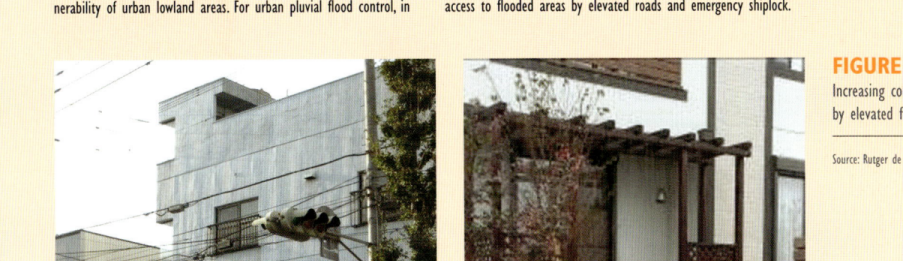

FIGURE 1-10
Increasing coping capacity by elevated floor levels in Japan

Source: Rutger de Graaf, 2008.

a quasi-stationary process (the past reflects the future). Flood disasters in recent years have contributed to the realisation that the future is inherently uncertain.

Many definitions of uncertainty co-exist. Laplace^1 argued that the world was completely deterministic: that uncertainty is a consequence of incomplete knowledge. This doctrine of scientific determinism remained the standard assumption for over 100 years until the early twentieth century when a sequence of discoveries in physics proved science can only partially reduce uncertainty. Knight^2 defined uncertainty as being clearly distinct from the notion of risk, as follows: 'It will appear that a measurable uncertainty, or "risk" proper, as we shall use the term, is so far different from an unmeasurable one that it is not in effect an uncertainty at all.' The U.S. Water Resources Council (WRC) (1983) defined risk and uncertainty in relation to the ability to describe potential outcomes in probabilistic terms. Within this context, the WRC states that: 'situations of risk are detained [defined] as those in which the potential outcome can be described in reasonably well-known probability distributions such as the probability of particular flood events' and that 'situations of uncertainty are defined as those in which potential outcomes cannot be described in objectively known probability distribution.'

^{1} Pierre-Simon, marquis de Laplace (1749–1827) was a French mathematician and astronomer whose work was pivotal to the development of mathematical astronomy and statistics. Laplace's law of succession states that, if before we observed any events we thought all values of (unknown probability) p were equally likely, then after observing r events out of n opportunities a good estimate of p is $p = (r + 1)/(n + 2)$.

^{2} Frank H. Knight (1885 - 1972), an American economist, made in his work Risk, Uncertainty and Profit (1921) a distinction between "risk" (randomness with knowable probabilities) and "uncertainty" (randomness with unknowable probabilities). Knightian uncertainty is risk that is immeasurable, not possible to calculate.

1.4.2 Types of uncertainty

Uncertainty comprises different and distinct components such as statistical variation, measurement errors, ignorance and indeterminacy, which all have one feature in common: uncertainty reduces the strength of confidence in the estimated cause and effect chain. If complexity cannot be explained by scientific methods, uncertainty increases. But even simple relationships may be associated with high degrees of uncertainty if either the knowledge base is missing or the effect is stochastic by its own nature.

Similar to the co-existence of many definitions of uncertainty, there are many different ways of ordering different types of uncertainty. For example, Walker *et al.* (2004) distinguish three dimensions of uncertainty—specific to model-based decision support:

- Location of uncertainty – where the uncertainty manifests itself within the model;
- Level of uncertainty – where the uncertainty manifests itself along the spectrum of different levels of knowledge between determinism and total ignorance: statistical uncertainty, scenario uncertainty and recognised uncertainty;
- Nature of uncertainty – whether the uncertainty is due to imperfection of knowledge or is due to the inherent variability of the phenomena being described.

The various dimensions of uncertainty are described in Table 1-3.

TABLE 1-3

Three dimensions of uncertainty (after Walker *et al.*, 2004)

Dimension	Forms	Description
Location	Context uncertainty	Ambiguity in the definition of the boundaries of the system to be modelled
	Model uncertainty	Uncertainty about both the conceptual model itself (i.e. competing interpretations of the cause-effect relationships within the system) and the computer model (i.e. software and hardware errors)
	Input uncertainty	Uncertainty associated with the data that describe the reference (base case) system and with the external driving forces that influence the system and its performance
	Parameter uncertainty	Uncertainty about the constants in the model, supposedly invariant within the chosen context and scenario
	Model outcome uncertainty	The accumulated uncertainty caused by the uncertainties in all of the above locations (context, model, inputs, and parameters) that are propagated through the model
Level	Statistical uncertainty	Any uncertainty that can be described adequately in statistical terms
	Scenario uncertainty	Any uncertainty that can be described by a range of possible outcomes, where the allocation of likelihood is not possible
	Recognised ignorance	Fundamental uncertainty about the mechanisms and functional relationships, and about the statistical properties
Nature	Epistemic uncertainty	Imperfection of knowledge—due to limited or unreliable data, incomplete knowledge, imperfect models, ambiguities, ignorance, etc. Uncertainty that comes from imperfection of knowledge can be reduced by more research or empirical efforts
	Variability uncertainty	The inherent uncertainty or randomness induced by variation in space and time, which is especially applicable in complex systems. They express a non-linear and sometime chaotic behaviour, what makes them impossible to be predicted.

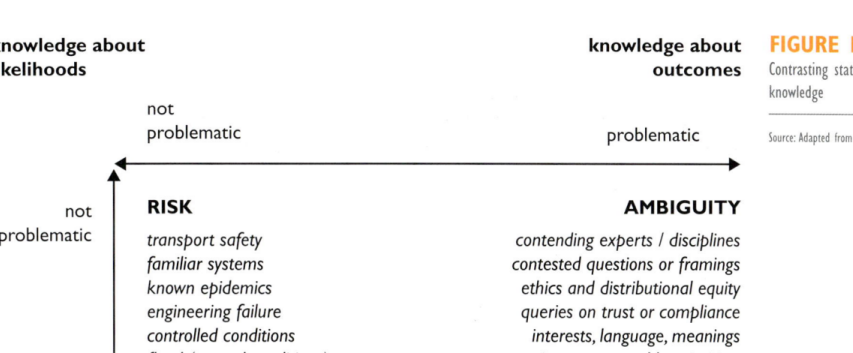

FIGURE 1-11
Contrasting states of incomplete knowledge

Source: Adapted from Stirling, 2003.

Based on the distinction of uncertainty by its nature, there are two types of uncertainty: epistemic uncertainty (incomplete knowledge) and variability uncertainty (unpredictability).

Epistemic uncertainty, characterising the behaviour of natural systems, implies that, in principle, the knowledge base could be improved by carrying out more research, or collecting more or better data. To this end, science and the scientific method can gradually work towards reducing uncertainties about a problem, or at least how to deal with it. In this context, approaches can be used that allow the evaluation and quantification of the effects of uncertainty in models, such as sensitivity, uncertainty and scenario analyses.

Variability uncertainty implies that the unpredictability of the system is accepted. With this type of uncertainty, additional science does not yield an improvement in the knowledge base and model outcomes. Different approaches for dealing with unpredictability, as needed for the managed flooding system, are discussed in Section 1.5.

1.4.3 Methods for dealing with uncertainty

Based on Stirling's framework (e.g. Stirling, 2003), Figure 1-12 gives a general overview of methodological responses to different forms of incertitude. The simplest approach for dealing with uncertainty is *sensitivity analysis*. Sensitivity analysis can be used to understand model behaviour by measuring the sensitivity of model outputs to changes in model inputs, and involves varying input factors (e.g. parameters, input data) that are not known with certainty. Another—more demanding—commonly used approach of risk analysis is termed *uncertainty propagation* in models. It is concerned with estimating the risk associated with the outcome and the (probability) distribution of output variables, given

1.4 TYPES OF UNCERTAINTY

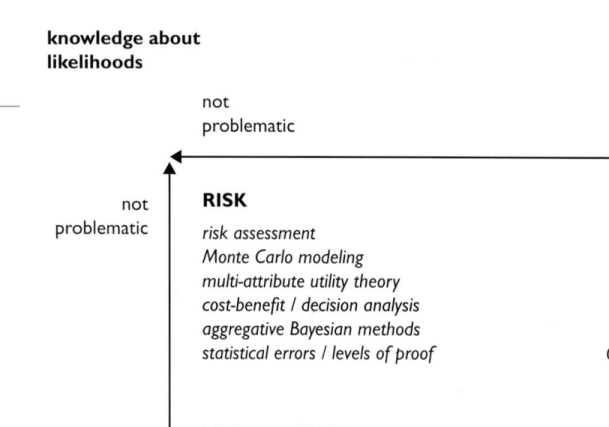

FIGURE 1-12
Methodologies for dealing with different forms of incertitude

Source: Adapted from Stirling, 2003.

the uncertainty associated with parameters or input data. The simplest implementation of propagation of uncertainty involves specifying a joint distribution on selected input factors and then propagating this uncertainty through the model to the output. A more complex implementation uses modelling of certain variables as stochastic processes. For computational purposes, uncertainty propagation usually involves sampling from a joint distribution using the Monte Carlo method or—if this is computationally too expensive—reduced Monte Carlo simulations based, for instance, on Latin Hypercube sampling.

In water management, new approaches for dealing with future uncertainties have been recently introduced. For example, the development of flood insurance, flood-risk mapping systems and general risk management approaches that specifically address the likelihood of certain future trends are commonly used in spatial planning research and are gaining increasing attention in water management (see Chapter 7).

Scenario analysis is an approach to understand the effects of incertitude on the model outcome. It aims to simulate a range of plausible future scenarios, each of which embeds a coherent and internally consistent set of assumptions about the key relationships and drivers. Figure 1-13 shows the scenarios used in the UK's Future Flooding study and how climate change scenarios related to the IPCC Special Report on Emission Scenarios were mapped on to the socio-economic scenarios; which comprise governance in the vertical axis and social values on the horizontal axis. Note that these scenarios are not predictions of the future they are simply logical and consistently defined visions of potential long-term futures.

Dealing with ignorance has two different components: dealing with imperfection of knowledge and dealing with unpredictability. Epistemic uncertainty, characterising the behaviour of natural systems, implies that, in principle, the knowledge base could be improved by carrying out more research, or collecting more or better data. To this end,

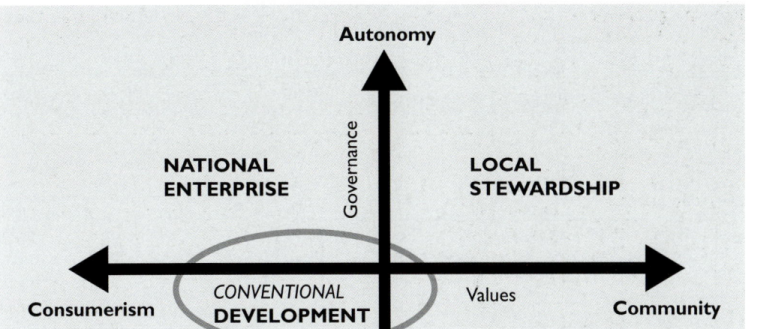

FIGURE 1-13
The scenarios used in the Foresight futures project

Source: Foresight, 2004.

science and the scientific method can gradually work towards reducing uncertainties about a problem, or at least how to deal with it. In this context, approaches can be used that allow the evaluation and quantification of the effects of uncertainty in models, such as sensitivity, uncertainty and scenario analyses. However, more additional science does not yield improvement in the knowledge base and model outcomes when the outcomes of a system are unpredictable by character. In Table 1-3, this type of uncertainty is defined as variability uncertainty. There are at least three basic strategies for dealing with this type of uncertainty:

- *Uncertainty control*, as by demand management;
- *Passive protection*, as by robust strategies that could perform well under a large range of (variation in) future conditions;
- *Active protection*, as by adaptable strategies that could easily be modified to fit the (changing) future conditions.

In terms of protecting the system, passive dealing with uncertainties, which will function without any significant management decisions, is the first strategy that comes to mind. It implies making the system robust—in short, 'future-proofing' decisions against unknown future parameters or changing assumptions; i.e. preparing for the worst but hoping it will never happen. The principal mechanisms that can provide robustness include, but are not limited to: reliability, sensitivity, rapidity, excess capacity, reserves, functional diversity, redundancy, autonomy, interdependence and modularity. Robustness refers to the ability to satisfactorily endure foreseen changes. It is arguable that when a system is robust it becomes insensitive or more tolerant of expected, unforeseen states of nature (see Section 1.5). Robustness (i.e. the ability to regain stability) can be measured as the degree to which a solution is affected (in terms of cost, performance or any other attribute) by unknown future parameters or changing assumptions. It differs from resilience (i.e. the ability to survive) as this focuses on the ability to recover after an event (see Chapter 7).

FIGURE 1-14
Comparison of protection strategies to dealing with variability uncertainty

Source: Defra, 2004.

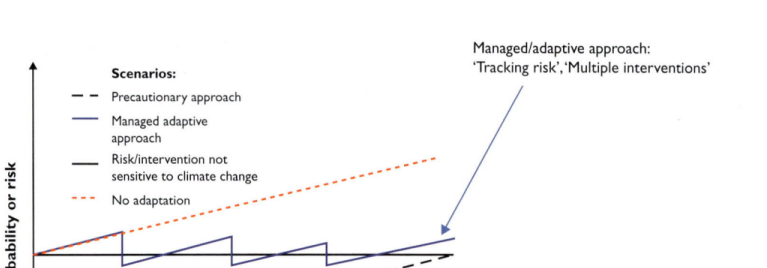

Alternatively, actively dealing with uncertainties enables decision-makers to modify the configuration of the system depending on the (changing) future conditions. It is about introducing flexibility^1 to adaptively respond to foreseen or expected changes. Adaptability (flexibility) is the inherent capability to successfully respond to foreseen (unforeseen) changes, i.e. uncertainty. The possibility of building options into the design of engineering systems offers a means to introduce flexibility to manage uncertainties (at least cost). It enhances the ability to preserve alternative options without having to abandon existing arrangements. Embedding options in the design creates flexibility which can be used in the future to accommodate a range of other possibilities, informed by new knowledge insights and understandings. Indeed, it creates possibilities for incorporating active learning and accommodating uncertainty throughout the process rather than a one-time intervention that then precludes other courses of desirable action. Depending on the available information, a decision is made to adapt to the future parameters in the best possible way.

¹ Note the difference between the concept of redundancy and flexibility. Both redundancy and flexibility refer to the idea that some components should be designed assuming that the design is not optimised and that the system will develop and not remain static. Redundancy refers to more than just sufficient design elements to serve the same functional requirements; while flexible elements may not serve the same functions as some currently existing components— indeed they may serve a new function.

Figure 1-14 illustrates the comparison of passive and active responses to managing future changes, which includes coping with uncertainty. The graph of probability or risk and associated efficiency shows the resulting saw-tooth effect as part of taking an adaptable strategy.

1.5 ADOPTION OF A SYSTEM APPROACH

This section applies a systems-oriented perspective to the management of the flooding system. The system approach is also used to formulate different strategies for dealing with uncertainty.

1.5.1 The managed flooding system

Many of the policy and management problems in the driver–response process for flood-risk management arise from causes that are both physico-environmental and socio-economic. Managing the flooding system is difficult due to multiple scale system dynamics, increasing uncertainty and the reflexivity of societal action. Therefore, flood-risk management requires the adoption of a more dynamic, systems-oriented perspective—where the system comprises the whole of the physico-environmental subsystem (PES) and socio-economic subsystem (SES) (see Textbox 1.5).

The managed flooding system is the complex, dynamic risk-producing–risk-response system that adapts to changes in drivers for flood risks. This system can be depicted

TEXTBOX 1-5
The 'managed' flooding system, PES and SES

The 'managed' flooding system is the complex, dynamic risk-producing–risk-response system that adapts to changes in drivers for flood risk. It comprises the **Physico-Environmental Subsystem (PES)** and the **Socio-Economic Subsystem (SES)** in mutual interactions. The PES includes the combination of physical structures and the environment. The SES includes the economy, actors, plus management. Management is thought of as an interface between the PES and the rest of the SES. It seeks to build flexibility and reversibility into the PES as needed for maintaining adaptive resiliency in the face of change. Management can, for example, expand or alter the system design if that is required to meet acceptable levels of functioning.

To effectively capture the dynamic nature of the linkages between the PES and SES, Figure 1-15 gives a dynamic conceptualisation of the 'managed' flooding system. In this conceptual model, dynamics is understood as the state of dynamic stability, known as social-ecological resilience (Folke, 2006). This is described by relevant thresholds where imposed acceptable levels of functioning (related to policy objectives) are exceeded. As an example, the likelihood of flooding or the risk is an important boundary condition for flood management.

The dynamics are represented by interactions inside each of the two subsystems as well as between the PES and SES (including management). Interactions result in physico-environmental and socio-economic change with both positive and negative effects on the linked system (PES+SES), and this is related to impacts.

Responses are diverse adaptations to PES and SES dynamics. There are four ways to do this (consistent with the different components of vulnerability): (1) changing the resistance/threshold capacity, (2) changing the coping capacity, (3) changing the recovery capacity, and (4) changing the adaptive capacity.

Drivers are formed by interdependent physico-environmental and socio-economic processes responsible for any type of change (impact) to the PES and SES, either external or internal. Pressures are then variables of such processes interacting within the two subsystems. It should be noted that the development of drivers and pressures only becomes relevant for adaptation decision-making if it would lead to alternate system states. That is, the trigger for adaptation is not future change in itself, but it is failing to meet the policy objectives that determine the acceptable levels of functioning.

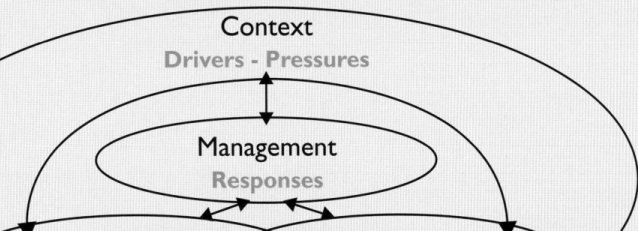

FIGURE 1-15
A conceptual model of the 'managed' flooding system

Source: Carl Folke, 2006.

FIGURE 1-16
Systems model for the managed water system

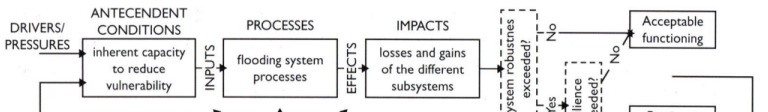

using a systems-theoretical model. The basic input-output systems model with feedback loops consists of: the antecedent conditions that provide the inputs to processes operating within the system; the processes in the system; and the effects of those processes that have an impact on the subsystems which feedback to alter the conditions of the system. External processes can act as drivers that alter the system conditions as well.

1.5.1.1 Antecedent conditions

Antecedent conditions are a product of place-specific processes that occur within and between the physico-environmental subsystem and the socio-economic subsystem. Antecedent conditions include the inherent characteristics that allow the system to absorb (adverse) flood impacts. These antecedent conditions can be viewed as a static state.

1.5.1.2 Flooding system processes

The flooding system processes and the relationships among them can be represented as the causal linkage between the sources of risk, exposed receptors and the pathways linking them. Antecedent conditions interact with the characteristics of the flood system to produce the immediate effects. The immediate effects are reduced or increased by the presence or absence of predetermined mitigation responses, which themselves are a function of antecedent conditions. After these mitigation responses are implemented, the impacts are realised.

1.5.1.3 System impacts

The effects of the flood system processes can have impacts on the physico-environmental subsystem and the socio-economic subsystem. The total impact is the cumulative effect of the antecedent conditions, event characteristics, and predetermined mitigation responses.

1.5.1.4 System functioning

The resulting system functioning is largely dependent on two systemic properties: robustness and adaptive resilience. These systemic properties emerge from a particular combination of the following capacities of the system to decrease its vulnerability: threshold capacity, coping capacity, recovery capacity and adaptive capacity. The four capacities have been summarised in Section 1.3. The following provides an overview of the systemic properties and the consequences in terms of system functioning.

Robustness refers to the physico-environmental and socio-economic properties of the system that provide the ability to resist perturbations and change, without re-organisation. If robustness occurs, the immediate effects are either avoided or reduced, and accordingly there is a greater likelihood that the system continues to function at at least acceptable levels. Dysfunction occurs only if the system's robustness is exceeded.

Adaptive resilience refers to the self-organising properties of the system that provide the ability to re-organise and adjust its structures or dynamics in the face of change. Resilience does not exclude dysfunction. However, if the system's robustness is exceeded and adaptive resilience occurs, dysfunction is likely to be transient, followed by a trajectory of at least acceptable functioning. If robustness is exceeded and resilience does not occur, persistent dysfunction may result. This is illustrated in Figure 1-12 with the 'No' arrow following resilience.

1.5.1.5 Feedbacks and drivers

Drivers alter the system conditions. They include both external and internal factors. The factors external to the system include physico-environmental changes, such as climate change and infrastructure deterioration; and socio-economic changes, such as population increases, economic change and political changes. Internal changes refer to the re-organisation of the physico-environmental subsystem or socio-economic subsystem induced by the system management. Over the short term, change drivers may be treated as a constant, while the sources are treated as variables, which cause perturbations in the system. However, over the longer term, the drivers become variables and lead to a change in the time-averaged state of the flooding system, which results in a change in the corresponding level of risk.

1.5.2 A system-oriented perspective of the managed flooding system

Within the systems approach it is possible to distinguish three categories of strategies to manage uncertain risks: strategies directed to the development of drivers; strategies directed to the system; and strategies directed to the system effects.

1.5.2.1 Influencing the development of drivers

While it is not possible to control the development of natural drivers to any considerable degree, although mitigation does set out to do this, it is still possible to affect much of the uncertainty caused by human disturbances, such as land-cover and land-use change or other socio-economic changes. Controlling, or at least attempting to manage, processes of land conversion whether in or beyond urban areas or providing incentives for sustainable land management are, among others, possible ways of controlling for uncertainty through managing demand. The adoption of uniform approaches and standards to control the sources of flooding can help regulate the changes in runoff rates (see Chapter 7). This in turn can control how the potential impact of urban development plans may increase risks and uncertainty. For example, the targets set in England (Interim Code of Practice for SuDS) require the maintenance of green field runoff rates for two-year return period events and the maintenance of green field runoff rates and volumes for 100-year return period events for all new and redevelopments. Another example of controlling for uncertainty in urban drainage management is the use of building codes to regulate construction and minimum safety levels.

1.5.2.2 Influencing the system

Two alternative or combined risk management strategies are needed for adapting the flood system: to consider a robustness-focused approach or to adopt a more resilience-focused approach. The robustness-focused approach mainly concerns increasing the insensitivity of the physical and social properties of the system to a range of uncertain risks, whereas the resilience-focused approach aims to implement flexibility and reversibility in the whole system so that it can deal with unknown or uncertain risks. Without attempting to be exhaustive, Table 1-4 provides a summary of the two approaches. The following provides a brief explanation of this.

A robustness-focused approach aims to avoid the risks associated with under-protection and under-adaptation, by improving the ability of the system to resist perturbations and changes, respectively. This type of approach is insensitive to a wide range of projected futures, in the sense that it gives acceptable functioning of the system in the presence of uncertainty. Consequently, a fully robust system design must be able to cope with the worst-case projection of the future. Mechanisms to build robustness include: incorporating headroom allowance, i.e. some degree of excess capacity above the design capacity to allow for the realisation of future uncertainties; introducing redundancy or rather dual-use functions that offer societal benefits even when the expected impacts do not occur; reducing the susceptibility of the exposed receptors; and diversifying protection levels for receptors according to their vulnerability. It should be noted that a robustness-focused approach is precautionary with respect to under-adaptation errors.

A resilience-focused approach aims to improve the self-organising ability of the system to respond to changes as they arrive and to restore the system functioning under the impact of perturbations. This implies managing for the system as a whole, rather than for its separated subsystems in isolation. In this type of approach, the management is seen as an integral part of the planning, design and implementation of the physical system. The management seeks to build real options into the physical system that enable the re-organisation of the system as needed to reduce risks or to capitalise on opportunities associated with inherent uncertainty. Real options in physical systems are created by changing the system design as uncertainty is resolved. These options offer a means to create technological flexibility and reversibility that the management can use to react to uncertainties. As such, the possibility of building real options into the physical system constitutes an important element of dealing with unknown or uncertain risks, because it allows learning about future parameters that were uncertain by only restricted errors towards over- and under-adaptation. Other mechanisms available to management that can enhance resilience include: commitment to monitoring and translating the system feedback into knowledge and action; and diversifying

TABLE 1-4

Uncertainty management strategies directed to the system

Approach	Robustness-focused approach	Resilience-focused approach
Aim	Reducing under-adaptation errors	Reducing both under-adaptation and over-adaptation errors
Mechanisms	Enhancing threshold and coping capacity through: - Some degree of excess capacity - Redundancy or dual-use functions - Reducing susceptibility - Diversification of protection levels	Enhancing recovery and adaptive capacity through: - Tight system coupling - Learning - Technological flexibility and reversibility - Diversification of measures
Valuation instruments	Net present value analysis	Real options analysis

TEXTBOX 1-6
Lessons learned from Japan

After World War II Japanese urbanisation exhibited an explosive growth, reflecting the onset of industrialisation. The population migrated to cities in the downstream alluvial plains, forming mega-cities such as Tokyo and Osaka. The occurrence of floods was accelerated by the conversion of rice paddies into residential areas. The dike systems constructed in the past resulted in high peak flows in major rivers. As a result, major river dike breaches occurred along large rivers and vast urban areas, causing flooding in Tokyo, Osaka and Nagoya. A historical example was caused by typhoon Kathleen in September 1947. This flood affected 1.6 million people, predominantly in the Tokyo Metropolitan Area. Urban floods were destructive not only in terms of economic damage, but also in terms of fatalities. After 1960, the national government launched a river rehabilitation plan, including major river improvement works. The progress of 'flood control works' and changes in social behaviour resulted in a reduction of the number of fatalities. Better weather forecasting and early warning allowed the population to prepare for floods. In the three consecutive years after 1974, dike breaches of the major rivers Tama, Ishikari and Nagaramajor resulted again in devastating

flood disasters. Realising that infrastructural measures (flood defences) alone could not completely overcome the floods, an integrated approach was launched in 1977 (Ando and Takahasi, 1997). Structural measures, such as regulating storage and retardation facilities, dams and embankments, were constructed in major river basins. Non-structural measures, such as land-use zoning, flood proofing and flood-risk mapping, were also implemented. In addition, a discharge suppression strategy was implemented for rainfall storage and infiltration facilities at individual buildings and the use of public spaces, such as parks for flood retardation. There were recurrent floods, and river improvement works continued in the 1980s and 1990s. Widening the river or raising higher levees was impracticable because the urban areas along the river Kanda are densely built up; and engineering solutions had almost reached their limits.

This situation lead to the construction of the spectacular Kanda River Loop 7 Underground Regulation pond, which was started in 1988 and the first phase (4.5 km long) was completed in 1997. This is a new method for an underground river—a multi-billion euro project, consisting of a pipe with a diameter of 10 to 12.5 m and

dimensioned to give protection against an hourly rainfall of about 75 mm, which has a return period of once in 15 years. This has resulted in a significant decrease in the number of fatalities since the 1960s. Such efforts have greatly contributed to the decreasing trend of the economic damage (in terms of economic/GDP ratio). Recent heavy rain events hit the Tokyo region, causing the collapse of river embankments, which affected urban functions such as transportation networks. These extreme floods were characterised as record breaking, as they have not been observed since the modernised rain gauge observations began. Although sufficient evidence is lacking of its actual manifestation, these flood events revealed Japan's vulnerability to climate change. In Japan, there is now a growing recognition that risk changes over time and that it is important to identify the nature of these changes and how global warming will effect these changes. Variability and uncertainty are becoming the key issues of future Japanese flood management. Thinking in static terms of risk has shifted towards an approach where variability needs to be managed.

Source: Ikeda, 2004.

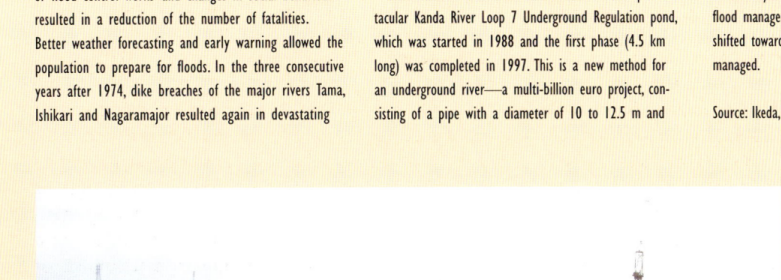

FIGURE 1-17
Super Levee Edogawa City

Source: Chris Zevenbergen, 2008.

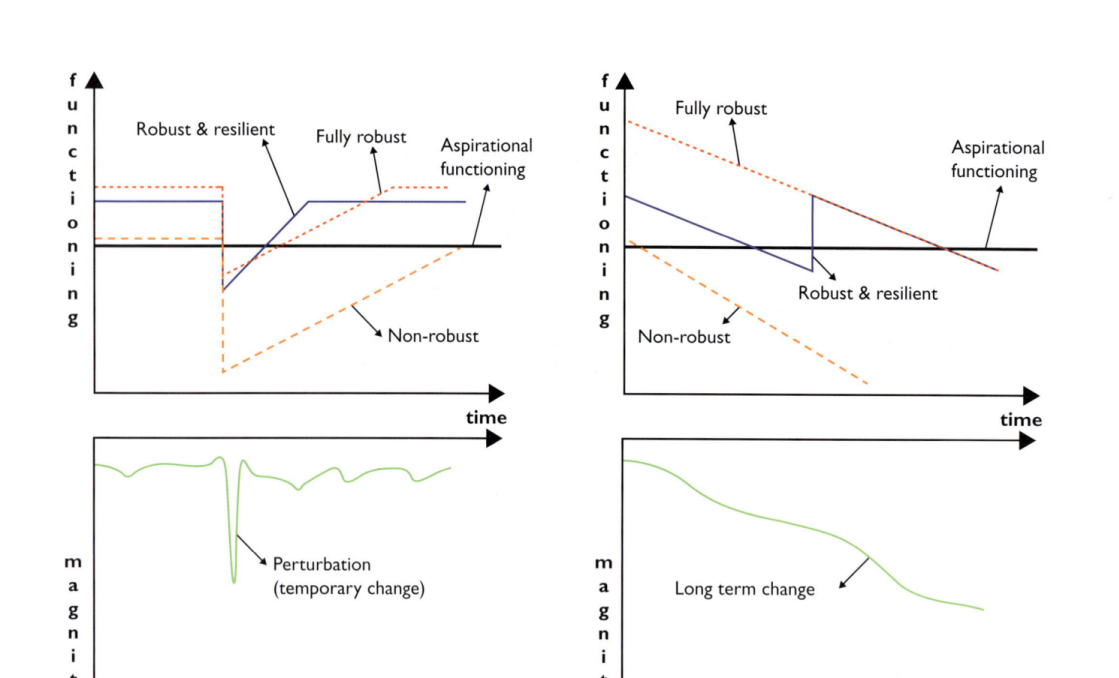

FIGURE 1-18
Functioning of the managed water system over time

Source: Berry Gersonius, 2009.

risk reduction measures. It should be noted that a resilience-focused approach is precautionary with respect errors associated with both under-adaptation and over-adaptation.

Figure 1-18 shows the consequential effects on the functioning of the system under the impact of perturbations and changes, as part of taking a robustness-focused—and a resilience-focused approach. It should be noted that both approaches acknowledge the uncertainty regarding the risk events, but that these differ in the way the impacts are managed over time.

1.5.3 Mitigating the adverse effects of perturbations and change

A third strategy to deal with uncertainty is to mitigate the adverse effects of perturbations and change. This strategy concerns mechanisms such as risk sharing, including flood insurance (see Chapter 10). Another example is to compensate for flood losses, for instance by setting up disaster funds. The availability of reconstruction plans and effective communication are other examples of responses enhancing the capacity of the system to recover quickly and effectively from the impacts of a flood.

1.5.4 Intervening in the spatial and temporal dimension

The highly interconnected and interdependent Physico-Environmental Subsystem (PES) and the Socio-Economic Subsystem (SES) (see Textbox 1-5) introduce significant challenges for risk management. These interdependencies, which span multiple spatial and temporal dimensions, make the system more vulnerable to floods. It is therefore prudent to acknowledge and address the different causal intra- and interconnected and interdependent relationships. For example, Figure 1-19 illustrates that cities can be divided by sub-systems such as the streetscape and allotment scale and that the urban system itself is part of a supra-system. Interactions take place within and between different scale levels. Within the streetscape scale (see also Section 1.3), for example, rainfall runoff interacts with sewage networks, which could cause damage at allotment scale in the case of flooding. Recognition of spatial scales can help to incorporate synergy effects that influence higher of lower scale levels; measures at allotment scale can contribute to problems and/or solutions at streetscape scale. For example, the paving of private gardens has a higher-level impact on the water storage capacity of quarters. Also, the lack of maintenance of flood defence systems at regional level can increase a city's vulnerability to floods.

Cities are the physical manifestation of networks with large numbers of interactions between agents (persons and organisations) in a relatively small area. Many processes in a city are not centrally steered, but self-organised. As a consequence, system behaviour at the macro level cannot be explained by the aggregate of elements at the micro level. This phenomenon is often referred to as *emergent* behaviour. Synergy from emergent behaviour creates high levels of robustness and stability of system states, but also leads to the inability to adjust running processes in, for example, urban development. In order to deal with changing conditions and the slow dynamics of climate change and its effects on cities, long-term planning and policy are required. Urban development and re-development should therefore not only meet present requirements, but should also take into account future change. Long-term visions of the system are crucial for optimal timing of investments in order to establish maximum cost efficiency. Insights in differences in the dynamics of change of the physical environment and climate change can be used to adapt cities to climate change. For example, to use the autonomous process of urban renewal to adapt to changed climate conditions or to speed up urban renewal by reducing the life cycles of buildings.

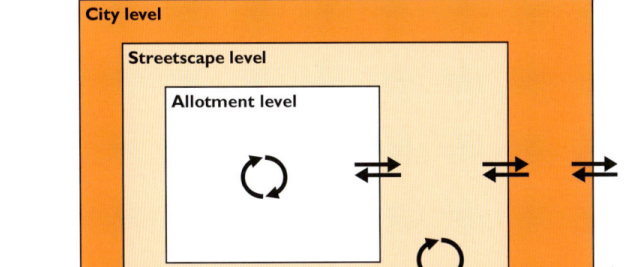

FIGURE 1-19
Spatial scales and interaction within and between sub and supra systems

Source: Chris Zevenbergen, 2008.

TEXTBOX 1-7

The advice of the Dutch Delta Commission 2008

In September 2007, the Government of the Netherlands assigned a Delta Commission to create an integral vision for flood protection and spatial development to prepare the low-lying county for the impacts of climate change and to prevent flood disasters in the coming centuries (2100–2200). A year later in 2008, the Delta Commission presented its advice that not only enhanced structural and non-structural flood protection measures, but also the physical and institutional design of the country. The advice was based on a systems approach that integrated different spatial and temporal dimensions in relation to water safety, living, working, agriculture, nature, landscape infrastructure and energy systems in the Netherlands. A risk mitigation approach that reduces the vulnerability of the Netherlands to the changing climate by reducing both its exposure and sensitivity was used as a starting point. To reduce exposure, protection and storage of water resources and a tenfold increase of the flood protection level by higher and stronger flood defence structures is proposed. To reduce sensitivity to climate change, especially the potential damage of flood disasters, measures such as regulation for spatial planning, evacuation routes, shelters and evacuation plans are proposed.

Although the advice aims to adapt to long-term changes, it does not neglect the near future. It proposes to gradually implement measures in line with climate change and other ecological processes in order to create cost efficiency and social value in the short term.

Key Questions

Cities as special cases

1. Why are cities losing their capacity to adapt spatial and demographic changes and how does this affect their ability to anticipate and deal with floods?
2. Historically, natural disasters were viewed as 'Acts of God'. How does this attitude or belief affect flood management practices?
3. Major flood events in the past century have acted as catalysts for changing policies towards floods. What are the most important changes in flood-risk management strategies?

Cities as living organisms

1. What difference does viewing cities as Complex Adaptive Systems imply for flood management?
2. What are the different timescales that affect urban dynamics and development?
3. Does policy follow development or development follow policy?

Vulnerability of urban areas

1. What are the four components of vulnerability and what is the main purpose of each component?
2. Why is it necessary to develop and implement a comprehensive strategy in which all four components are addressed?
3. What is the result of a flood control strategy that is only focused on threshold capacity?

Uncertainty

1. How would you describe the difference between risk and uncertainty?
2. What alternatives are there to engineering solutions to cope with flooding?
3. In what way could Social Learning contribute to flood-risk management?

FURTHER READING

Ando & Takahasi. (1997). Water International, 22(4), 245–251.

CRED (2010). The International Disaster database EMDAT, Centre for Research on the Epidemiology of Disasters.

de Bruijn, K.M. (2004). Resilience indicators for flood risk management systems of lowland rivers. *International Journal of River Basin Management*, 2(3), 199–210.

de Graaf, R.E., van de Giesen, N.C. & van de Ven, F.H.M. (2007). The closed city as a strategy to reduce vulnerability of urban areas for climate change. *Water Science & Technology, 56*, 165–173.

de Neufville, R. (2003). Real options: Dealing with uncertainty in systems planning and design. *Integrated Assessment*, 4(1), 26–34.

Folke, C. (2006). Resilience: The emergence of a perspective for social–ecological systems analyses. Global Environmental Change, 16, 253–267.

Ikeda, T. (2004). Flood management under the climate variability and its future perspective in Japan. *Int. Conf. Climate Change: a challenge or a threat for water management*. IWA. September 27–29, Amsterdam.

Klinke, A. & Renn, O. (2001). Precautionary principle and discursive strategies: Classifying and managing risks. *J Risk Res.*, 4, 159–173.

Kundzewicz, Z., Mata, L., Arnell, N., Doll, P., Jimenez, B., Miller, K., Oki, T., Sen, Z. & Shiklomanov, I. (2008). The implications of projected climate change for freshwater resources and their management. *Hydrological Sciences Journal/Journal des Sciences Hydrologiques*, 53, 3–10.

Milly, P.C.D., Betancourt, J., Falkenmark, M., Hirsch, R.M., Kundzewicz, Z.W., Lettenmaier, D.P. & Stouffer, R.J. (2008). Climate change: Stationarity is dead: Whither water management? *Science*, 319, 573.

Mitchell, J.K. (ed.) (1999). *Crucibles of Hazard: Mega-cities and Disasters in Transition*. Tokyo: United Nations University Press.

Sage, A. & Armstrong, J. (2000). *Introduction to Systems Engineering*. New York: Wiley.

Sheppard, S.C. (2007). *Infill and the Microstructures of Urban Expansion*. Homer Hoyt Advanced Studies Institute, January 12.

Stirling, A. (2003). Risk, uncertainty and precaution: some instrumental implications from the social sciences. In: *Negotiating Environmental Change*, F. Berkhout, M. Leach, I. Scoones (eds), 33–76. Cheltenham, UK: Edward Elgar.

Tucci, C.E.M. (2008). *Urban Flood Management*. WMO Cap-Net publication, APFM.

UN-Habitat (2009). *State of the World's Cities 2008/2009 – Harmonious Cities*. UN-HABITAT. ISBN: 978-92-1-132010-7.

van Asselt, M.B.A. (2000). *Perspectives on Uncertainty and Risk*. Dordrecht: Kluwer Academic Publishers.

Walker, B., Holling, C.S., Carpenter, S.R. & Kinzig, A. (2004). Resilience, adaptability and transformability in social-ecological systems. *Ecology and Society*, 9(2), 5.

Zevenbergen, C., Veerbeek, W., Gersonius, B. & van Herk, S. (2008). Challenges in urban flood management: Traveling across spatial and temporal scales. *Journal of Flood Risk Management*, 1(2), 81–88.

Urban floods

Learning Outcomes

Chapter 1 provided an overview of Integrated Urban Flood Management and suggested that by thinking of urban areas as living organisms we will be better placed to understand the different ways in which they respond to extreme events such as flooding. It also suggested that urban areas will respond in different ways depending on prior knowledge and their particular characteristics. In this chapter we consider in more detail some of the features that are associated with different forms of flooding in urban areas. So the attention now moves away from the urban area to the characterisation of the types of flooding they can experience. In the forthcoming pages you will learn about the following aspects of Urban Flooding:

- the influence of climate on the various causes of flooding and the different modes of flooding;
- the different types of urban flooding and their characterisation;
- how historical hydrological data is not a good guide to future urban hydrology and how climate change can be accommodated.

2.1 THE INFLUENCE OF CLIMATE AND OTHER FACTORS

Climate can be considered as the result of the functioning in the hydrosphere and atmosphere of a complex thermodynamic 'machine'. Uneven heating of the earth's crust and oceans via solar radiation gives rise to temperature gradients on all areas of the planet, depending upon location and natural conditions. The uneven heating of the atmosphere contributes to the generation of convective movements and horizontal gradients producing the winds, the displacement of air masses and the atmospheric circulation.

Once the global atmospheric dynamic is established as the first phase of climatic formation (through its variations in the distribution of energy levels), it is the more localised low pressure areas which cause surges in lakes, seas and oceans and which through resonance can give rise to seiches in inland seas or coastal areas. At the same time, wind energy produces swells, set-ups and currents, resulting in the localised and transient rise of sea level. These sea level rises interact with those caused by solar and lunar gravitation (tides), and can lead to coastal flooding under certain circumstances as well as inducing rises in water levels in river estuaries that can give rise to inland flooding.

2.1 THE INFLUENCE OF CLIMATE AND OTHER FACTORS

Climate weather patterns are unpredictable, and are constantly changing over time and across the surface of the globe, as changes in the water cycle and atmospheric and oceanic dynamics (as well as seas and lakes) are the 'mechanisms' by which thermal balance is regained through an interactive process driven by uneven solar radiation. However this spatial–temporal variability is inherent to a stationary climate. Global climate changes themselves, showing changes in this variability, may be considered a natural occurrence in the history of the planet, both before and after the Central American isthmus closed around 10 million years ago and the current climatic pattern is considered to have been established. Current human influence, however, has the potential to effect changes in the climate.

The first worldwide consequence of climate change that normally comes to mind is sea level rise caused by the addition of large volumes of water previously held in polar icecaps. Sometimes a connection is extended to its impact on floods and droughts, though not based on Arrhenius's^1 arguments (1896) precisely. But the changes in the hydrosphere that accompany climate go beyond sea level and hydrological phenomena, even if the first to be observed, and the most significant for a long time in determining climate changes, was the fluctuating sea level. As our knowledge of the nature of climate and of its changes, along with its causes, processes and consequences, continues to advance, other observable parameters of increasing significance are also becoming better known.

¹Svante August Arrhenius (1859–1927) was a Swedish scientist, originally a physicist, but often referred to as a chemist, and one of the founders of the science of physical chemistry. The Arrhenius equation is a simple, but remarkably accurate, formula for the temperature dependence of the rate constant, and therefore, rate of a chemical reaction.

Thermodynamic changes in the global water cycle allow for the humidification of the atmosphere through the evaporation of surface water, and its precipitation on the surface of any area of the planet through condensation in the form of rain, snow or ice. Atmospheric dynamics determine, in turn, the spatial localisation of this precipitation as masses of air varying in humidity and temperature are displaced until processes of condensation are generated.

FIGURE 2-1
Flooding in Venice

Source: P.A. Piazelli, 2009.

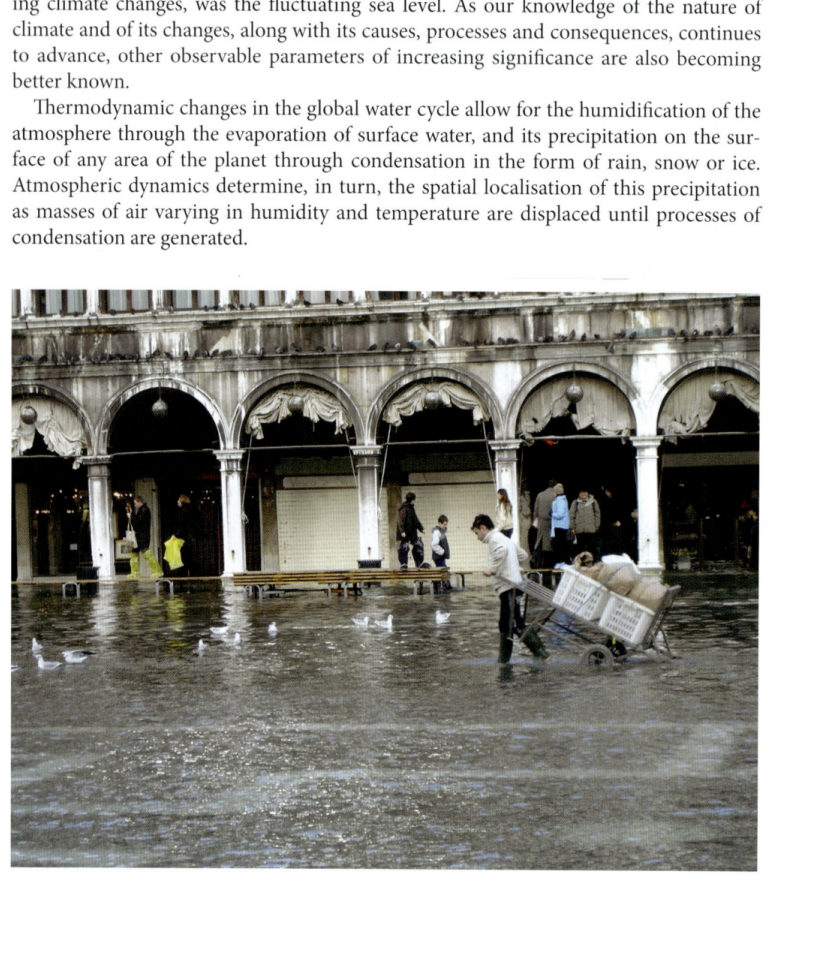

Continental precipitations start the hydrologic cycle, resulting in soil infiltration and excesses in the form of surface runoff. These actions together lead to the formation of a catchment drainage network which in turn gives way to processes of erosion and sediment transport, creating a landscape erosion or inverted relief. The river's slope decreases over the final stretch, slowing the flow of water so allowing the sedimentation of suspended solids, generating floodplains. In near coastal areas, rivers build littoral plains in which the subterranean flow of water can reach values comparable to the sub-aerial one, most strictly 'fluvial'. Levels near the sea and estuaries allow sedimentation of solids, producing alluvial and coastal plains in which the flow of subterranean water can reach levels comparable to strictly backshore rivers.

Climate also influences water levels in the seas and oceans, affecting the sea levels in coastal areas. Coastal plains may flood, altering sub-surface topography and estuaries, and break or realign berms and other coastline features. The sea (or lake) level, controlling the mouth of the main channel of the basin's drainage system, is ultimately what governs estuarine water levels, the structure of the channels, the extent of floodplains and the evolution of the river's profile.

In the Tropics especially, intense precipitation occurs frequently under atmospheric conditions of relative low pressure and/or with winds from the sea (hurricanes, extra-tropical cyclones, monsoons). This also raises sea levels in the form of storm surges, and the wind blowing from the sea or accompanying the Coriolis force causes additional set-up (i.e. tides). Under these conditions, flows from the landward side resulting from the intense precipitation events are exacerbated and accentuated by the rise of the drainage flow base level.

On a much longer timescale the structure and dynamics of plate tectonics of the Earth and its crust come into play. Processes such as isostatic compensation resulting from the release of pressure of ice caps on continental land masses and subduction processes at the margins of crustal plates may influence sea–land levels. Climate-induced changes and movements of the crust affecting the relative sea level—the base level—causes river profiles to be in a state of ongoing adjustment. A temporary rise in sea level may act as a dam at the mouth, blocking river drainage and causing a rise in the water level on the river, its floodplain and the surrounding areas.

2.1.1 Floods as natural phenomena

The continuous movement of water above, over and below the surface of the Earth is generally referred to as the 'Hydrological Cycle' and is governed by the mass–balance principle. The changes that characterise the hydrological cycle that are of particular relevance to flooding may be described as follows:

Precipitation (P) = Runoff (Q) + Evaporation (EVA) +/– Storage (S)

In this formulation floods may be defined as natural physical phenomena. Runoff and floods are themselves influenced by geomorphological, spatial and social factors. Floods are associated with natural abnormal events often referred to as extreme events that are area specific, whether the area is highly localised or extensive. The common factor of such areas is that they are associated with catchment areas and river basins. Such areas can be rural, urban or a combination of both. The idea of floods as disasters though is an anthropogenic concept, they are a disaster because of the adverse and unlooked for impacts that they have on the functioning of human society. Floods are therefore a natural phenomena and it is a human construct to label flooding as being acceptable or not.

One of the most important factors affecting this is the discharge rate of open channels such as rivers and streams and is correlated with the maximum discharge that can be routed through without causing out of channel flow within a catchment basin. The flow of water within a river catchment basin is mainly dominated by the infiltration ratio— the amount of water absorbed by the soil and which, as a consequence, acts as a barrier for any further downward water movement within the soil. Several factors determine the infiltration ratio:

- soil saturation: correlated with the height of the water table in the soil;
- vegetation cover: the type and extent of cover may act as a natural flow barrier, influencing infiltration and transpiration as well as surface flow velocities;
- soil type: affects the absorptive capacity, clay soils for example are generally impermeable;
- frozen ground: only small amounts of water can penetrate frozen ground;
- human construction: can 'harden' natural surfaces, preventing infiltration and increasing runoff.

Within this overall hydrological cycle flooding can have different causes:

- Fluvial flooding caused by an increase in the concentration of runoff from a catchment in a river system such that the discharge capacity of the river channels is exceeded, necessitating an expansion of the river channel (e.g. by overflow into floodplains). In urban areas this can be intensified by the occupation of the floodplain or the increased rigidity of the channel. A consequence being the possible transference and intensification of the problem downstream (see Section 5.3).
- Pluvial flooding caused principally by rain water precipitation of an intensity and duration that exceeds the infiltration (including drainage capacity), interception and evaporation capacity of the affected area to absorb that rainfall, which can be intensified in urban areas because of an increase in runoff compared to filtration (see Section 5.4).
- Flooding caused by the emergence of underground and sub-basin waters (groundwater flooding) resulting from the saturation of the soil and subsoil from intense and prolonged periods of infiltration without sufficient drainage. In urban areas this can be accentuated by drainage infrastructures. Another typical example is groundwater flooding due to infiltration of flood water from the river into the underground. This infiltration produce groundwater flood waves, especially in the hinterland of dykes, which can cause substantial flooding of the hinterland.
- Flooding caused by the elevation of the base drainage level, typical of sea and lake coastal zones where tides, surges or other causes can lead to increased flooding. In those areas which have been urbanised, such flooding can be accentuated by the reduction of the filtration capacity of the natural soil.
- Flooding caused by the overtopping or failure of flood defence, protection or attenuation structures (see Section 5.5).

Two or more types of floods may occur simultaneously, which is known as joint flooding. An example is monsoonal flooding in the Indian sub-continent, where rains on the extensive coastal plains, (in structurally deltaic–coastal areas such as Kolkata) intensify

FIGURE 2-2
People fleeing floods in Sri Lanka

Source: Trokilnochchi, 2008.

the flooding caused by melting snow from the Himalayas. Drainage is influenced by the relatively low sea level, increased by the same winds and low pressure that cause the seasonal rains. Other examples are other low-pressure storms such as hurricanes and typhoons on other coastal areas of the world.

Urban flooding can vary greatly, depending on the location of the urban area:

- Urban areas located in coastal areas could be subjected to any one or a combination of the different modes of flooding outlined above, depending on the combination of climatic, atmospheric and maritime conditions.
- The situation of urban areas located on lakeshores, depending on the size of the lake, may be subject to similar modes of flooding as coastal areas but with less variability (e.g. Chicago and Venice).
- On a river channel but with an attenuated influence of sea level. The circumstances and conditions of flooding depend on topographical factors, such as flattish terrain, which leads to flooding primarily as a result of overflowing and the inundation of property and people. In the case of steep hillsides, damage to property and washing away may occur. In both cases debris and material being swept along introduces an additional risk factor.
- In poorly drained or impermeable basins, flooding results from direct rainfall or from a rise in the water table. This is influenced by the degree of urbanisation, the remaining percentage of natural surface, and the dimensions and distribution of buildings that affect the runoff-damming ratio.

The type and extent of flood damage will vary depending on the location and characteristics of the urban area, the nature and duration of the events leading up to and contributing to the flooding, the degree of preparedness and responses to the events.

TEXTBOX 2-1

Case of study: The effects of maritime climate on pluvial and fluvial flooding: Hurricanes Ike and Katrina

The effects of Hurricane Katrina (2004) were extraordinarily devastating in New Orleans, located on the Gulf of Mexico at the mouth of the Mississippi River. The eye of the hurricane virtually passed right over the middle of the city, making the storm surge greatest facing the coast and at the very mouth of the river. Relatively speaking, however, the storm surge was not as significant as it was 100 miles to the east where set-up was strongest (Figure 2.3). All things considered, due to winds that had begun days before, wave run-up was particularly relevant, hitting the city's coastline and Lake Pontchartrain with maximum force. Flooding occurred when levees overflowed and were breached.

Although the effects of Ike (2009) were also devastating, the New Orleans area was affected to a lesser degree. It made landfall more to the west, affecting Galveston, making it the Texas coastal area which suffered the greatest storm surge, which was less serious in New Orleans and at the mouth of the Mississippi River, and weaker than under Katrina. But the greater wave set-up and run-up were experienced here, making the water level at the mouth of the Mississippi higher and more sustained under Ike.

Storm surges at the mouth of the river generate a higher profile of river balance where water moves upstream in the form of a backwater effect, in such a way that it affects the balance profiles of tributaries and their drainage networks. Both hurricanes rapidly lost strength as they made landfall, but they each fused with extra-tropical cyclones over the Mississippi Valley, sustaining their storm activity up to the Great Lakes region as storm patterns typical of these latitudes. The heavy rains throughout the valley made it more difficult for the tributaries to drain into the great river. The flooding in its basin was due, then, to both this drainage difficulty and to the actual rain and runoff. This effect was more significant after the passing of Hurricane Ike, whose flooding in cities like Cincinnati and Pittsburgh, located on the Ohio River, and Columbus, on one of its tributaries, was extraordinary

FIGURE 2-3
Hurricanes Katrina and Ike

Source: Nifanion, 2005.

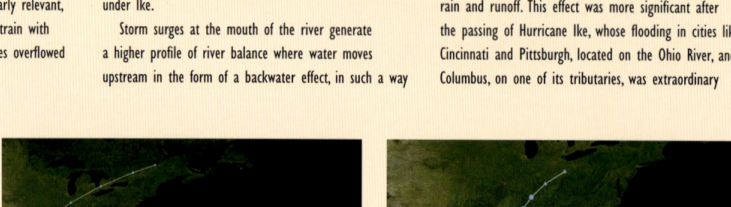

2.2 TYPES OF FLOODING

Urban areas are found in a wide range of geographic locations and because of this they are subject to all conceivable forms of flooding. We take as the starting point that there are two basic types of flooding:

1. inland, which is driven by rainfall;
2. coastal, which is driven by a combination of the sea and/or rainfall, but which only occurs at times of extremely elevated sea levels. At all other times flooding in coastal areas is caused by inland flooding.

Inland flooding can be sub-divided into categories which are dependent on the reaction between rainfall and the earth; these categories can interact with each other, building up a complex picture of causes of flooding and flood risk. These are shown in Table 2-1. However, the different types of flooding rarely act entirely independently from each other and there are many combinations in which they can act.

For inland areas a typology of urban rivers and watercourses can be suggested, each of which causes different problems associated with flooding:

- major rivers along which cities, towns and other settled areas have grown up;
- upland rivers which have relatively steep watercourses as they enter urban areas;

TABLE 2-1
Types and causes of flooding (inland and coastal)

System	Flood type	Component	Comment
Inland	Overland	Rural green space	Rural green space draining away from urban area
		Peri-urban green space	Rural green space draining into urban area
		Intra-urban green space	Green space within urban area
		Developed urban surface	Buildings, structures, highways etc.
	Groundwater	Interstitial	Saturation of topsoil and shallow drift material
		Deep	Water from rock aquifers and deep drift material
	Fluvial	Stream capacity exceedance	Small watercourses with rapid response times
		River capacity exceedance	Large watercourses with slow response times
	Major system incidents	Groundwater pumping discontinuation	Failure of groundwater pumping
		Reservoir dam breaches structure	Dam structure fails
		Reservoir dam breaches spillway	Dam spillway fails
		Canal breaches	Canal bank fails or overtops
		Land drainage PS failure	Failure of land drainage pumping
		Channel blockage	Blockage in stream or river channel
		Channel collapse	Collapse of stream or river bank
		Bridge or culvert collapse	Collapse of structure over stream or river
		Bridge or culvert blockage	Blockage within structure over stream or river
		Bridge or culvert capacity	Capacity of stream or river structure capacity
		Defence collapse	Defence fails
	Minor system	Combined sewer capacity exceedance	Flooding of urban surfaces during extreme rainfall
		Surface water sewer capacity exceedance	Flooding of urban surfaces during extreme rainfall

2.2 TYPES OF FLOODING

		SUDS/Source control capacity exceedance	Flooding of urban surfaces during extreme rainfall
		Foul sewer capacity exceedance	Overloading due to wrong connection of surface water
		Pipe drain capacity exceedance	Flooding of urban surfaces during extreme rainfall
		Open drain capacity exceedance	Flooding of urban surfaces during extreme rainfall
	Minor system incidents	Combined sewer blockage	Blockage resulting in reduced capacity
		Combined sewer collapse	Collapse resulting in reduced capacity
		Surface water sewer or drain blockage	Blockage resulting in reduced capacity
		Surface water sewer or drain collapse	Collapse resulting in reduced capacity
		Foul sewer blockage	Blockage resulting in reduced capacity
		Foul sewer collapse	Collapse resulting in reduced capacity
		Sewer pump station failure	Failure of sewer pumping
		Sewage treatment works	Failure within or overloading of SWT
		Water main burst	Water main burst causes flooding and/or failure of other minor system components
Coastal	Estuarial/Delta	Sea level	Global temperature driven
		Tide	Astronomically driven
		Surge	Weather driven
	Open sea	Sea level	Global temperature driven
		Tide	Astronomically driven
		Surge	Weather driven
	Incidents	Tsunami	Earthquake
		Defence collapse	Defence fails

- urban watercourses which are now entirely within urban conurbations;
- artificial urban drainage networks including sewers some of which might formerly have been natural watercourses.

All of these situations can give rise to flooding in rural areas as well as urban ones, but other causes of rural flooding which can affect urban areas include:

- surcharges in water levels due to natural or man-made obstructions in the flood path (bridges, gated spillways, weirs);
- dam failure;

FIGURE 2-4
Flooding in Slovenia (2007)

Source: Mitja Brilly, 2007.

- landslides and mud flows;
- inappropriate urban development (excessive encroachment in the floodway);
- ice jams;
- rapid snowmelt;
- deforestation of catchment areas.

Whilst rural floods tend to be 'river basin' events, urban floods can be both of localised origin as well as area wide. Furthermore, as catchment areas become urbanised and increasingly managed through 'hard engineering', so the number of flash floods tends to increase. Furthermore, whilst water pollution is a common problem, it tends to be more serious in urban areas, which have more diverse sources of pollution.

Overflowing rivers are not the only cause of urban flooding; they can also be caused by intense rainfall events over the urbanised areas combined with inappropriate sewer systems and diverse urban land cover. Generally though, flooding here originates from extremely high flows in major rivers as a result of severe regional meteorological events, or from local high intensity thunderstorms. Thus, management of urban floods requires both the knowledge of the physical characteristics of specific flood events and an understanding of urban hydro-meteorological issues. Urban flooding therefore requires special attention due to its sheer complexity, as it incorporates a host of social, economic, institutional and technical factors within both rural and urban environments.

The various components of flow, shown in Figure 2-5 are discussed in the following sections.

2.2.1 Surface runoff

In an urban area the management of surface runoff has been highlighted as a key driver for flood risk. Surface runoff is not only generated by flow that is in excess of the rate of infiltration, but also on saturated topsoil layers and on water bodies (lakes, rivers and streams) as saturation flow. Subsurface runoff can be generated by the rapid through flow of newly infiltrated storm water within macro-pores and soil pipes feeding directly into the stream flow. In the case of saturation of the soil matrix, pre-storm soil water returns to the surface. This return runoff mainly occurs close to rivers and is significant on concave hill slopes and in wide, flat river valleys.

2.2.2 Subsurface flow

Interflow forms in the unsaturated soil layer and flows more or less directly through this soil layer to the river. In urban areas, interflow is reduced by various man-made environmental changes. First, because the sealing of the surface reduces the infiltration of the precipitation, less free water is available in the unsaturated soil layer. Furthermore, much of the original natural soil layer may have been removed or have been consolidated by construction activities. As a result of their origin and uses, urban soils may contain pollutants that can find their way into interflow and groundwater and thus into rivers.

2.2.3 Groundwater flow

Groundwater is subsurface water which totally fills the pores and cavities of the saturated zone. Between 60 and 80 per cent of the long-term flow volume in rivers is a result

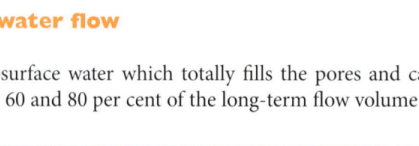

FIGURE 2-5
A stream hydrograph

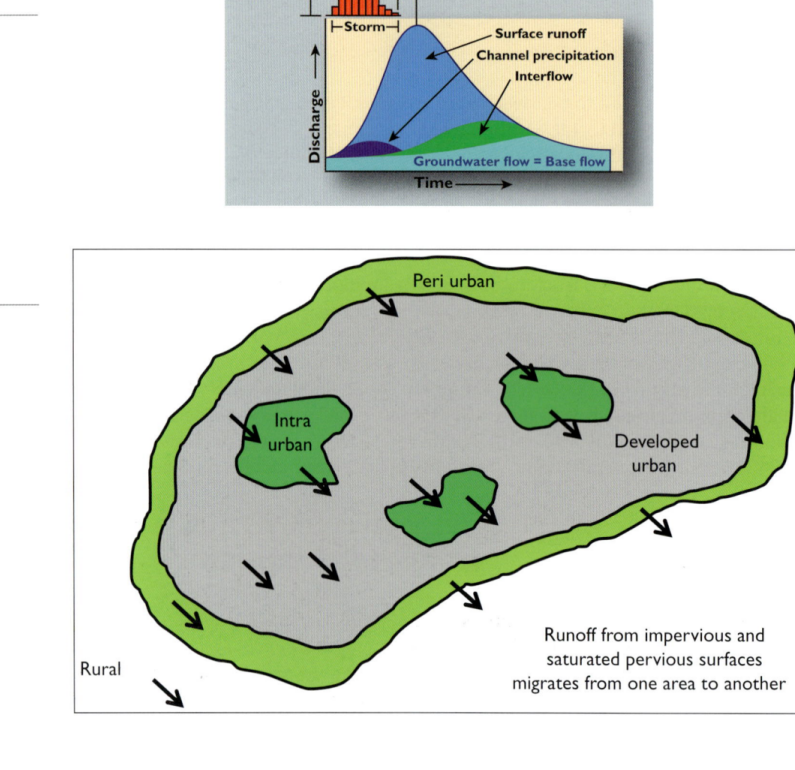

FIGURE 2-6
Flooding from surface runoff

of groundwater discharge. Compared to interflow and surface runoff, this flow process is much slower and it can take years before the infiltrated precipitation water can reach the river. The outflow of groundwater into rivers persists even during long periods of no rain. During floods, rapid changes in groundwater outflow can be observed, which are not due to lateral flow processes through the aquifer. In confined aquifers a distant rise of the groundwater table through infiltrated precipitation water can increase the groundwater pressure close to the river with the new water pushing out the old water. Old groundwater outflows along the river, while new groundwater enters distant aquifer regions.

Probably the main characteristic of groundwater flooding is long duration (weeks and months), which is controlled by the groundwater hydraulic conditions and aquifer properties. Here, two types of aquifers can be distinguished: aquifers formed in intergranular porosity (e.g. in sands or gravels) and aquifers formed in rock fractures, fissures and caverns (e.g. carbonate rocks, such as limestone). The latter represents an extremely complex system of conduits, where the interaction of recharge process and discharge is very difficult to describe physically. Generally, the rise of groundwater level in fractures is significantly faster than it is for intergranular porosity. At the surface, water is released through springs, which in the vertical plane represent an intersection point of the rock fracture and the surface.

The aquifer recharge process is of great importance in understanding how floods are generated. Natural recharge mainly occurs through percolation of water from the surface through the unsaturated zone of the soil, or/and by the leakage through semi-pervious soil layers between two aquifers. In urban areas, an additional source of aquifer recharge occurs through leakage from water distribution and sewage systems.

The inherent uncertainty of groundwater distribution and a time and space scale of groundwater flow and recharge processes make prediction modelling of groundwater floods extremely difficult. Such models should integrate all relevant processes, such as flow in saturated and unsaturated media, interaction of surface streams and groundwater, runoff generation, surface flow, groundwater interaction with sewage and water distributions systems, etc. Recently, a few models have been proposed for the simulation of groundwater regimes in urban areas with those capabilities.

Like interflow, groundwater in urban areas is affected by land surface changes. Surface sealing reduces groundwater recharge. Thus the overall amount of groundwater in urban watersheds declines and the groundwater table is lowered, reducing the groundwater discharge to rivers. In dry periods, rivers draining urban catchments have lower runoff than rural streams of comparable size and at times may become totally dry. The characteristics of groundwater base flow may have a significant influence on the quality of urban surface watercourses. Transport of dissolved phase contaminants from the aquifer to the river will take place across the groundwater/surface water interface where processes are governed by the rapid change in physical and chemical conditions.

2.2.4 Flow processes

Rainfall runoff finds its way into water bodies and watercourses such as rivers, pipes and channels. Within these elements that transport water from one location to another, the flow that is runoff is subject to translation and retention. Translation corresponds to the travel time of the water, which is dependent on the flow velocity and the length of the watercourse. The flow velocity is influenced by the gradient of the bed of the watercourse, the water depth and the flow resistance.

Retention is caused by temporary storage of water within the watercourse. This occurs either on the rising stage, when the river itself is filled with water, or when

FIGURE 2-7
Schematic illustration of groundwater flooding occurrence for two types of aquifer porosity: a) intergranular porosity, and b) aquifer formed in fractures and caverns

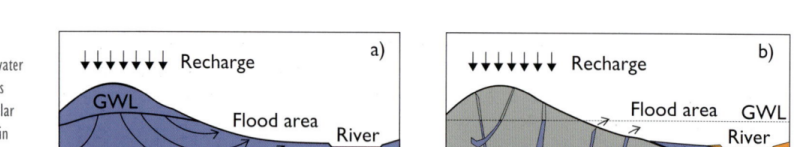

FIGURE 2-8
Flooding from groundwater

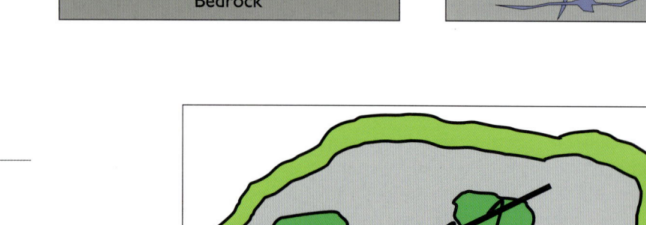

overbank flow occurs, inundating flood plains. Especially in natural rivers, the retention has a strong effect on the propagation of the flood wave. Here wide floodplains, meandering riverbeds and wooden vegetation cause high flow resistance and vast areas of inundation. They have a positive effect on the retention at flood by dampening the flood wave and attenuating its peak. However, they are very complex and not fully understood.

Retention is only of minor relevance in stormwater pipes and drainage channels, where compact cross sections produce a nearly constant flow velocity. In addition, the travel time of the water is reduced by a straight channel and a smooth surface along the wetted perimeter. This acceleration of flow reduces the translation time with the effect that flood waves, especially in stormwater pipe networks and channels, reach the downstream end of their system without attenuation and much faster than flood waves in natural rivers. The more catchments are developed and drained by sewers and channels the more the flood waves will overlap with the negative effect of increasing the flood peaks in the central draining rivers. One of the differences between flooding from natural watersheds and urban areas is that in urban areas they overload the pipe networks, flooding streets, cellars and the low-lying parts of urban areas. The flooded area, however, is restricted to the area surrounding the pipes and seldom comprises extensive urban areas. In natural watersheds, long-lasting rain events with high total precipitation volumes but relatively low intensity cause the most critical floods. The larger the river basin, the longer the flood wave lasts. For large rivers, the floods may last several weeks.

Due to these distinctive differences in the flood waves, their accumulation in the river is hard to determine. Therefore it is difficult to assess the effect on flood attenuation in the receiving urban rivers of retention measures in urban drainage networks, such as detention in green roofs, ponds and reservoirs, and infiltration to the subsurface through drains, porous pavements and ponds. In extreme floods, only reservoirs that retain large volumes of floodwater clearly exhibit an impact on flood magnitude.

2.2.5 Causes of flooding

The different types of watercourses, runoff regimes, flow processes and land uses all combine with physical geography to create various causes of flooding. Broadly speaking, flood events affecting urban areas stem from six main causes:

1. in mountain areas, the effects of heavy, warm spring rains on winter snow causing sudden melting;
2. regional weather systems that are blocked by high pressure systems and produce widespread heavy rain over large sections of major river catchments;

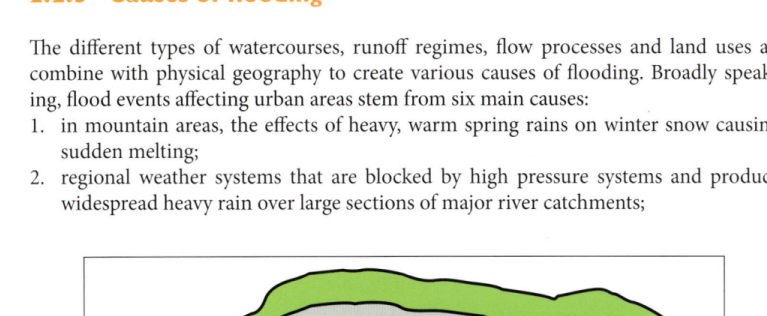

FIGURE 2-9
Flooding from small watercourses

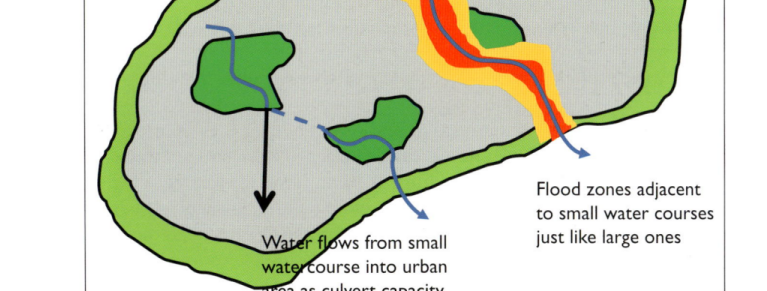

FIGURE 2-10
Flooding from minor drainage systems

3. flash floods in hilly and mountainous regions;
4. short duration, high intensity thunderstorm-driven local flooding on small streams entirely within the urban areas;
5. sewer flooding associated with blocked sewer overflows to larger rivers or surcharging through manhole covers;
6. groundwater flooding can be a result of prolonged heavy rainfall in certain geological conditions.

Regional widespread flooding may result from prolonged heavy rain, saturating the ground and reducing its permeability. Under these conditions, any significant rainfall event will give rise to immediate increases in runoff and can lead to widespread flooding which will also affect urban areas. Streams that drain water from upland areas and then descend into urban areas often cause serious damage to urban infrastructure. However, flash floods are not evenly distributed in time and space. Heavy rains and long duration rainfall may cause landslides and flash floods and the runoff from small catchments may exceed periodic maximum discharges.

Cities experience the impacts of occasional extremely high intensity thunderstorm rains. These events are predicted to grow in frequency due to increased development and the intensification of storm events associated with the effects of climate change. They are also especially difficult to defend against effectively, as they can occur within a very short time period and there is little time to issue warnings or enact predefined management strategies.

Flooding from sewers occurs when the rivers into which storm waters overflow are so high that the stormwater cannot escape or when the sewers are not large enough to cope with all the water flowing into them from upstream. Quite commonly, low-lying urban areas close to rivers suffer from the overflowing of sewers, rather than the overtopping of riverbanks.

Under certain geological conditions, groundwater flooding may be a real problem, especially where groundwater levels fluctuate widely. Unlike surface water flooding events, which are usually of relatively short duration, groundwater flooding may last several weeks or even months. Such flooding may follow an unusually wet period when water tables rise so much that flows from normally dry springs and winterbournes inundate roads and overwhelm all drainage systems.

2.2.6 Causes and effects of urban pluvial flooding

Pluvial flooding results from intense, localised, often short duration rainfall events and consequent surface water runoff over impermeable surfaces which overload the hydraulic capacity of the urban drainage system. The underlying geology and soil type affect the quantity and the rate of infiltration. Urbanisation changes the natural drainage regime, thus increasing the risk of flooding in depressions and other areas where surface water runoff naturally collects.

Surface water runoff is affected in three ways, by:

- increasing runoff volumes due to reduced rainwater infiltration and evapotranspiration;
- increasing the speed of runoff, due to hydraulic improvements of conveyance channels, and;
- reducing the catchment response time by restricting absorption and interception, and thereby increasing the maximum rainfall intensity causing the peak discharge.

TEXTBOX 2-2
Case study: The hydrological and flood situation in Heywood, Greater Manchester, UK

This case study covers pluvial flooding in the Heywood area of Rochdale, an old industrial town to the north-west of the Greater Manchester conurbation, UK. Located at 53° 35' N and 2° 17' W and at an elevation of around 130 metres above mean sea level, Heywood is in a storm-prone area immediately south and west of the Pennine Hills (the first uplands in England in line with westerly air-streams from across the Atlantic Ocean). The undulating terrain on fluvio-glacial deposits left by the retreating Devensian ice-sheets is drained by two streams flowing into the River Roch from south to north across an urbanised catchment of about 8 km^2 (Figure 2-11). Although the main urban development took place between 1750 and 1900, since 1960 many brownfield sites, both within the town and on it southern margins, have been occupied by new housing and new low-rise, large warehouses at a new distribution centre (Figure 2-12). Heywood is still primarily drained by combined sewers, the utility company having inherited an antiquated system on privatisation in 1989. Many of the privately owned sewers in Heywood are still unknown to the utility company and Rochdale Metropolitan Borough Council (MBC) reports that illegal sewage attachments may still be being made from new developments to old sewers.

Climate change in North West England will result in increased temperatures, more winter rainfall, higher wind speeds, fewer winter frosts, perhaps more variable weather, higher sea levels and perhaps more stormy weather and higher wave heights (Shackley *et al.*, 1998). Seemingly paradoxically, computer models predict an increase in *intense* summer rainfall with global warming across northern Europe and this is likely to result in major pluvial floods in urban areas. Instead of the reduced amount of rain being spread across summer months, there will be a tendency for this precipitation to clump into extreme weather events (Christensen and Christensen, 2003). At present, the Heywood area experiences heavy rainfall from two types of rain event, large widespread storms that produced prolonged rain for many hours, at rates of around 10 mm per hour, particularly in winter, from westerly depressions that move across the Atlantic into western Britain; and short duration, high intensity, summer thunderstorms that may produce intensities of the order of 20 mm in 15 minutes and 50 mm in 1 to 2 hours. However, Heywood had no previous record of flooding affecting property until summer storms on 3 August 2004 and 2 July 2006, which resulted in severe sewer flooding in six distinct and identical areas on both occasions. The 2006 event resulted in over 200 homes being flooded with up to 900 mm of sewage-contaminated water for up to 3 hours and around 90 properties had to be evacuated for varying periods whilst renovation was taking place. All six areas that experienced flooding are located along the two streams which were, to a large part, turned into culverts soon after the area was developed (Figure 2-13) and some reaches of these streams are still to this day part of a combined sewer system. Neither of these streams is included in the EA's *Register of Critical Ordinary Watercourses*, which means that the flood events are outside the remit of the EA, and fall under the auspices of various departments within the Local Authority, the utility company and riparian landowners.

This case study demonstrates that the risk of urban pluvial flooding is intensified by reducing natural infiltration, turning local streams into culverts, climate change and fragmented institutional responsibility.

FIGURE 2-11
Terrain map of Heywood

Source: Nigel Lawson, 2009.

2.2 TYPES OF FLOODING

FIGURE 2-12
Land-use change (infill) in the Wrigley Brook catchment Heywood, 1968 (black) to 2007 (red)

Source: Nigel Lawson, 2009.

FIGURE 2-13
Wrigley Brook and Millers Brook Heywood and location of 2004 and 2006 floods

Source: Nigel Lawson, 2009.

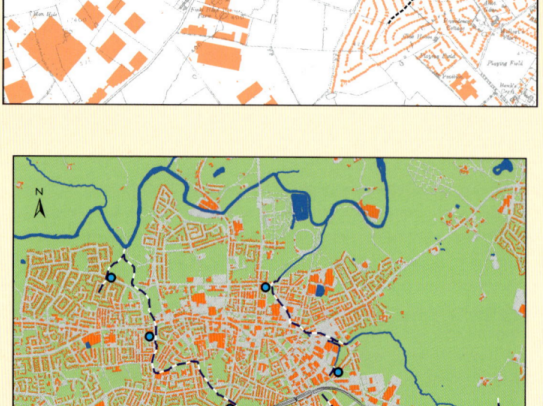

The key element is the sealing of the ground surface, or the extent of the impermeable area. Nearly 100 per cent of net precipitation (that not lost by evaporation) will runoff from impermeable areas and the extent of sealing of the surface varies considerably within urban areas. A further contributor to urban pluvial flooding in many cities is turning urban streams into culverts, which then become part of the sewerage system. Creating culverts removes the possibility of floodwaters spilling on to natural floodplains. Riordan estimates that urbanisation increases the mean annual flood from 1.8 to 8 times, and the 100-year flood from 1.8 to 3.8 times. Typically, the economic cost in renovating a house after an internal sewer-infected flood is between 10,000 and 50,000 Euros (see also Chapter 6).

TEXTBOX 2-3

Groundwater flood-prone areas: an example of Pančevački rit region in Serbia

Pančevački rit is a region north of the Serbian capital Belgrade, on the left bank of the river Danube. The region is surrounded by water bodies: the Danube on the west and south, the river Tamiš on the east and the channel Karaš on the north side. The region is about 40,000 ha, and it is partially urbanised (30 per cent), including the north suburbs of Belgrade. The remaining area is agricultural. The whole area is relatively flat, and is surrounded by 90 km of levees.

The aquifer of interest comprises layers of gravels and sandy gravels (up to 5 m thick), sands (10–25 m thick, that reach the surface in several places) and silty sands with clay (up to 10 m thick). The aquifer is mainly confined by a surface layer of silt and clay

(up to 5 m thick). Since it is hydraulically connected with rivers, the groundwater regime depends highly on the river stage. The Danube water level is mainly influenced by the downstream Djerdap dam. Its effect on the water-level duration curve in comparison with natural conditions is remarkable, as shown in (Figure 2-14). About 30 per cent of the area is currently below the average water level of the Danube.

Groundwater and runoff is collected by a huge system of drainage channels, 816 km in length, or 25 m/ha. Collected water is pumped into rivers at five location, with a total installed capacity of 34 m^3/s. Nevertheless, groundwater flooding occurs frequently. Moreover, water management becomes even greater challenge because of rapid, mainly uncontrolled, urbanisation.

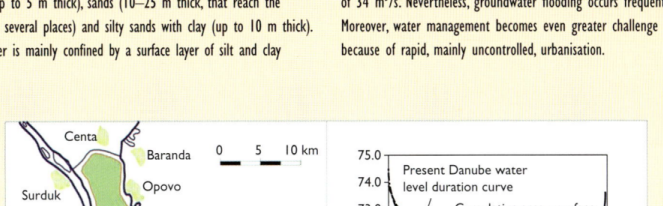

FIGURE 2-14
a) Pančevački rit region location (green), b) diagram showing relation between Danube water level duration and cumulative surface level curve, for natural conditions and after Djerdap dam construction

Source: N. Jaćimović, 2009.

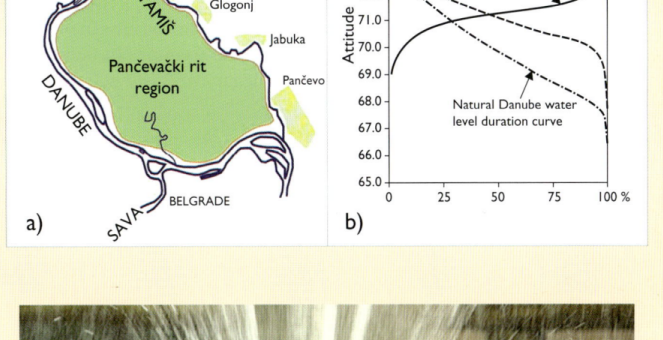

FIGURE 2-15
Piezometer overflow at the south of Pančevački rit region, implying groundwater head above the surface level

Source: Marcin Stępień, 2008.

2.3 PITFALLS IN USING THE HISTORICAL RECORD (OR 'STATIONARITY IS DEAD')

A key issue at present is how to predict future flood drivers under the uncertainty that exists about future climate change and variability. Stationarity is the central principle around which the analysis of hydrological time series has been built, but this is now being questioned. In simple terms it means that the statistical properties of a part of a time series does not differ from those of another part—in other words, properties do not change with time. No time series in the natural world is stationary in the strict sense. For example a rainfall time series will definitely show a seasonal change of properties and in some climates a diurnal cycle as well. However, taken over a number of years (usually a century or longer) natural time series like rainfall and temperature were assumed to be stationary. There are long-range trends, but in many historical time series these were not very pronounced could be removed without major impacts on the analysis of the results. However, due to the rapid change of the climatic system of the earth due to anthropogenic influences like global warming, the trends appearing on the forcing variables of urban hydrology (e.g. rainfall, temperature, incident solar radiation) can no longer be neglected.

Marked non-stationarities that have already appeared in the hydrological time series and the fact that these are likely to get larger in the future has forced urban hydrologists to consider revising the traditional analytical techniques.

In practice, time series encountered in hydrology and in geophysics in general are made of two types of variation, namely deterministic and stochastic. The deterministic variation comprises periodic and trend components. For example, in the case of meteorological variables like rainfall or temperature, diurnal cycles are periodic. Stationary time series have negligible trends.

2.3.1 Ergodicity assumption

All the principles of time series analysis are strictly valid only for an ensemble of all possible realisations of a process. In practice this is impossible—hydrologists often have a single realisation of a process (e.g. a rainfall time series). Ergodicity is the principle

---

TEXTBOX 2-4

Theory probability density function

If the probability density function, $P(X = x) = f(x)$, can be used to calculate the probability of the variable X taking values between x and $x + \Delta x$ as

$$P(x < X < x + \Delta x) = \int_x^{x+\Delta x} f(x) dx \tag{2-1}$$

The variable x is given the special name a process, if its value varies with another (independent) variable, e.g. $x(t)$ is a time series process where t is time. Then the probability density definition becomes $P(X = x(t)) = f(x(t))$. For a stationary process, the underlying distributions are not dependant on time, or:

$f(x(t_1)) = f(x(t_2)) = f(x(t_3)) = .. = f(x(t_n)) = f(x)$, implying that the statistics like

mean and variance are not functions of time. Under some assumptions, we use the following results as a proxy for stationarity.

$$\mu(t) = \mu$$
$$\sigma(t) = \sigma$$
$$\text{cov}(t_1, t_2) = \text{cov}(t_2 - t_1) \tag{2-2}$$
$$\rho(t_1, t_2) = \rho(t_2 - t_1)$$

Where μ, σ, cov and ρ are the mean, standard deviation (), auto-covariance and autocorrelation, respectively.

FIGURE 2-16
Using stationary theories in a non-stationary future

Source: Assela Pathirana, 2009.

that assumes the properties (e.g. mean, standard deviation, autocorrelation) obtained from a realisation are equivalent to those of the ensemble realisations of the underlying process. This is equivalent to the common assumption in statistics that the population statistics are represented by their sample estimates. The ergodicity principle has important implications for dealing with non-stationarities.

2.3.2 Climatic non-stationarity and dealing with it

Non-stationarity is nothing new for hydrologists and geophysicists. Removal of non-stationarity is a standard step in the process of data analysis. When the random variability of data is much larger in magnitude compared to the trends, the latter can be removed without major implications. The traditional point of view was that the effects of non-stationarities over the planning horizons do not warrant specific models of climatic change, for the other uncertainties like forecasts of populations and water demands could outweigh those of climate. However, faced with the swift changes that are taking place in the earth-atmosphere system due to changes in anthropogenic forcing, it is not possible to ignore the trends in the statistics, at least in the long-range analyses. Trends are not limited to the mean, but could be observed in other statistics like variance. For example, it is possible for the variability of rainfall to increase much more rapidly compared to the increase in mean in a climatic region. Then the latter could be more relevant for the study of extreme events like floods compared to the former.

In the case of small-scale, urban floods, the runoff statistics computations in urban drainage are primarily based on the Intensity-Duration-Frequency (IDF) curves for rainfall observation points (see also Chapter 6.2). IDF curves are essentially the conditional cumulative distributions of extreme rainfall intensity: Each curve is conditioned on frequency of occurrence, often expressed as a 'return period'. The practice is to construct the curves based on annual maximum statistics of a long rainfall series, fitted to an extreme probability model like Gumbel distribution (see Textbox 2-5). The central principle of the construction of IDF curves is that the rainfall observed over a long period of time is stationary in a statistical sense.

In order to apply such a principle to future rainfall modified by climate change, which is by definition non-stationary, a different approach to model fitting is required: In the place of the stationarity of a long time-series, a statistical ensemble of extreme values generated by impact models can be used without violating the mathematical principles on which IDF curves are based. As an example, if the (stochastic) nature of rainfall of a given locality as a result of a scenario at a given point of time in the future can be established, these statistics can be used to derive the IDF curve for that point in time. In theory, the future climatic scenarios provided by methods like GCM simulations with suitable downscaling techniques can provide the statistical basis for achieving this. However, there are many practical issues with this approach that contribute in major ways to increase the uncertainty of the end results, namely the design guidelines specified by IDF curves.

2.3.2.1 Dealing with non-stationarity

In spite of the obvious problems in assuming stationarity in hydro-meteorological analysis in the view of imminent climate change, there are few alternatives for the central principles of hydrologic design like the IDF curve available at the present. Therefore it is important to understand the limitations of traditional theories based on stationarity assumption and to attempt to circumvent the problems imposed by non-stationarity. While the removal of trends is always possible, this defeats the purpose of incorporating climate change effects in the long-range analysis.

Figure 2-16 shows a general framework that can be used to apply stationarity-based techniques (e.g. IDF curves) in a non-negligible non-stationarity. Even if a continuous, high time resolution estimate is available (which is hardly ever the case) spanning over the design horizon of urban floods, the pronounce trends expected in such data makes it quite unsuitable to remove the stationarity. (s in Figure 2-16 is such an estimate.)

Historical data (period a to b) can be used to establish the present estimates for such techniques (IDF curves to be used for current state), provided that the degree of non-stationarity observed in the data is negligible. This requirement essentially restricts the duration of data that can be used in the analysis. The statistical robustness of the estimates will be at risk due to insufficient data—but we'll visit this problem in a moment.

The prediction of the future state invariably involves modelling (either physical or statistical) to predict the behaviour of the measure (rainfall in this example) in the future. (The specific means of doing so is beyond the scope of this chapter, but an illustrative example will be given later.) A window (c to d) in the future can be used to compute the future estimates for the techniques (future IDF curves). This estimate could also suffer

TEXTBOX 2-5
Gumbel and extreme value theory

Extreme value theory is a branch of statistics dealing with the extreme deviations from the median of probability distributions. The *Gumbel* distribution (named after Emil Julius Gumbel (1891–1966)) is used to model the distribution of the maximum (or the minimum) of a number of samples of various distributions. Extreme value theory is important for assessing risk for highly unusual events, such as extreme earthquake, flood or other natural disasters.

from similar drawbacks as the historical analysis, due to insufficient duration of data. The next section discuss a method that can be used to circumvent this problem.

2.3.2.2 Robust statistics by using space for time

The ergodicity assumption (discussed in the theory section) is central to any type of geophysical time series analysis. This is essentially similar to the procedure of selection of a sample to estimate population statistics. Keeping this similarity in mind, it can be argued that the loss of robustness due to shortening the period of analysis (say 10 years instead of 100 years of rainfall record) to avoid non-stationarity, can be overcome by using more than one realisation of a time series. In the case of historical rainfall data this could be a number of rain gauge stations in the close spatial proximity. Often this was not easy to achieve in the past, due to the fact that the rain gauge networks are carefully designed to minimise the probability of having multiple gauges in a same microclimatic zone.

However, today there are very viable means of overcoming this issue of lack of data. For the present case (historical data), there are numerous sources of space–time estimates (e.g. radar precipitation estimates in the case of rainfall) that provide better ways of robustly estimating statistics from a shorter period. For example, time series extracted from twenty hydro-meteorologically similar locations of a space–time dataset (e.g. radar) for a period of five years is statistically equivalent in terms of robustness of estimates to a hundred year long rainfall time series. (Series p_1, p_1, … p_n from an ensemble e_p) But, the former has the important advantage of having a negligible non-stationarity.

For the future case, it is possible to use high-resolution limited area atmospheric models under forcings provided by the climate models to produce space–time estimates of future variables. Once available such space–time data can be used in a similar way to past observations to produce robust estimates of the future situation. (Series f_1, f_1, … f_n from an ensemble e_f).

At this stage it should be emphasised that the 'robustness' referred to in the above paragraphs is strictly in a statistical sense. It does not imply that the technologies or models used to estimate present and future time series realisations (e.g. radar, limited

TEXTBOX 2-6

Case study—Urban heat islands and localised rainfall enhancement

Urban heat islands (UHI) are the increase in temperature in large urban areas relative to the surrounding rural or natural landscape due to urban-heat build-up due to anthropogenic activities. The reasons could be both direct heat generating activities (e.g. residential heating, vehicles) and indirect effects like retardation of latent heat flow to the atmosphere due the reduction of evapo-transpiration. It has been long understood and observed that such differential heat build-up can cause changes in the local atmospheric circulation and has significant effects on the urban rainfall.

There have been a number of studies on the qualitative nature of the UHI and rainfall relationship and also the possibilities of large historical urban rainfall being related to UHI. However, there is little if any literature on the quantification of rainfall enhancement due to UHI as a result of future urbanisation. The study presented here, takes an initial step in this direction.

An atmospheric model (WRF) was used to conduct a series of controlled experiments where the degree of urbanisation was changed while keeping all other forcing variables (e.g. terrain shape, lateral boundary conditions including winds, moisture, pressures) exactly the same. A WRF model can produce rainfall estimations based on atmospheric physics. The rainfall change that is resulting from the changes in the urbanisation (changing the land use to more urbanised situation) was estimated. The statistics of rainfall was estimated and used to demonstrate possible changes that can happen in the design of IDF curves. The system was applied to a number of large cities in the world and in many cases a large increase (about 2 times) of urban sprawl, results in a very significant increase of rainfall and marked changes in the IDF curves. The following figures show the results for Dhaka city.

FIGURE 2-17
Rainfall change in Dhaka city due to future urbanisation

Source: Assela Pathirana, 2009.

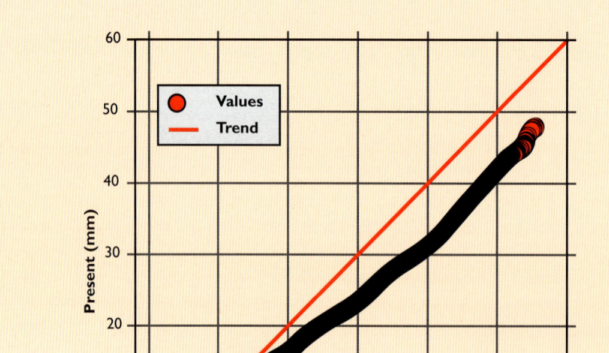

FIGURE 2-18
IDF curves before and after urbanisation for Dhaka city

Source: Assela Pathirana, 2009.

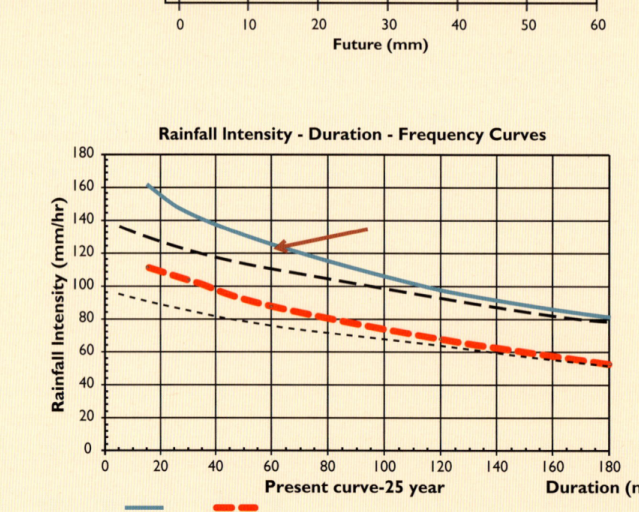

area atmospheric models) are as accurate as directly measured rain gauge data. The uncertainties arising from these are important, but out of the scope of this chapter.

There is no question about the fact that the stationarity assumption used in developing the bulk of the hydrological time series analysis techniques is no longer valid for long-range planning purposes. Climate change has rendered it useless in this direct application. However, this does not imply that the stationarity-based techniques are completely useless for future applications. Using innovative approaches to circumvent the problem of errors introduced by large non-stationarity, it is possible to use these techniques in a rapidly changing hydro-climatic environment. Therefore it fits better to say 'Stationarity is dead. Long-live stationarity!'

Key Questions

The influence of climate

1. How will climate change affect urban pluvial flooding?
2. Could you give examples of joint flooding?

Types of flooding

1. What are the implications of land-use planning designed to reduce urban pluvial flooding on the urban form?
2. Could you describe the different types of flooding?
3. How does urban development influence the groundwater regime?

Pitfalls of historical data

1. Describe how hydrological time series data can be analysed.
2. What are the implications of stationarity and non-stationarity for flood modelling?
3. What approaches are applicable for dealing with the impact of climate variability and climate change on hydrological modelling?

FURTHER READING

Bates, B.C., Kundzewicz, Z.W., Wu, S. & Palutikof, J.P. (eds) (2008). *Climate Change and Water*. Technical Paper of the Intergovernmental Panel on Climate Change. Geneva: IPCC Secretariat.

Burry, K.V. (1975). *Statistical Methods in Applied Science*. John Wiley & Sons.

Castillo, E. (1988). *Extreme Value Theory in Engineering*. New York: Academic Press, Inc.

Dankers, R. & Feyen, L. (2008). Climate change impact on flood hazard in Europe: An assessment based on high-resolution climate simulations, *J. Geophys. Res., 113*, D19105, doi:10.1029/2007JD009719.

Dankers, R. & Feyen, L. (2009). Flood hazard in Europe in an ensemble of regional climate scenarios, *J. Geophys. Res., 114*, D16108, doi:10.1029/2008JD011523.

Feyen, L., Dankers, R. & Bödis, K. (2009). *Evaluating the Benefits of Adapting to Changing Flood Hazard in Europe*. Institute of Physics Publishing.

FlOODSite (2009). *Development of Framework for the Influence and Impact of Uncertainty*. Report No T20-07-03. Available from: http://www.floodsite.net.

Gumbel, E.J. (1958). *Statistics of Extremes*. Columbia University Press.

Milly, P.C.D., Betancourt, J., Falkenmark, M., Hirsch, R.M., Kundzewicz, Z.W., Lettenmaier, D.P. & Stouffer, R.J. (2008). Stationarity Is Dead: Whither Water Management? *Science, 319(5863)*, 573–574 DOI: 10.1126/science.1151915.

Milly, P.C.D., Wetherald, R.T., Dunne, K.A. & Delworth, T.L. (2002). Increasing risk of great floods in a changing climate. *Nature, 415*, 514–517.

O'Connor, J.E. & Costa, J.E. (2004). *The World's Largest Floods, Past and Present: Their Causes and Magnitudes* [Circular 1254]. Washington, DC: U.S. Department of the Interior, U.S. Geological Survey.

II

Drivers for change

In Part I the idea that urban areas are special and distinct was introduced along with an understanding of what might constitute flooding and the circumstances giving rise to flooding for urbanised areas. In Part II those concepts are expanded and looked at in more detail. Urbanisation is a multi-faceted phenomena but one of the most important features of urbanisation is the distinctive ways in which land is used, the purposes for which it is used and their relationships with each other. The very idea of land use as a singular attribute may not be that helpful as it fails to take into account changes over time and the possibility that a single use may have multiple effects. Urban typologies provide some insight, though this is a developing area especially with respect to flood management. We have also acknowledged that flooding, and not just urban flooding, may be profoundly impacted by climate change. It is therefore important to understand what is meant by climate change, its variability and the forces driving it, and how the changes being experienced and projected differ from those of the geological past. Part II provides the reader with much to consider concerning the process of urbanisation, climate change and their relationship with urban flooding and management.

Extreme precipitation in the catchments of Europe's rivers Rhine and Meuse caused peak flows in January 1995. The floods of 1995 urged to speed up flood defence reinforcements and other flood protection programs in The Netherlands and Germany and an international flood action plan for the Rhine came into force.

Source: Rob Lengkeek, 1995.

Urbanisation

Learning outcomes

In the previous chapters you learned that urbanisation increases the magnitude and frequency of floods in the following ways:

- the creation of impervious surfaces such as roofs and pavements, and roads reduces infiltration so that a higher proportion of storm rainfall generates runoff;
- hydraulically smooth urban surfaces serviced with a network of surface drains and underground sewers, deliver water more rapidly to the channel;
- the natural river channel is often squeezed by the intrusion of bridge supports or riverside facilities, thus reducing its carrying capacity. This increases the frequency with which high flow overtops the embankments.

In this chapter you will learn about the following key concepts:

- Basic concepts and principles of land-use planning.
- Cities have flourished and declined over the centuries but now many are growing at record pace.
- Cities exhibit different spatial configurations and patterns of urban growth that may be classified in urban typologies.
- Urbanisation increases the magnitude and frequency of flooding but also its impact.
- The flooding characteristics and impacts depend on and are closely linked to the spatial configurations and patterns of urban growths, respectively.
- Human settlements, including many large cities, are also concentrated near or on coastlines, and a large proportion of global economic productivity derives from coastal areas.

3.1 PRINCIPLES OF LAND-USE PLANNING

When travelling across cities, suburbs and countryside, you might ask yourself: "Why is this factory located over here?" or "Why are they building this new suburb left of the highway instead of on the right side?" Historically, much of the development of human-related land use was unplanned. Cities started as small settlements along strategically located areas (e.g. close to fertile land often located in delta areas) based on a loose organisation. This organisation became rationalised by both strategic defence considerations (fortified camps/cities) and commercial development (trade). The rapid expansion of cities caused by industrialisation in the nineteenth century induced a more formal planning process of activities. Issues like sanitation, water supply and solid waste management transformed urban development from short term 'organic' planning into a rationalised process with a long-term horizon. This transformation took place on various scale levels, ultimately resulting in a spatial planning discipline in which land-use planning became the dominant means for organising society within spatiotemporal dimensions. Consequently, *land-use planning is a catch-all term focusing on the strategy and methodology for the organisation of human-related activities in space and time.*

As such, land-use planning operates within the context of societal, governmental and technological trends as well as within a theoretical and methodological framework. By means of legal regulations (planning laws), social conventions and rules, land-use planning sets in motion social processes of decision-making and consensus building concerning the functional (different types of activities) and legal (e.g. private, communal or public areas) differentiation of land. Ideally land-use planning should be forward looking, anticipating and guiding future uses with the goal of achieving a state of pareto-optimality. In principle, a differentiation is made between sectoral and technical planning (e.g. transportation planning or the planning of water resources) and planning which overlaps sectors or is partially integrative. Basically, the technical planning process provides the boundary conditions for the sectoral planning process, since it emphasises physical characteristics and the limitations/potentials these provide for specific activities.

By nature, methods and available tools within the planning process differ depending on the specific task and scale level: local, town, district and regional, national and (in some cases) international (see Figure 3-1). This suggests a coherent and consistent system of nested land-use plans across scales into a general planning framework; local and urban plans are guided by regional plans which are in turn informed by national planning objectives. In an ideal world these are all seamlessly integrated with each other, although in practice this is not necessarily so. The greater the interaction between the levels of planning, the better will be the complementarity between the respective goals and objectives of such plans and the means used to realise them.

3.1.1 National level

At the national level, planning is concerned with national goals and the identification of resources. In many cases, national land-use planning does not involve the actual allocation of land for different uses, but the establishment of priorities for district-level projects. A national land-use plan may cover:

- land-use policy: balancing the competing demands for land among different sectors of the economy food production, export crops, tourism, wildlife conservation, housing and public amenities, roads, industry;
- national development plans and budget: project identification and the allocation of resources for development;
- coordination of sectoral agencies involved in land use;
- legislation on such subjects as land tenure, forest clearance and water rights.

National goals are complex since policy decisions, legislation and fiscal measures affect many people and wide areas. Decision makers cannot possibly be specialists in all facets of land use, so the planners' responsibility is to present the relevant information in terms that the decision makers can both comprehend and act on.

3.1.2 Regional/district level

This is not necessarily an administrative unit but relates to areas that fall between national and local levels. Development projects are often at this level, where planning first comes to grips with the diversity of the land-related issues and suitability to meet project goals. The kinds of issues tackled at this level include:

- the siting of development projects and regional development schemes;
- the improvement of infrastructure such as water supply, transport systems;
- the definition of the urban boundaries (envelopes) in which cities can develop building activities.

3.1.3 Urban/local level

At a local level, land-use planning focuses on parcel division, functional zoning, building type regulations and many other requirements necessary to ensure the allocation of activities, culminating in a 'zoning plan' that serves as the blueprint for many cities.

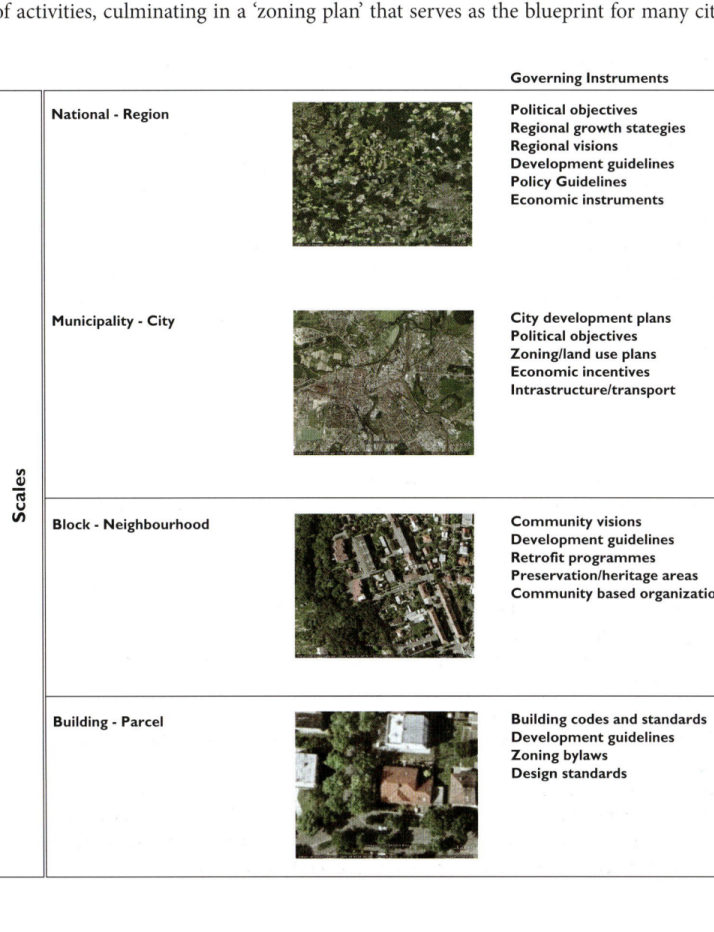

FIGURE 3-1
Scales and governing instruments

At this level, the zoning plan is bounded by numerous restrictions (e.g. the stormwater drainage network or land ownership distribution) and is therefore much more 'informed' than land-use plans on higher scale levels. Typically a zoning plan includes:

- division of functional zones (e.g. housing, industry, offices, recreation);
- land parcel division;
- maximum building heights;
- additional regulations (e.g. amount of vegetation, parking facilities).

Furthermore, the zoning plan is embedded in a complex legal framework that sets it apart from the more policy-related regional or national land-use plans. It is not unusual for building activities to stop because of non-compliance with the zoning laws.

3.1.4 Stakeholder involvement

Although in practically every country, land-use planning is executed by governmental institutions, a range of stakeholders is involved in the process; it is a multi-party competition over a community's future land-use pattern. Depending on the scale level of the land-use plan, these stakeholder groups differ. At the top of the nested hierarchy of planning, say at the national or even international level, the time and other resources required of stakeholders to be able to participate is such that only a few can afford or be able to participate. Often such stakeholders consist of institutional stakeholders, businesses, interest groups and NGOs. These parties can mobilise or access resources to enable them to participate. They do so in response to aspects of plans or planning which are of particular interest to them or have an impact on them. The degree of saliency, knowledge and experience, and perceptions of the ability to influence decisions are among the factors that influence stakeholders in the decision of whether or not to participate. The other factor is the degree of openness of the planning process to allow participation. At finer planning scales not only does the nature of what planning addresses but also

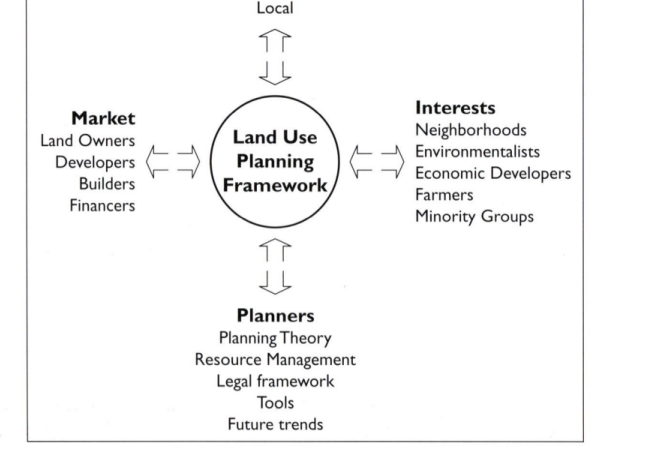

FIGURE 3-2
Land-use planning: a multi-stakeholder perspective

the interests become increasingly localised. This opens opportunities for other stakeholders to become engaged as the issues are more aligned to their interests and abilities. Conversely some stakeholders may absent themselves from the process for similar reasons—the issues are not aligned with their interests. The ideal is that all interested and affected parties should have a say in the planning process, given that by its nature planning is undertaken in the interests of a society.

Land-use planning not only involves the competition over land by different stakeholders; it also facilitates choices about competing functions, favouring one function over another. From an Urban Flood Management perspective this might include the planning of residential areas in the flood-prone areas for the sake of providing attractive urban environments for new urban dwellers. Similarly, cities might want to increase densities by integrating high-rise buildings into empty lots while compromising their drainage capacity. This indicates that land-use planning is not always aiming towards long-term sustainability. Short- or middle-term economic development often dominates the urban agenda. This is especially the case in developing economies, where building pressure prevails over rational land-use planning based on current and future allocation of resources, environmental quality and resilience towards natural hazard impact (including water). Often these risks are neglected by relying on future technological innovations (e.g. more efficient urban drainage systems) while these might not necessarily be available. This places a burden on future generations in which urban development might be focused on 'repairing' mistakes made in the past. To make matters worse, comprehensive land-use planning isn't even institutionalised in many developing countries. Currently, the United Nations (UN-Habitat, 2007) estimates that 95 per cent of the urban development in the world's expanding cities in developing countries is 'unplanned'. These include fast-growing slums often located in flood-prone areas but also excesses of wild ground speculation outside existing zoning laws.

Because of the increasing pressures of urban growth (see Sections 1.2 and 3.2) rational and long-term land-use planning is a vital instrument in keeping natural hazard impacts within bounds. Urban growth often concentrates more people and assets in floodplains, thus increasing the sensitivity towards flood damages and casualties. The potential effects of climate change only increase this sensitivity.

3.2 URBAN TYPOLOGIES: FROM CENTRAL SQUARE TO EDGE CITY

Cities can be considered as 'patterns in time' and act as focal points for a wider complex of economic, social, political and environmental linkages and flows of people, power, energy and information. Even in ancient history, cities provided the economic, social and cultural backbones within nations. Modern urban analysis focuses on the interplay between physical characteristics and the complex dynamic processes that govern these characteristics. Before such an analysis can take place, it is vital to provide a classification framework in which the physical structure and composition of a city can be identified.

3.2.1 Urban typologies

Cities differ. The sprawling neighbourhoods of Los Angeles might seem like villages when compared to the overly dense centres of Mumbai. Also, the regular grids that form the urban block and street network of Barcelona might seem a relief after endlessly strolling through the narrow streets of the historical centre in Athens. In other words: cities are composed of different urban typologies. These can be considered as the spatial composition of areas in relation to the interplay between building types, infrastructure and other forms of public space. The application of urban typologies is very much a product of changing ideas on urbanism and of local conditions. During the industrial revolution for instance, the dense historic city centres were extended by open garden cities. Later on, in post-Second World War Europe, technological advances in combination with a large housing demand resulted in new areas filled with multi-storey apartment blocks surrounded by communal gardens and parking space. In many cities, different urban typologies live side-by-side and provide heterogeneity and identity.

Although such a descriptive approach provides vital clues about the expected ground coverage and spatial characteristics, the urban flood manager might be served better by a classification that is quantitatively grounded and relates to the observed densities in built-up areas. In practice, such a perspective connects nicely to the application of land-use and land-cover data used for urban analysis and the identification of natural drainage zones. Angel *et al.* (2005) define a set of five major urban typologies including the metrics for identification of these typologies (see Figure 3-3):

- *Main urban core*: contiguous group of built-up pixels in which at least 50 per cent of the surrounding neighbourhood within an area of 1 km^2 is built-up. Typically, this area represents the central business district or the historical centre of the city.
- *Secondary urban core*: pixels not belonging to the main urban core with 1 km^2 neighbourhoods consisting of at least 50 per cent built-up area. These areas represent new urban centres or centres formerly belonging to towns that were swallowed by the urban expansion of the main urban core.
- *Urban fringe*: pixels with 1 km^2 neighbourhoods that are 30–50 per cent built-up. The urban fringe often hosts the majority of residential areas built during a process of expansion (suburbs, sprawl).
- *Ribbon development*: semi-contiguous strands of built-up pixels that are less than 100 metres wide and have 1 km^2 neighbourhoods that are less than 30 per cent built-up. Generally, major infrastructural connections (rivers, highways) act as attractors for settlements causing linear, strip-like development.
- *Scattered development*: built-up pixels that have neighbourhoods that are less than 30 per cent built-up and not belonging to the ribbon development. This category describes settlements outside the direct proximity of the city. Their location is often determined by specific landscape features (e.g. a lake) and the often lower ground prices. In time these areas can be absorbed by the urban expansion of the main city.

While this density-driven classification method is able to identify some of the urban characteristics, it does not provide the means to analyse the structure of a city; analysis of the urban centre of Las Vegas, which is based on a wide grid filled with large buildings, results in a similar main urban core characterisation as the centre of Paris, which developed more organically into an area filled with small buildings and curved narrow streets. To identify such differences, Hillier (1996) developed a series of metrics that apply a more spatial perspective: Space Syntax. These include the isovist or viewshed analysis of convex urban spaces, axial lines and ultimately axial maps. After application to a city, the metrics and resulting statistics tell us something about the distribution of the size of spaces and their interconnectedness. Cities showing a more global structure (e.g. grids characterised by long streets) are classified differently from those characterised by local structures (e.g. intricate street patterns combined with small buildings).

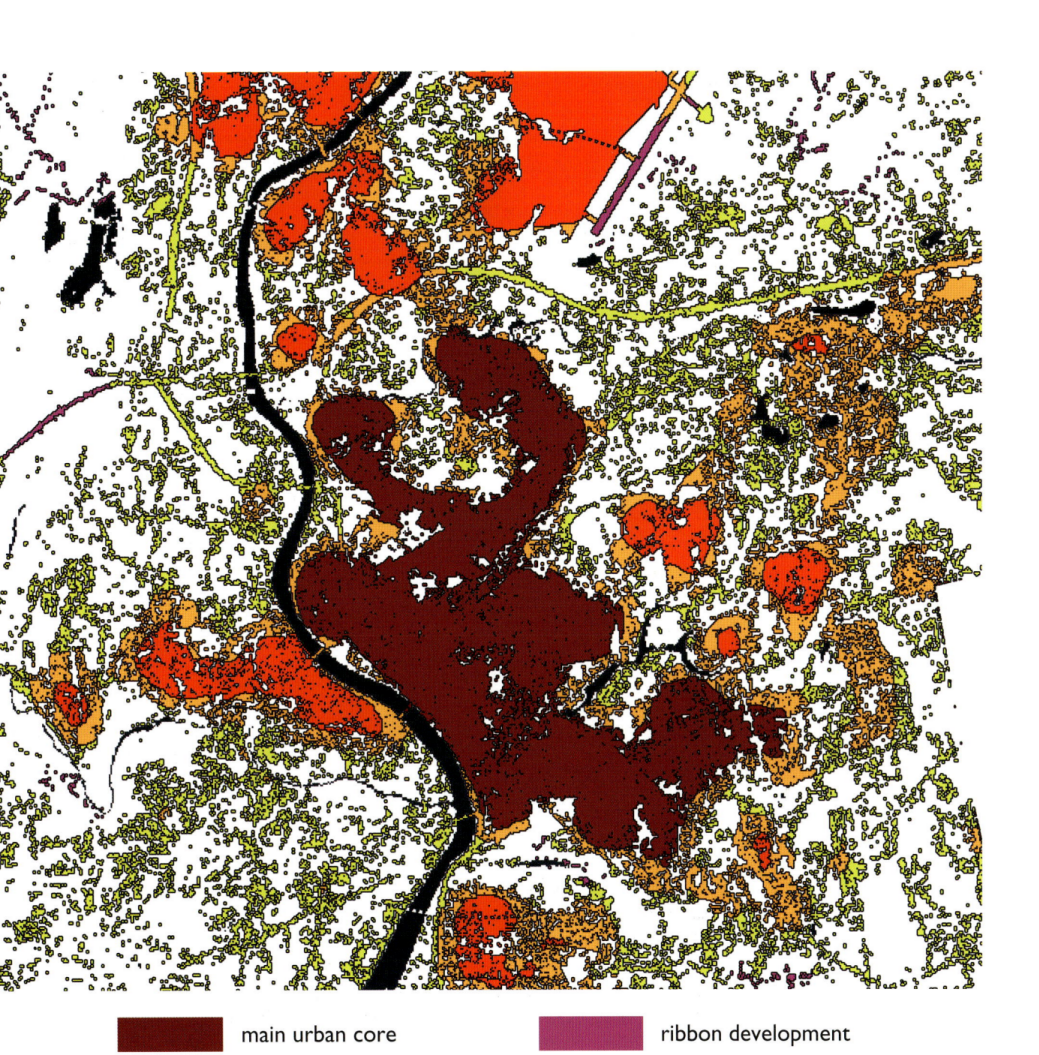

FIGURE 3-3

Spatial distribution of urban typologies for Springfield, Ma, 1989 (data courtesy of Center for Landuse Research and Education, 2009)

Source: Center for Landuse Research and Education, 2009.

Apart from differences in density and open spaces, cities can be analysed from a variety of other perspectives that all provide clues for flood vulnerability assessment or flood management strategies. These include:

- *Functional distribution.* Basically, such a view implies a land-use classification in which different functional zones (e.g. retail, business, industry, housing) are identified. Apart from the spatial distribution of different functions, land-use characteristics also provide information on the heterogeneity of a city; mono-functional cities that

3.2 URBAN TYPOLOGIES: FROM CENTRAL SQUARE TO EDGE CITY

FIGURE 3-4
Tokyo's streetscape. Metropolitans like Tokyo, Yokahama, Kyoto, and Osake are the most densily populated cities of the world.

Source: Paul Regouin, 2002.

developed because of single industrial sectors (or even individual factories) differ substantially from multi-functional cities that provide a multitude of functional zones.

- *Temporal distribution of neighbourhoods.* Since neighbourhoods are often developed at the same time, this perspective provides information on the differences in age between different areas (e.g. historical centres, post-war rapidly developed residential districts).
- *Ownership distribution.* Scattered individual ownership distributions can potentially hinder redevelopment of areas since dispossession procedures can take a long time. Furthermore, ownership distribution tells us something about the power-relations within a city, since owners of large areas can be identified as major stakeholders within development or redevelopment.
- *Value distribution.* From a materialistic perspective, cities are concentrations of value in the shape of assets. Especially when assessing economic vulnerability towards flooding, it is vital to analyse the spatial (and temporal) distribution of value 'hotspots'.
- *Social distribution.* Through the spatial distribution demographic, the location of fragile social groups can be identified as well as a range of other factors.

The application of these perspectives, classification schemes and metrics depends very much on the problem at hand. Traditionally, the urban flood management framework mostly focuses on the physical characteristics of our cities (e.g. the distribution of impervious land-cover). Yet, when a more integrated flood management policy is applied, the range of possible variables that influence flood vulnerability is substantially larger and thus requires a more comprehensive urban analysis. However, there is still a long way to go to fully exploit these perspectives as we are barely scratching the surface of this emerging domain of urban analysis and its relevance for urban flood management.

3.3 GROWING AND SHRINKING: DENSITY ISSUES INDUCED BY GLOBALISATION

3.3.1 Introduction

The year 2007 marked a turning point in the process of rapid urbanisation of the world's population: more than 50 per cent of the world's population lived in urban areas. This is striking since a little more than 100 years ago the world population consisted predominantly of rural inhabitants. In 1800, only 3 per cent of the world's population lived in cities. In the coming years, our urban population will further skyrocket. In developing countries alone, by 2030 the urban population is expected to rise to almost 4 billion inhabitants, a 100 per cent growth within about 30 years. For industrialised countries, this growth is only expected to be 11 per cent.

This process of urbanisation is not new; during the nineteenth century industrialisation, cities expanded rapidly. The causes for urban growth are relatively straightforward:

- *Rural to urban migration.* The majority of urban growth is caused by farmers and villagers moving towards the more prosperous cities which offer better facilities and most of all an increased potential for personal economic growth.
- *Autonomous urban growth.* Better economic conditions in the cities often lead to a household and population growth.
- *Intra-urban migration.* Increasing competition between cities and subsequent economic attractiveness generate an increasing intra-urban migration. Furthermore, an overall relaxation of immigration laws makes workforces more mobile than ever before (e.g. a substantial amount of Philippine women work as au pairs abroad).

Although the world's urban population is exploding, paradoxically, the average densities of many urban areas are declining. The World Bank (2005) estimates an annual decline of urban densities of 1.7 per cent for developing countries, resulting in a built-up area of 600.000 square kilometres by 2030. To put this in perspective: urban agglomerations in 2030 will have tripled occupation space with about a 160 square metre transformation of non-urban to urban per new resident. Within the industrialised world, these figures are less dramatic. Here, the urban population is expected to rise by 11 per cent within the next thirty years to about 1 billion inhabitants. Occupied land is expected to increase 2.5 times with an annual decline in density of 2.2 per cent. Individual occupation is substantially higher though; every new resident is expected to convert on average 500 square metres of non-urban into urban land. This confirms the much higher amount of used square metres per capita. Furthermore, declining densities are in some cases also caused by a different phenomenon: urban shrinking. In particular, mono-functional cities, that often depended on a single industry, can suffer this problem. Many cities in, for instance, former East Germany are now shrinking because of intra-urban migration; poor economic prospects make the young people move towards the more prosperous Western part of the country. Note that this will have a multiplier effect, since the increasingly old population will not provide any prospects for autonomous urban growth.

3.3 GROWING AND SHRINKING: DENSITY ISSUES INDUCED BY GLOBALISATION

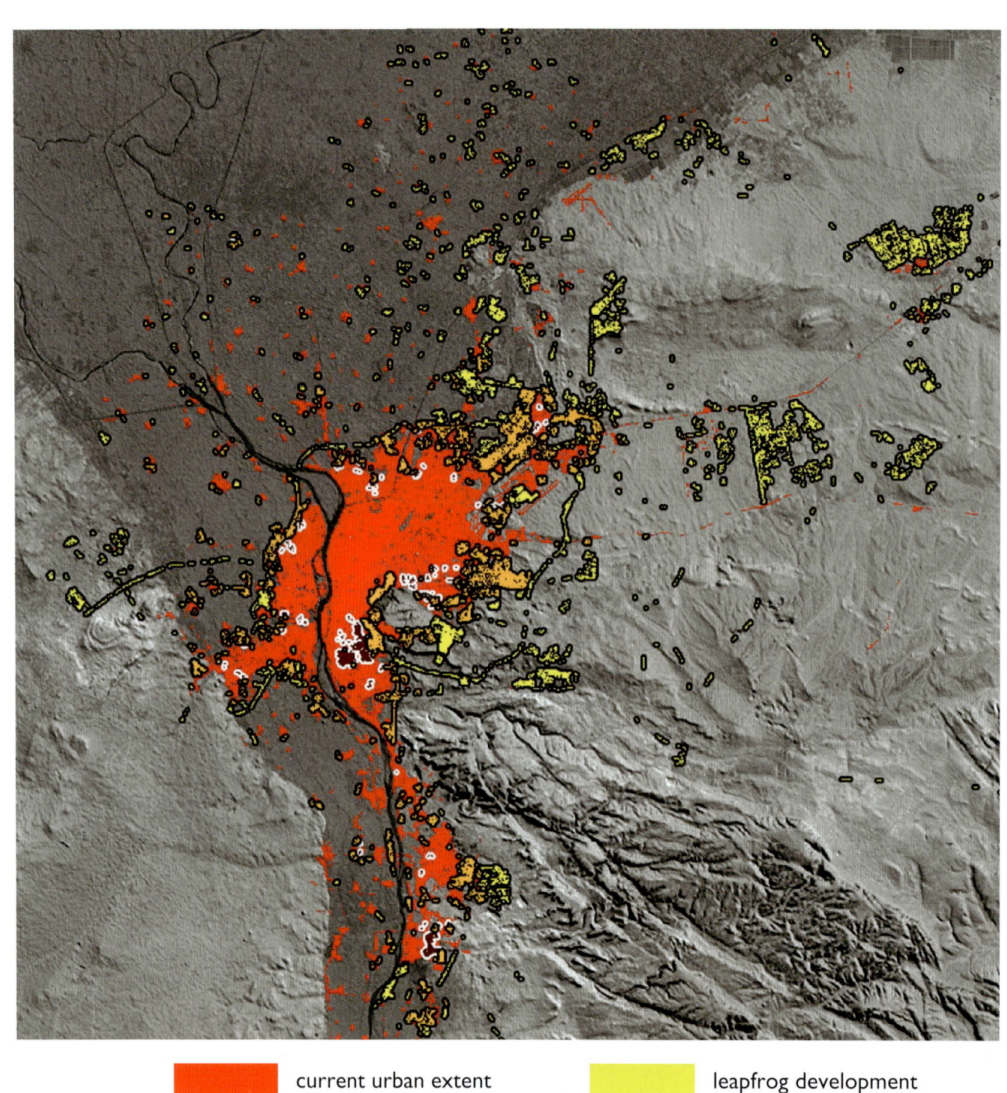

FIGURE 3-5
Urban growth types for Cairo between 1984 and 2000 (data courtesy of Center for Landuse Research and Education, 2009)

Source: Center for Landuse Research and Education, 2009.

3.3.2 Urban growth characteristics

One of the major manifestations of urban growth in the past decades is the ongoing process of urban sprawl. Conceptually, the notion of urban sprawl has not been clearly defined, although urban densities are good indicators for identifying urban sprawl. Especially in the United States, sprawl has dominated the urban growth process of many cities over the last decades. This low-density development also creates a relatively large footprint per capita, which is further amplified by an increased necessity for infrastructural and utility-lifeline networks. Apart from the 'planned' examples of urban sprawl, many cities in developing countries experience unplanned sprawl in the shape of slum areas. Such unplanned and fast urban expansion can lead to many problems, including an increased susceptibility to the impacts of natural disasters due to a concentration of inhabitants and assets in hazard-prone areas.

The physical characteristics of urban growth can be separated into three categories that explain something about how new urban development relates spatially to the existing city. These urban growth categories are:

- *Infill*: new development within remaining open spaces in already built-up areas. Infill generally leads to higher levels of density and increases the contiguity of the main urban core.
- *Extension*: new non-infill development extending the urban footprint in an outward direction.
- *Leapfrog development*: new development not intersecting the urban footprint leading to scattered development.

Together, urban extension and leapfrog development are responsible for the majority of observed urban growth, ultimately leading to sprawling cities like Los Angeles, where the majority of the city consists of detached housing on small plots

Currently, there is a lot of attention on the development of spatial urban growth models that can provide future predictions on urban development. Within the domain of flood vulnerability assessment and urban flood management in general, this could provide valuable information to assess the impact of urban growth on the overall risk distribution towards flooding.

3.4 MEGACITIES IN THE DELTA

More than two-thirds of the world's large cities are vulnerable to rising sea levels, exposing millions of people to the risk of extreme floods and storms. At present, the population densities in coastal areas are three times the global mean. Within the coming 30 years, the United Nations project that the number of people living in cities in coastal areas (within 100 km of the coast) will increase to 60 per cent of the world's population, resulting in even more people living in highly exposed areas. Hence, socioeconomic trends further amplify the possible consequences from future floods as more people move towards urban coastal areas and capital is continuously invested in ports, industrial centres and financial businesses in flood-prone areas.

A historical review of the populations of cities reveals that the world's largest cities have flourished and declined over the centuries. Since the trade and transport networks consisted primarily of water-based infrastructures for a long time, the 50 largest cities to date are all located in deltaic regions, along rivers, seacoasts or lakes. These cities are

TEXTBOX 3-1

Delta cities

Although delta cities and other ports share many characteristics, delta cities are unique in at least three ways:

- their locations are exceptionally unstable due to both shifts in the river and the encroachment of the sea;
- they are exceptionally isolated from their proximate hinterland (because the width of the delta typically hinders cross-delta transportation and communication); and
- they are exceptionally vulnerable to flooding due to subsidence and low-lying land.

Delta cities have always been and still are the locus of intense engineering efforts and expenditures. Delta cities have a topography and hydrology which are artificial. The first settlers constructed flood defence structures and raised and stabilised natural banks for flood control and drained the wetlands for agriculture using irrigation canals to continuously adjust the water levels. Later they relied on the invention of first the windmills and later on moderns pump, levee systems and pilings for building foundations.

TABLE 3-1

World's typical delta cities and their rivers

Delta Cities	
City	**River**
Marseilles, France	Rhone River
Amsterdam, The Netherlands	Amstel
Rotterdam, The Netherlands	Rhine
Kolkata, India	Ganges
Shanghai, China	Pearl
Port Harcourt, Nigeria	Niger
Bangkok, Thailand	Chao Phraya
Phnom Penh, Combodia	Mekong
Basra, Egypt	Tiber
Alexandria, Egypt	Nile
Sacramanto, California	Sacramento-San Joaquin River
Yangon, Myanmar (Burma)	Ayeyarwady
St. Petersburg, Russia	Neva

also defined as the metropolitans of metropolitan areas. In the literature the term metropolitan area is often defined as a city with more than 1 million inhabitants. A metropolitan area typically is a city with a global impact, an area where global relationships dominate over local ones.

FIGURE 3-6
Delta City, Shanghai

Source: Johan vande Pol, 2010.

TEXTBOX 3-2

Role of watercourses and canals in city planning in delta cities

In low-lying areas of The Netherlands and Japan the drainage and irrigation system of the landscape, the polder allotment, offered a master plan for the cities just growing over these patterns. Natural watercourses and man-made canals in the cities have for a long time functioned as discharge, storage, drainage, drinking water, transport and sewer. Water also gained a military function when it was used as protection in the shape of moats around fortifications. The function of water has remained a determining factor in the structure of the urban environment, for example in Amsterdam. The canals have been, up until recently, crucial in fulfilling the many functions mentioned above. They form the central axes in the city and shape the urban structure. The trade people, for whom the expansion has been built, are reflected in the careful detailing of the canals. Nice bridges and rules for the buildings, the rows of trees along the canals all made sure it would be an elegant city district representing the social status.

3.4 MEGACITIES IN THE DELTA

FIGURE 3-7
Historical map of Amsterdam, The Netherlands: watercourses and canals are shaping the city lay-out of Amsterdam

Source: Stadsarchief Amsterdam, 2001.

TEXTBOX 3-3
Top 20 ranked cities

The scale of the disruption to urban populations and economies by extreme weather events in coastal areas in recent years highlights their vulnerabilities. Worldwide, there has been a rapid growth in the number of people killed or seriously impacted by storms and floods, and also in the amount of economic damage caused.

A global screening has made a first estimate of the current (2005) and future (2070) exposure of the world's large port cities ($n = 136$) to coastal flooding due to extreme water levels.

The results reveal that socioeconomic changes are the most important driver of the overall increase in both population and asset exposure and that climate change has the potential to exacerbate this effect. The concentration of future exposure to sea level rise and storm surge exhibit a steep increase in rapidly growing cities in developing countries in Asia, Africa and to a lesser extent in Latin America. Given the widespread distribution of and the threat to the highly vulnerable large port cities, large-scale city flooding may remain frequent even if all cities are well protected against extreme events.

3. URBANISATION

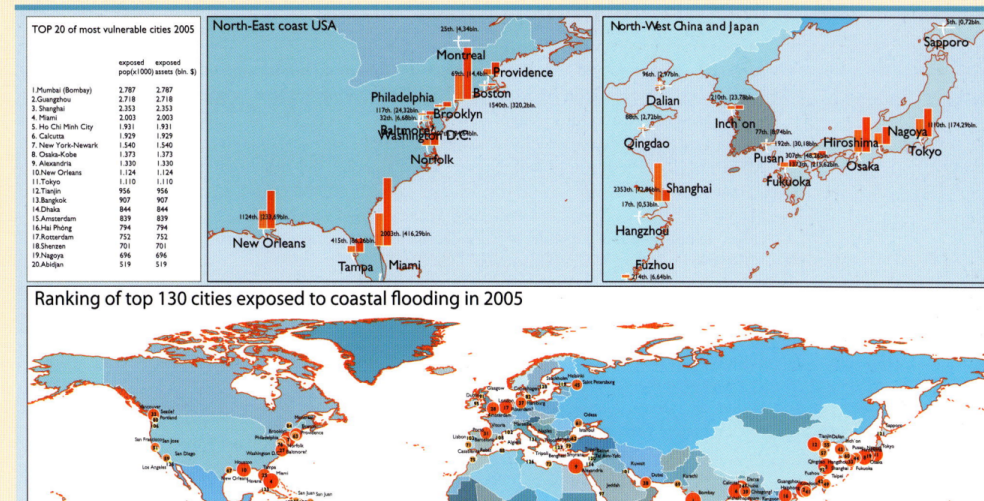

FIGURE 3-8
Top 136 cities exposed to coastal flooding in 2005

Source: William Verbeek (based on Nicholls et al.), 2007.

Key Questions

Urban typologies: from central square to edge city

1. How would you describe your own (or capital) city in terms of urban typologies using the five classes?
2. What are the strengths and weaknesses of the classification scheme developed by Angel?

Urban growth characteristics

1. How do the different growth types relate to urban flooding in terms of susceptibility?

2. Does urban growth have a larger impact on flood risk than climate change (give reasons for your answer)?

Megacities in deltas

1. What does a metropolitan area look like? And what are the physical characteristics?
2. Explain why large cities in developing countries in Africa and Asia are particularly vulnerable to flooding?

FURTHER READING

- Angel, S., Sheppard, S.C. & Civco, D.L. (2005). *The Dynamics of Global Urban Expansion*. Washington, DC: The World Bank. Related online version available from: http://www.williams.edu/Economics/UrbanGrowth/WorkingPapers_files/WorldBankReportSept2005City MapPortrait.pdf [Accessed 27th November 2008].
- Atkinson, G. & Oleson, T. (1996). Urban sprawl as a path dependent process, *Journal of Economic Issues, 30(2)*, 609–615.
- Bacon, E.N. (1974). *Design of Cities*. New York: Viking Press, rev. edn.
- Cities Alliance (2006). *Guide to City Development Strategies: Improving Urban Performance*. Washington, DC: Cities Alliance, p. 11.
- Ericson, J.P., Vorosmarty, C.J., Dingman, S.L., Ward, L.G. & Meybeck, M. (2006). Effective sea level rise and deltas: causes of change and human dimension implications, *Global Planet Change, 50*, 63–82.
- Harris, C.D. & Ullman, E.L. (1945). The Nature of Cities. *Annals of the American Academy of Political and Social Science, 242*, 7–17.
- Hillier, B. (1996). *Space is the Machine: A Configurational Theory of Architecture*. Cambridge: Cambridge University Press.
- Nicholls, R.J. (2004). Coastal flooding and wetland loss in the 21st century: changes under the SRES climate and socio-economic scenarios. *Global Environmental Change, 14*, 69–86.
- Nicholls, R.J., Hanson, S., Herwijer, C., Patmore, N., Hallegatte, S., Corfee-Morlot, J., Chateau, J. & Muir-Wood, R. (2007). *Ranking Port Cities with High Exposure and Vulnerability to Climate Extremes—Exposure Estimates*. Environment Working Papers No. 1, Organisation for Economic Cooperation and Development, Paris. Available from: http://appli1.oecd.org/olis/2007doc.nsf/ linkto/env-wkp.
- UN (2007). *The Millennium Development Goals Report*. New York: United Nations. Available from: http://www.un.org/milleniumgoals/pdf/mdg2007.pdf.
- UN-Habitat (2007). *Sustainable Urbanization: Local Actions for Urban Poverty Reduction, Emphasis on Finance and Planning*. 21st Session of the Governance Council, 16–20 April, Nairobi, Kenya.

Climate change: key uncertainties and robust findings

Learning outcomes

In this chapter you will learn that:

- Climate change is already happening and is a long-term risk issue, though clearly extreme climatic events can occur at any time.
- There is uncertainty about the nature and magnitude of climate change and its impacts. There is also uncertainty about how these impacts should be valued.
- Most of the urban fabric has been designed on the premise that the future climate will be similar to that experience in the past. Cities are therefore particularly vulnerable to climate change.

4.1 A REVIEW OF THE PAST

4.1.1 Introduction

Today the planet is experiencing a significant warming. The predictions of what the future might be is the subject of much debate and argument, often based more on ideological opinion than on scientific facts. For example, for a brief period in the 1970s there was a supposition that the Earth was entering a period of cooling, which caught the public's imagination with stories of glaciers advancing on Western Europe. The observed warming trend and the preceding cold spell known as the Little Ice Age, which some have suggested lasted approximately between the 1300s and 1800s have had precedents. These periods of climatic fluctuation have been linked to changes in sea level and some have speculated that the collective myths and memories of universal floods, Atlantis and sunken continents are an echo of these events. This is one reason why the observation and investigation of signs associated with sea-level change has been a source of reference indicators for estimating past climates. It was the obvious localised signs of former sea levels, in the form of raised beaches and fossilised dunes, along with evidence of submerged or elevated remains of human settlements that served to highlight past climate changes. They were assumed to have been caused by sea-level changes in recent geological time, known as the Plio-Quaternary period of about 2 million years ago.

As the history of the Earth was progressively established and elaborated on, other types of change in sea levels became apparent, such as those resulting from vertical movements of the crust, modifications of sea basins due to continents fracturing, drifting and coming together, volcanic activity or to different types of climate pattern.

These changes in temperature, climate and sea level were the result of the interplay of natural forces, in which human activity had no global influence. That changed with the increase in human activity associated with industrialisation and the burning of fossil fuels. The discovery and more importantly the increasing use of fossil fuels, which allowed man to escape the tyranny of the climate, has had a destabilising effect when combined with the ensuing qualitative increase in population. The resultant impact makes the present trend of climate change possibly unique in history. The implications and consequences of climate change for the water sphere go far beyond just changes in sea level and offer another clue as to the nature of climate itself and its changes, causes, processes and consequences, allowing for more thorough analyses.

4.1.2 Current warming

Whenever climate change is talked about there are two things that catch the attention of the general public: sea level rise (SLR) and greenhouse gas emissions (such as CO_2). In 1896, the Swedish scientist Arrhenius developed a theory which linked changes in atmospheric carbon dioxide with changes in surface temperatures through a greenhouse effect. But it has been later observations of changes in global temperatures during the twentieth century that has been responsible for much of the scientific and public interest. The concerns over trends in global temperatures are due, primarily, to the use of the three scenarios posed by Hicks (1973) and presented by Hoffman (1984) and the work of Mann, Bradley and Hughes (1998), whose 'hockey stick' graph was featured in the

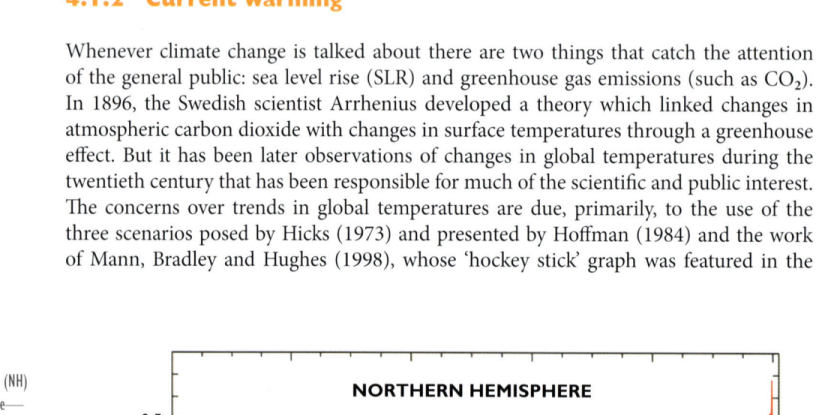

FIGURE 4-1
Millennial Northern Hemisphere (NH) temperature reconstruction (blue—tree rings, corals, ice cores, historical records) and instrumental data (red) from AD 1000 to 1999. A smoother version (black), and two standard error limits (grey) are shown. Source: IPCC Third Assessment Report (2001). Over the years, the 'hockey stick' chart has gradually become a symbol of man's impact on global climate in the post-industrial age in spite of the debates over the technical correctness and implications for global warming

Source: IPCC Third Assessment Report, 2001.

2001 United Nations Intergovernmental Panel on Climate Change report among others. In reality, most of their future forecasts, including subsequent ones by the IPCC (2000), employ a generalised version of those three scenarios.

The current warming period started after the Little Ice Age (LIA), with many placing its origins, or at least its intensification, in the mid-nineteenth century, associating it with the production of greenhouse gases arising from the industrial revolution. There still remains, though, difficulty in accurately determining the chronology of the LIA, as the available evidence from different geographic locations differs in terms of types, accuracy and completeness of parameters. Thus estimates for the times for the beginning and end of the LIA differ probably due to the asynchronous influences of global and regional processes as well as both short- and long-term trends. The following examples highlight this variation. The maritime climate of the east coast of the United States had reversed the trend by the end of the seventeenth century, whilst the generalised river floods along the Mediterranean coasts seemed to start at around the same time, only to end by the latter half of the eighteenth century. Until recently, there was said to be an obvious link between history, geography, politics and the climate in which the population, and its productivity, became climate independent precisely because of the use of energy resources. According to this relationship, the last third of the eighteenth century seems to indicate better than any other period of time the change in climate trends. All estimates for a climate minimum are subjective in part due to a lack of agreement as to what constitutes 'normal' climate, and the data in each field of knowledge are related by succession rather than by the concurrence of the phenomena they study.

The mechanisms causing the planet to warm are based on the often-misunderstood greenhouse effect. An understanding of the phenomenon required the conceptual development of quantum resonance, and an understanding of its effects on sea levels requires a complete insight into climate–atmosphere–hydrosphere interactions. All of this had already been presented globally in 1896 by Arrhenius^1, who established a logical chain which goes beyond the mere impact of the greenhouse effect on temperature:

1. Solar radiation is the cause of 'ambient' planetary warming through the pre-heating and the subordinate radiation reflected from its surface. This warming is produced by the quantum resonant capture of this radiation by 'greenhouse' substances, mainly water vapour (H_2O), carbon dioxide (CO_2) and methane (CH_4).
2. This 'ambient' temperature melts continental surface ice on the planet, increasing the mass of water in the sea (today we would have to qualify that it also dilates it), and, as a direct result, the level of its basins.
3. Consequently, all other factors being equal, the burning of fossil fuels demanded by the industrial revolution had to result in a rise in average relative sea levels, with the appropriate local deviations resulting from the accentuation or compensation of this rise.

This clear statement was overlooked for almost seventy years in general terms, and did not appear until it was taken up with the same clarity in the second half of the twentieth century, when phenomena such as the regression of coral reefs in the Pacific or generalised beach erosion, among others, triggered the alarm which the 1970 report by the Club of Rome transformed into a formal warning against the burning of fossil fuels. Since the late twentieth century there has been an increasing body of work devoted to the topic of climate change and its impact on the Earth. Climate change from greenhouse gases was mentioned in 1968 by Paul Ehrlich and in 1976 the World Meteorological Organisation issued a warning that '*a very significant warming of global climate*' was probable. At the same time other factors, such as eustatic changes (Fairbridge, 1983) and the effects of changes in the geoid along with the resurrection

^{1}Svante August Arrhenius (1859–1927) was a Swedish physicist and one of the founders of the science of physical chemistry. The Arrhenius equation is a simple, but remarkably accurate, formula for the temperature dependence of the rate constant, and therefore, rate of a chemical reaction.

of Milankovitch's on the effect of the Earth's planetary movements on climate have all contributed to a better understanding of the influences affecting climate.

4.1.3 Climate changes

Milankovitch (1930) developed a theory in which a significant role was ascribed to solar activity through the relative movement of the Earth's surface with respect to the sun and of its modulating effects on the three components of the Earth's planetary movement. This enabled him to explain the four great waves of quaternary glaciations and to predict many of the smaller ones comprising them, which have left their mark on ancient sea levels. The theory was capable of explaining with surprising accuracy not only the four glacial cycles generally referenced, but also the other significant oscillations within them. However, it cannot explain all or most of the stoppages and reversals of their associated processes, as evidenced by the unleashing of secondary effects such as the Younger Dryas (~11000 years ago), which saw a return of European glaciers in the midst of a global warming trend. Glaciologists know that our experience with the behaviour of common ice cannot be extended to the formation and distribution of glaciers. The genesis of glaciers requires processes under unique pressure and temperature conditions, which are particularly affected by the rhythms and durations of precipitation and temperature over annual and hyper-annual cycles. As the Pleistocene progressed, deglaciations occurred more rapidly than glaciations, occurring as they did in random saw-tooth patterns.

All of these variations are part of the natural components of current climate change. It is changes in climate or sea level, on a time scale that ranges from decades to centuries

FIGURE 4-2

Thermal and pressure gradients are generated in the atmosphere and oceans by solar radiation

Source: Chummu Hugh, 2010.

that are going to establish the environmental conditions of the future, giving more importance still to a precise knowledge of its origins and possible recurrence. It is ironic, therefore, that these are the changes with which we are least familiar. The question of climate change, then, centres on the nature of changes in radiative forcing, which are of two types: those involving the sun's radiant activity and of its position relative to the planet; and those involving the atmospheric greenhouse. Today the nature of solar activity is better understood and data are available to deduce its influence on climatic cycles over minor periods. These are perhaps not given the level of attention they deserve by expert panels. It seems we are also reluctant to fully recognise the influence of the hydrosphere on climate, or the fact that the role of oceans has not been completely incorporated into coupled models of global climate. In them, climate phenomena are seen as consequential, when to a great extent they are determining and causal factors. In reality, climate must be understood as the outcome of the functioning of the atmospheric–oceanic thermal machine.

4.1.4 A global thermal machine

The atmosphere and oceans as a whole comprise a double system of interactive fluids in which great thermal and pressure gradients are established due to the energy provided by solar radiation. The fact that atmospheric warming is produced from the planet's surfaces leads to inversions and gradients that, through the Earth's rotation (Coriolis) and the distribution of the continents, results in their motion. These drive the movements of the oceans, which in turn are influenced by their thermohaline gradients as well. As a consequence of this coupling of atmosphere and oceanic masses, and also of the superimposing of the variability of the cycles of the Earth's planetary rotation and oscillation together with its topology, a complex double circulation, atmospheric and oceanic, results, which drives the climate. The climate, then, is made changeable as a result of the intrinsically variable character of those interactive phenomena, but not by the nature of its own pattern.

The constancy of global climate patterns since the closing of the Atlantic Ocean (~10 million years ago) does not, however, ensure the absence of significant climate modifications. This can occur when certain changes, some resulting from the alteration of the essentially turbulent climate system, exceed certain thresholds. This has happened with for example the glaciations, the Younger Dryas period, or as occurs with phenomena associated with El Niño (ENSO) events over much shorter time periods.

The effect of carbon dioxide has perhaps been overestimated, in part because the greenhouse effect is not due exclusively to CO_2. It is also because we do not have a complete understanding of all the mechanisms and processes that regulate the proportion of greenhouse gases in the atmosphere, such as the exchange mechanism between the atmosphere and the sea. The greatest, and almost only, effort to understand the problem has involved the 'greenhouse effect', and has focused on the evolution and forecasting of atmospheric carbon dioxide. Hence the question of carbon dioxide merits special attention.

For a long time, climate models practically ignored the role of the oceans in the extraction of CO_2 from the atmosphere through the equilibrium process dictated by Raoult's Law. The role of oceans has not even been used to reject the hypothesis that the increase of atmospheric CO_2 could be due to its exclusion from the top layers of superheated oceans. Nor does there seem to be a way to incorporate the role of carbon dioxide in the oceans in the genesis of calcareous rocks, the planet's great storehouse of CO_2.

The solubility in water of carbonates in the form of bicarbonates is low but significant, and the oceanic volume is vast, meaning the amount of carbon dioxide that could be stored to the point of saturation is enormous. It is also constantly being renewed by the biogeochemical 'sedimentary thermal machine' and transformed into calcareous deposits, a complex and ongoing process that takes CO_2 out of the atmosphere in order to create calcareous sediments. However, the ability of oceans to absorb CO_2 decreases with increasing water temperature, slowing the ocean's response to emissions. Warming of waters can also lead to an increase in ocean thermal stratification, isolating the surface ocean from deeper waters and changes in the ocean's thermohaline circulation may act to decrease the transport of dissolved CO_2 into the deep ocean. However, the magnitude of these processes is still uncertain and so prevents long-term estimates of their effects on atmospheric carbon dioxide levels.

4.1.5 Records of sea level

The process of understanding the various mechanisms that link climate with sea levels and the role of global temperatures is advancing relatively slowly. Eustasy initially only accounted for the change of state from water to ice. It took time to realise that there was also a steric factor; a volumetric change in water caused by a thermal or saline effect on its density. In a strict sense, both temperature and salinity affect density, and not always in the same way, since other local factors are also at work, some of them also climatic and different from temperature. It took even longer to consider the effects of plate tectonics, isostasy and geoid deformations on the capacity of sea basins through the modification of their shapes, which obviously affects levels. Most of these factors can be considered independent of each other, though they sometimes exhibit opposing or compensatory characteristics. Finally, glacial rebound or readjustment is not the only significant source of isostasic subsidence.

We must bear in mind that historical sea levels are relative to some level on a nearby coast and therefore localised. In other words some long-standing levels can differ notably from real average levels once changes are accounted for, and can lead to errors in the trend analysis. Furthermore, the global average sea level requires a statistical treatment of historical data on a secular level to allow for a trend analysis that will enable short-term (less than a year) cyclical oscillations to be filtered out as noise. The relevance of this point it that the average relative sea level has been an important piece of data for estimating the planet's climatic condition, but that it has also been tempered by the workings of complex subsidence processes.

A heating process gives rise to an increase in ocean surface level, but since that rise could only be determined in reference to a fixed point on the coast, any modification of the crust affects its relative position. Hence the importance of a measurement system in the future that can account for the impact of these crustal movements, both vertical ones that affect the elevation itself, as well as horizontal ones that, by modifying the sea basins, affect their water-holding capacity. Also important is the maximum achievable accuracy when analysing the various causes and types of crustal movements, and calculating their magnitudes. Records from different areas of the planet point to variations of the degree of accuracy of the available measurements and this can complicate the interpretation of results or estimates of sea levels.

The development of methods for the analysis of sea level records is allowing for a progressively more reliable revision of values that were estimated in the past, after having been a trigger about the effects of Climate Change. Critical reflection (Emery and Aubrey, 1991) has led to a more prudent attitude, one that, while not predominant, is still encouraging.

TEXTBOX 4-1
Impact of climate change in Hawaii

In Hawaii, sea-level rise resulting from global warming is a particular concern. Riding on the rising water are high waves, hurricanes and tsunami that will be able to penetrate further inland with every fraction of rising tide. Rahmstorf (2007) estimates twenty-first century sea-level change on the empirical relationship between twentieth century temperature changes and sea-level changes. The study establishes a proportionality constant of 3.3 cm of sea-level rise per decade per °C of global temperature warming. When applied to future warming scenarios of the Intergovernmental Panel on Climate Change, this relationship results in a projected sea-level rise in 2100 of 0.5 to 1.4 m above the 1990 level. On the basis of Rahmstorf's research, and the documented accelerations in melting of both the Greenland and Antarctic ice sheets, it seems highly likely that a sea level of approximately 1 m above present could be reached by the end of the twenty-first century. In Hawaii, as the ocean continues to rise, natural flooding occurs in low-lying regions during rains because storm sewers back up with saltwater, coastal erosion accelerates on our precious beaches and critical highways shut down due to marine flooding.

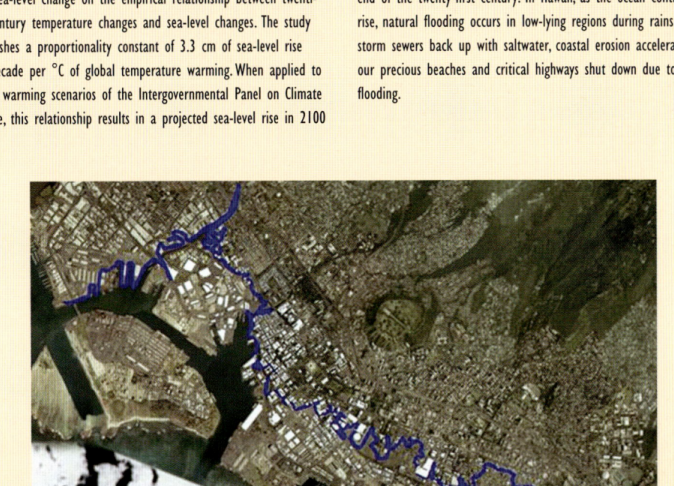

FIGURE 4-3
The blue line marks the 1 m contour above present high tide of the Hawaii's coastline

When the sea level rises 1 m above the present level, the land below the blue line will be highly vulnerable to coastal hazards.

Source: Victoria Munoz.

4.2 SIGNS OF CHANGE

Weather is what is happening outside today, it is the day-to-day conditions in a particular place. Climate is an area's long-term weather patterns characterised by certain regularities and seasonal changes such as average temperatures, rainfall intensity, type and timing, amount of sunshine, average wind speeds and direction, and other factors. From this it can be seen that there is an inherent *variability* in climate around some mean. Climate change, on the other hand, is a change in the average weather that an area experiences, persisting for an extended period (typically decades or longer) in other words it describes a change in the long-term average conditions, and it is still the case that there will be climate variability around that mean. However, climate change as it is currently being used is also taken to imply something about the rate and magnitude of that change as well as the variability of the climate patterns. This variability is usually associated with changes in the frequency and intensity of severe and extreme weather events, such as heat waves, floods, storms and droughts, that will become more frequent, more widespread or more intense. In some cases climate change is also being associated with consequent changes in nature such as the appearance of new pathogens and pests. Climate change may be due to natural internal processes or external forcing, or to persistent anthropogenic changes in the composition of the atmosphere or in land use.

Climate change is the result of a great many factors including the dynamic processes of the Earth itself, as discussed in the previous section; external forces, including variation in sunlight intensity; and more recently by human activities. External factors that can shape climate are often called climate forcings and include such processes as variations in solar radiation, deviations in the Earth's orbit and the level of greenhouse gas concentrations. The Earth's climate is governed by how much energy, in the form of heat, is stored mostly but not exclusively in the atmosphere. The processes that influence how this energy is stored and in what form can be either natural or manmade through pollution. Changes in the amount of greenhouse gases have occurred naturally and have resulted in previous changes, as discussed earlier. Changes in the way ocean water circulates have influenced climate through the way energy is stored in the oceans and transferred to and from the atmosphere. Variations in incident solar radiation—the energy input into the system—has also varied. Volcanic eruptions emitting particles into the atmosphere and affecting the amount of sunlight have also had an influence. And on a much longer timescale, processes driven by plate tectonics; mountain building, uplift and patterns of landmasses affect climate.

Human influences on climate change arise principally from the use of fossil fuels as an energy source and for the production of petroleum-based goods. The conversion of fossil fuels through combustion, for example, releases carbon and other gases that had previously been captured and stored by natural processes. The burning of fossil fuels can also produce aerosols, which are tiny particles that are suspended in the air. Other human activities that produce aerosols include industrial processes, slash-and-burn agriculture and the alteration of land surfaces by deforestation or by agricultural activity. Aerosols have an effect on climate by reflecting incident sunlight back into space and by changing the chemical composition of clouds. Livestock play a dual role in contributing to climate change. Firstly, their manure is a source of methane and nitrous oxide, both greenhouse gas. Secondly, they impact through land-use practices, the growing demand for livestock products (e.g. meat and milk) and the quantity of animals raised to satisfy the demand.

Some of the key signs of climate change for Europe are:

- Recent observations confirm that the global mean temperature has increased by 0.8°C. Europe has warmed more than the global average (1–1.2°C), especially in the south-west, the north-east and mountain areas.
- More frequent and more intense hot extremes and a decreasing number of cold extremes have occurred the past 50 years.
- Changes in precipitation show more spatially variable trends across Europe. The intensity of precipitation extremes such as heavy rain events has increased.
- Climate variability and change have contributed to an increase in ozone concentrations in central and south-western Europe.

4.3 EXPECTED CONSEQUENCES

In the previous sections you learned that climate change is a long-term risk issue, though clearly extreme climatic events can occur at any time. Most climate impacts are

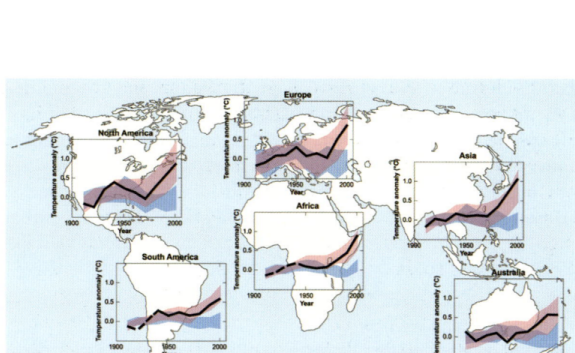

FIGURE 4-4

Comparison of observed continental- and global-scale changes in surface temperature with results simulated by climate models using either natural or both natural and anthropogenic forcings. Decadal averages of observations are shown for the period 1906–2005 (black line) plotted against the centre of the decade and relative to the corresponding average for the 1901–1950. Lines are dashed where spatial coverage is less than 50%. Blue shaded bands show the 5 to 95% range for 19 simulations from five climate models using only the natural forcings due to solar activity and volcanoes. Red shaded bands show the 5 to 95% range for 58 simulations from 14 climate models using both natural and anthropogenic forcings

Source: IPCC Synthesis Report, 2007.

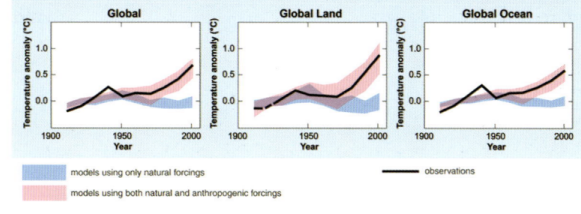

FIGURE 4-5

The modelled change in mean temperature over Europe between 1980–1999 and 2080–2099

Source: IPCC Synthesis Report, 2007.

Note: Left: annual; middle: winter (DJF); right: summer (JJA) changes in °C for the IPCC-SRES A1B emission scenario averaged over 21 models (MMD-A1B simulations).
Source: Christensen *et al.*, 2007. Published with the permission of the Intergovernmental Panel on Climate Change.

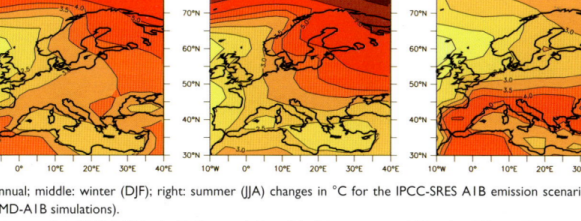

FIGURE 4-6

Modelled precipitation change between 1980–1999 and 2080–2099

Source: IPCC Synthesis Report, 2007.

Note: Left: annual; middle: winter (DJF); right: summer (JJA) changes % for the IPCC-SRES A1B emission scenario averaged over 21 models (MMD-A1B simulations).
Source: Christensen *et al.*, 2007. Published with the permission of the Intergovernmental Panel on Climate Change.

expected to intensify over the coming decades, as the climate continues to change. The predicted effects of climate change present a number of primary challenges for cities. These include an increase in the risk of flooding and winter storm damage as well as increased subsidence risk in subsidence-prone areas, water shortages and prolonged

4.3 EXPECTED CONSEQUENCES

drought. There is, however, uncertainty about the nature and magnitude of climate change and its impacts. There is also uncertainty about how these impacts should be valued, let alone how to respond to the anticipated impacts. It is important for decision makers to understand and manage this uncertainty (see Chapter 11).

Generally, climate change is likely to result in drier conditions for most regions of our globe. However, extreme rainfall events are likely to increase in frequency and

TABLE 4-1
Expected consequences of climate change

	High risks	**Extreme risks**
Water	Degradation and failure of water supply	Water shortages
	Degradation and failure of sewer piping	Stormwater drainage and flood damage
	Sewer spills to rivers and bays	
	Degradation and failure of drainage infrastructure	
Buildings	Degradation and failure of foundations	Increased storm and flood damage
	Degradation and failure of materials	Coastal storm surge and flooding
	Increased storm and flood damage to urban facilities	
	Coastal storm surge and flooding to urban facilities	
Transport	Road foundation degradation	Storm impacts on ports and coastal infrastructure
	Tunnel flooding	
	Storm damage to bridges	
	Extreme events impacts to airport operations	
Telecommunications	Storm damage to above ground transmission	
	Flooding of exchanges and underground pits, access holes and networks	
Power	Storm damage to above ground transmission	Increase in demand pressure shortages
	Substation flooding	
	Reduction in hydro-electricity generation	
	Reduction in coal electricity generation	
	Offshore infrastructure storm damage	

Source: CSIRO, (2006) State of Victoria, Department of Sustainability and Environment 2006 ISBN: 978-1-74152-864-0

intensity in most places, particularly at northern mid-high latitudes. An increasing frequency of extreme daily rainfall events would affect the capacity and maintenance of stormwater, drainage and sewer infrastructure. Significant damage costs and environmental spills are likely if these water systems are unable to cope with major downpours. Due to drier conditions, increased ground movement and changes in groundwater could accelerate the degradation of materials and the structural integrity of the water supply, sewer and stormwater pipelines. Lower rainfall is likely to lead to water shortages, exacerbated by higher temperatures and increased demand from a growing population.

Cities are particularly vulnerable to climate change since most of the urban fabric such as the infrastructure and buildings have been designed, built and maintained on the premise that the future climate will be similar to that experienced in the past. Recognition of the risks associated with climate change (see Table 4-1) is a first step towards better planning of new infrastructure investments and mitigating potential damage to existing infrastructure.

TEXTBOX 4-2
Case study Greater Manchester

Current and future flood risk in Greater Manchester in the Northwest Region of England. This example illustrates how the forecasting of future climates can be used in urban flood-risk management.

Current vulnerability to climate change

Current vulnerability to climate change should be assessed by combining a knowledge of past flood events in a given area and comparing them with the best available climate data over a period of time. The key changes in climate between 1961 and 2006 in the Northwest of England which affect vulnerability to flooding are summarised in Table 4-2.

The UK Climate Impacts Programme (UKCIP) also found a significant increase in the number of severe storms over the UK as a whole since the 1950s.

In Greater Manchester, pluvial flooding has become more prevalent and more severe than riverine flooding since the 1990s (Figure 4-7) and most pluvial flood events happen during the summer months (April–September). This increase in all flood events can be attributed to the overall increase in precipitation, and in particular to the considerable increase in winter precipitation since 1961. The increase in pluvial flooding since 1998 reflects the increase in severe storm events of recent times. Virtually every day when precipitation exceeded 38 mm at the Ringway weather station at the southern periphery of Greater Manchester (Figure 4-8), there was flooding, although many flood events happened when the daily precipitation at Ringway was in the region of 25–30 mm and even lower.

Future vulnerability to climate change

Forecasts of future climate are based on scenarios compared with the average climate from 1960 to 1990. Those covering the northwest region of England are modelled at a scale of 50 km^2 UKCIP from weather recordings at Ringway. The scenarios covering flood risk are summarised in Table 4-3.

TABLE 4-2
Changes in key climate variables in Northwest England (1961–2006) affecting vulnerability to flooding

	Spring	Summer	Autumn	Winter	Annual
Mean temperature (°C)	1.44	1.45	1.07	1.81	**1.40**
Days of rain >1 mm	0.4	–1.1	2.9	6.8	**7.5**
Total precipitation (% change)	6.3	–13.2	5.6	43.0	**8.8**

4.3 EXPECTED CONSEQUENCES

FIGURE 4-7

Riverine/pluvial flooding events in Greater Manchester

Source: UKCIP, 2009.

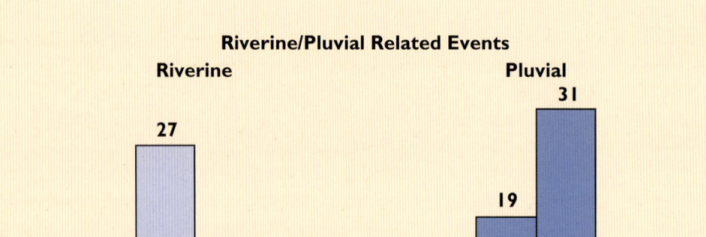

FIGURE 4-8

Greater Manchester: physical geography and principal weather stations

Source: UKCIP, 2009.

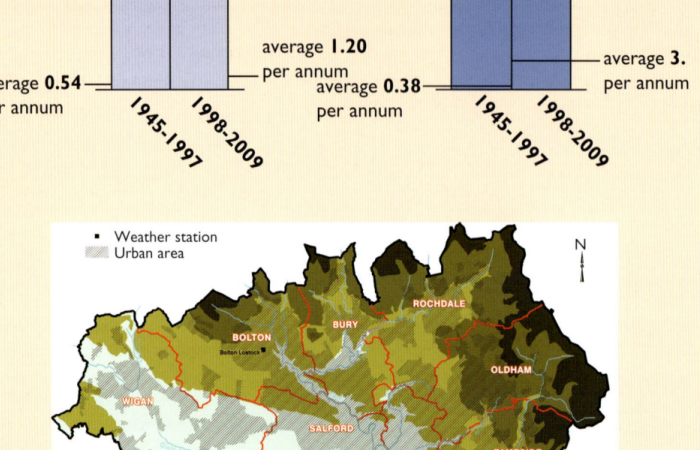

TABLE 4-3

UKCIP02 climate projections for Northwest England compared with the average for 1960 to 1990

	2020s (2011–2040)	**2050s (2041–2070)**	**2080s (2071–2100)**
Changes in winter rainfall	0 to 10% increase	0 to 20% increase	10 to 60% increase
Changes in annual snowfall	10 to 30% decrease	30 to 60% decrease	40 to 100% decrease
Changes in summer and autumn soil moisture content	Not available	Not available	20 to 50% decrease
Changes in sea level	Not available	7 to 36 cm	9 to 60 cm

Significant to urban flood risk, depressions moving across the UK from the North Atlantic are predicted by UKCIP to increase, thus leading to the possibility of more frequent and more intense storm events.

These climate scenarios, as well as the predicted increase in storm events, are most likely to significantly add to the vulnerability of urban conurbations such as Greater Manchester to the risk of flooding. To reduce this vulnerability, agencies concerned with urban drainage (the Environment Agency, Local Authority highways and drainage departments, the utility company) flood forecasting and mapping (the Met Office and the Environment Agency), emergency services (civil contingency departments, the Fire Service, the Police), land-use control (planning departments) and Building Control, need to act in a cohesive manner over the entirety of the conurbation. Local Authority planners will need to consider how best to improve the drainage system, increase natural infiltration rates and provide for Sustainable Urban Drainage Systems (SUDS) to reduce runoff. Furthermore, flood victims suffer considerable disruption and stress and this will require input by social and public health services.

In 2009, UKCIP published new probabilistic weather scenarios (i.e. including extremes and the level of uncertainty) at a 25 km^2 grid scale which will be downscaled to 5 km^2 using a weather generator tool. However, the varied geography of an urban area such as Greater Manchester means that these new scenarios will have to be topographically modelled because they will still be based on the Ringway weather station, the only available continuous climate data for the region (Figure 4-8).

To further reduce uncertainty, UKCIP generally advise that Local Authorities determine time frames for consideration of impacts and adaptation strategies and they suggest three broad time zones:

- Short term 2009–2015: projected by slightly extending current trends.
- Mid term 2015–2045: the 2009 scenarios, which reflect current emissions to the atmosphere, should be considered as being 'certain' and based on a single set of projections.
- Long term 2045–2075: to be based on emission scenarios because carbon usage is still uncertain.

TEXTBOX 4-3
Flooding in Colombia

Studies of Colombia have examined the consequences of a one-metre sea-level rise over the next 100 years. These have concluded that in addition to the erosion of beaches, marshes and mangroves, there could be permanent flooding of 1,900 square miles in low-lying coastal areas, affecting 1.4 million people, 85 per cent in urban areas. In two major cities on the Caribbean coast, Barranquilla and Cartagena, most manufacturing facilities are in highly vulnerable areas, as are 45 per cent of roads in the area (with another 23 per cent slightly vulnerable). In the tourist-oriented island of San Andrés, 17 per cent of the land would be flooded, especially along the northern and eastern shores where most economic activity and the richest natural resources are concentrated. In general, the Caribbean region of Colombia is projected to become drier while the southern region of the country becomes wetter. Since, however, some of the major rivers' headwaters are in the southern region of the country, heavier precipitation in the south may result in increased flooding in northern river deltas. River deltas also will be at risk of increased flooding from rising sea levels.

FIGURE 4-9
Flooding in the streets of Catagena, Colombia

Source: Victoria Bejarano (Dura Vermeer), 2010.

TEXTBOX 4-4
Climate change impacts on river flows and runoff

There are several hundred studies that have looked at the potential effect of climate change on river flows. Most of these have been confined to Europe, North America and Australasia. Few global-scale studies that have been conducted using both runoff simulated directly by climate models and hydrological models. Most show that runoff increases in high latitudes and the wet tropics, and decreases in mid-latitudes and some parts of the dry tropics. Runoff and flows reduce in southern Europe and the major rivers of the Middle East and Central America and increase in south-east Asia and in high latitudes. The 2008 IPCC Report on Climate Change and Water stated that warming would lead to changes in the seasonality of river

flows where much winter precipitation currently falls as snow, with spring flows decreasing because of the reduced or earlier snowmelt, and winter flows increasing. This was found for the European Alps, Scandinavia and around the Baltic, Russia, the Himalayas, and western, central and eastern North America. The effect was greatest at lower elevations, where snowfall is more marginal, with peak flows occurring at least a month earlier. In regions where changes in runoff are much more dependent on changes in rainfall than on changes in temperature there is a projected increase in the seasonality of flows, often with higher flows in the peak flow season and either lower flows during the low-flow season or extended dry periods.

FIGURE 4-10

Large-scale relative changes in annual runoff for the period 2090–2099, relative to 1980–1999. White areas are where less than 66% of the ensemble of 12 models agree on the sign of change, and hatched areas are where more than 90% of models agree on the sign of change

Source: Milly et al., 2005.

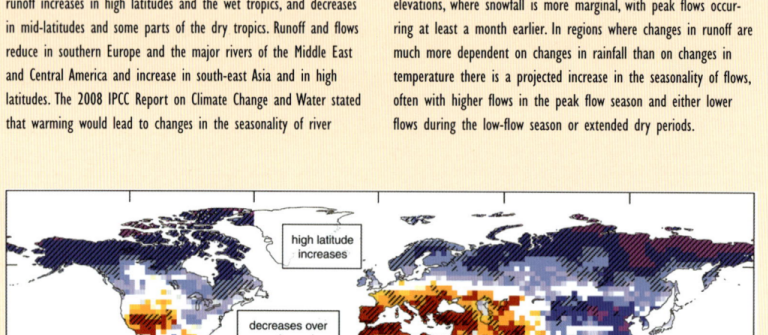

Key Questions

Climate change: a review of the past

1. What are the most important natural forces that change sea level?
2. Which mechanisms causing the planet to warm?
3. Couldn't a warmer climate be better?

Signs of change

1. What is a global climate model?
2. How reliable are predictions of future climate?

Expected consequences

1. Why should a few degrees of warming be a cause for concern?
2. What are the impacts to citizens from developing countries that we can expect from climate change?
3. In what respect do they differ from developed countries?

FURTHER READING

Dow, K. & Dowing, T.E. (2006). *The Atlas of Climate Change. Mapping the World's Greatest Challenge.* Earthscan. Myriad Editions Limited. ISBN 978-1-84407-522-5.

Intergovernmental Panel on Climate Change (2007). 'Summary for Policymakers'. In: Parry, M., Canziani, O.F., Palutikof, J.P., van der Linden P.J., & Hanson, C.E. (eds) *Climate Change 2007: Impacts, Adaptation and Vulnerability. Contribution of Working Group II to the Fourth Assessment Report of the Intergovernmental Panel on Climate Change.* Cambridge, UK: Cambridge University Press., pp. 7–22.

Mann, Michael E., Bradley, Raymond S. & Hughes, Malcolm K. (1998). Global-scale temperature patterns and climate forcing over the past six centuries, *Nature, 392,* 779–787.

Milankovitch, M. (1930). *Mathematische Klimalehre und Astronomische Theorie der Klimaschwankungen, Handbuch der Klimalogie Band 1.* Teil A Borntrager Berlin.

Milly, P.C.D., Dunne, K.A. & Vecchia, A.V. (2005). Global pattern of trends in streamflow and water availability in a changing climate. *Nature* 438: 347–350.

Rahmstorf, S. (2007). A semi-empirical approach to projecting sea level rise. *Science, 315,* pp. 368–370.

Roaf, S., Chrichton, D. & Nicol, F. (2005). *Adapting Buildings and Cities for Climate Change: A 21st Century Survival Guide.* Architectural Press, Elsevier.

Stern, N. (2007). *The Economics of Climate Change: The Stern Review.* Cambridge, UK: Cambridge University Press.

UK Climate Impacts Programme (2003). *Building Knowledge for a Changing Climate: The Impacts of Climate Change on the Built Environment.* Oxford, UK. Available from: http://www.ukcip.org.uk.

UK Climate Impacts Programme (2009). *UK Climate Projections 2009: Science report.* Oxford, UK.

III

Urban flood risk

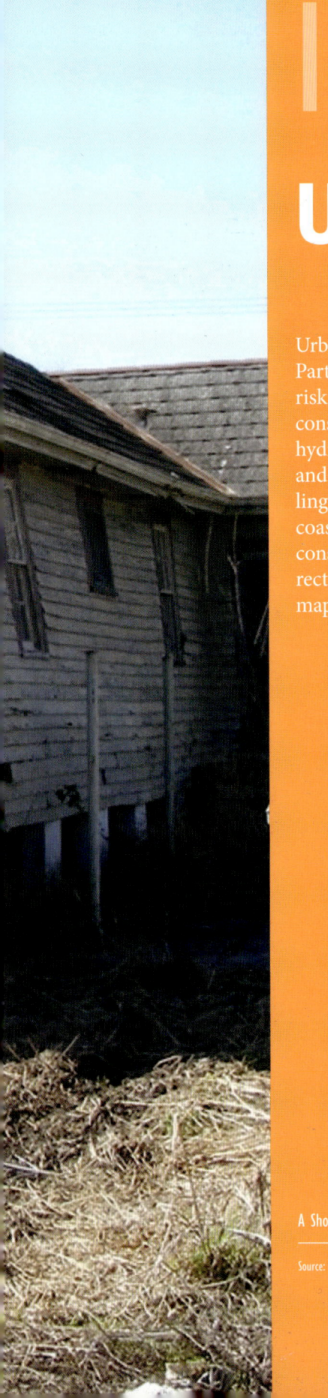

Urban floods have large impacts particularly in terms of economic and social losses. In Part III you will become familiar with the different components that constitute flood risk. Flood risk is commonly thought of as being a combination of the probability and consequences of flooding. To analyse flooding requires a basic understanding of the hydrology of cities. Therefore, Part III firstly describes the urban hydrological cycle and the processes within the cycle that result in flooding. It goes on to discuss modelling techniques for both surface runoff and water flows at different scale levels (fluvial, coastal and pluvial). Thirdly, it deals with the assessment of both the probability and the consequences of flooding. The consequences are discussed in terms of direct and indirect losses, and tangible and intangible losses. Part III ends with a chapter on flood-risk mapping.

A Shot Gun house abandoned after Hurricane Katrina (2005), Lower Ninth Ward, New Orleans.

Source: Chris Zevenbergen, 2009.

The hydrology of cities

Learning outcomes

This chapter provides an overview of the consequences of urbanisation on the hydrology of towns and cities. It introduces some of the modelling techniques that have been developed to provide the basis for flood-risk assessment and management and in particular, it looks at pluvial and coastal flooding. By the end of this chapter, you will have been introduced to:

- The impact of urban systems on water quality and the hydrology of built areas.
- The factors associated with land use in urban areas that influence how water flows.
- The different approaches that there are to urban hydrological modelling.
- How pluvial and coastal flooding events can be modelled.

5.1 THE HYDROLOGICAL CYCLE

5.1.1 The urban water system

Hydrology is the science that describes the occurrence and behaviour of water above, over and through the Earth. The continuous movement of water is called the hydrological cycle, because the amount of water remains fairly constant over time. The hydrological cycle is driven by solar energy and gravitation that cause continuous interrelated processes such as evaporation and transpiration, condensation, precipitation, infiltration and runoff. In contrast to the global hydrological cycle, in urban areas there is in practice almost never a closed hydrological cycle. Many cities depend on water resources from surrounding rural areas and discharge their wastewater into the sea or rivers outside the city, so that the urban water cycle is not being closed. However, since the start of the twenty-first century, various attempts have been made to create more sustainable urban development by closing the urban water cycle in order to reduce the environmental load of cities on their surroundings and reduce the dependency of cities on water resources form surrounding areas. Textbox 5-1 gives an example of such an attempt in the case of the Aurora development in the city of Melbourne, Australia.

In urban areas, five interrelated types of water can be determined: groundwater, surface water, stormwater, drinking water and wastewater. Figure 5-1 gives an overview of the urban water system and the relation between the different types of water. It shows that wastewater from households and industry is transported to a treatment facility by

TEXTBOX 5-1
Closing the water cycle in Aurora Estate

The Aurora Estate in Melbourne, Australia is a residential development under construction of approximately 650 ha and 10,000 lots. The estate is developed as a demonstration for sustainable urban design and includes water sensitive design features to reduce the environmental impact of the development and to reduce its vulnerability to drought. The developers aimed at to create a new benchmark for water sensitive urban design (WSUD) that considered water supply, wastewater and stormwater as integral streams in order to achieve the following goals:

- to minimise the amount of potable water imported to the site;
- to minimise the amount of wastewater generated from the development;
- to maximise the use of recycled wastewater;
- to maximise the use of rainwater for those end users not considered acceptable to be supplied by recycled water
- to minimise the stormwater pollutant loads discharge from the site;
- to implement mimicked natural stormwater flows

The built form was included in the strategic planning of the project and a wide range of WSUD techniques, including sewage water recycling, rainwater harvesting and retention of stormwater runoff, were applied at the estate level, streetscape level and allotment level. This has resulted in a reduction of 70% of import of potable water to Aurora Estate compared to conventional developments.

Source: McLean, 2004.

a sewer system, after which it is discharged into surface water outside the city. In the case of Figure 5-1, this is taking place in a combined sewer system that conveys both wastewater and stormwater runoff. However, during heavy rainstorms the capacity of the combined sewer systems could be exceeded, leading to combined sewer overflows (CSOs) taking place. This leads to the emission of diluted wastewater and sewage sludge to the urban surface water. It is now generally accepted that relatively clean stormwater should not be mixed with wastewater flows. Therefore, separate sewer systems that drain stormwater to the urban surface water and wastewater to the treatment plant are more or less standard in contemporary urban drainage systems. To reduce peak flows from surface runoff, process stormwater infiltration facilities and other Best Management Practices (MPPs), also called Sustainable Urban Drainage Systems (SUDS), have, since the late 1980s been increasingly implemented (see Chapter 9).

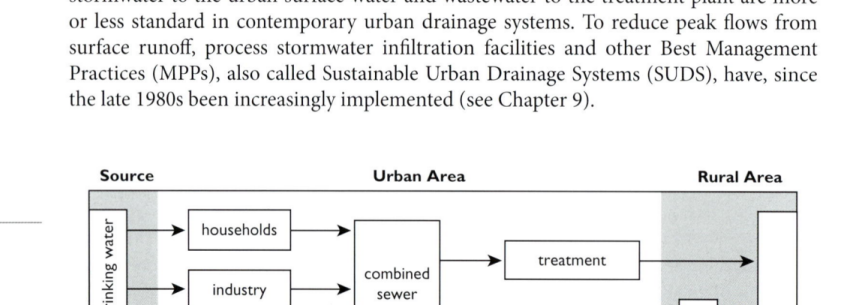

FIGURE 5-1
The urban water system

Source: Jeroen Rijke.

R runoff	D drainage	G groundwater flow	IR irrigation	L leakage
S supply	E evaporation	I interception	P precipitation	O overflow
C capillar flow	ET evapotranspiration	IN infiltration	PE percolation	

Depending on the land cover and climate of a certain area, precipitation is transformed into groundwater, stormwater runoff and vapour. Before precipitation reaches the soil, it can be intercepted by for example trees or buildings with flat roofs. This water generally evaporates. However, when precipitation does reach the soil, surface water does not only evaporate, it also runs off and/or infiltrates the saturated and unsaturated groundwater zones (see Figure 5-2). The percentage of paved area and the climate influence the share of water that evaporates and runs off. The permeability of the soil determines the percentage of water that infiltrates into the saturated and unsaturated zones. In moderate climates, precipitation on natural ground cover can for example result in approximately 40 per cent evapo-transpiration, 10 per cent runoff, 25 per cent shallow infiltration and 25 per cent deep infiltration (see Figure 5-2). In dense urban areas with up to 75 to 100 per cent impervious land cover, a completely different pattern occurs (30 per cent evapo-transpiration, 55 per cent runoff, 10 per cent shallow infiltration and 5 per cent deep infiltration).

It is of interest that:

- Only a limited part of the precipitation—much less than we tend to expect for an urban area—is discharged through the storm sewerage system.
- The subsurface drainage discharge is higher than generally would be expected, certainly on the almost completely paved parking lot.
- Evaporation from paved terrain is less than from unpaved terrain, but certainly cannot be ignored in the water balance.

An interesting addition to this water balance is provided by the import of drinking water and industrial water. Using our average daily consumption of drinking water per capita for the residential area would mean that drinking water input exceeds 70% of the total precipitation input.

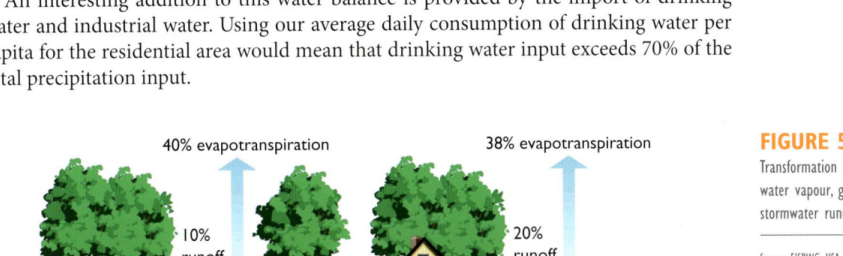

FIGURE 5-2
Transformation of precipitation to water vapour, groundwater and stormwater runoff

Source: FISRWG, USA, 1998.

TABLE 5-1

Water balance of two experimental basins in Lelystad, 1968–1980

	Residential area		**Parking lot**	
	mm	**%**	**mm**	**%**
Precipitation	698	87	739	88
Upward seepage	108	13	101	12
Total inflow	**806**	**100**	**838**	**100**
Discharge stormwater sewerage	159	20	376	45
Subsurface drainage discharge	320	40	337	40
Evapo-transpiration unpaved area	214	27		
Evaporation paved surface	75	9	112	13
Evaporation solitary trees	27	3	27	3
Change in groundwater storage	11	1	5	1

The size of the residential area is 4.5 ha of which 41% is paved; the parking lot is 0.7 ha with 99.7% paved. The amounts are expressed in mm per year over the total surface of the basin.

Source: Van de Ven, F.H.M. and Voerman, B.R. (1990) The water balance of an urban area; Experiences of two monitoring sites in Lelystad, The Netherlands (in Dutch), H_2O, 18(8), p. 170–176.

5.1.2 Polder systems

Many cities across the globe are located in low-lying delta regions. Polders are most commonly found, though not exclusively so, in river deltas, former fen lands and coastal areas.

Sometimes delta cities are located in polders. A polder is an area protected by dykes that are located below sea level. A polder forms an artificial hydrological entity, meaning it has no connection with outside water other than through manually operated devices. There are three types of polder:

- land reclaimed from a body of water, such as a lake or the sea bed;
- floodplains separated from the sea or river by a dyke;
- marshes separated from the surrounding water by a dyke and subsequently drained.

Due to subsidence, all polders will eventually be below the surrounding water level some or all of the time. Polder land made up of peat (former marshland) will show accelerated compression due to the peat decomposing in dry conditions.

Due to the difference in water levels outside and inside the polder, upward seepage occurs. This excess of water needs to be pumped out or drained by opening sluices at low tide. In polder systems, seepage is an additional source for floods and should be adequately dealt with by polder systems that have the capacity to retain and drain excess water. The seepage velocity depends on the differences in piezometric level and the resistance that the flow meets in the confining layers. Often a 'boezem' is located between the polder and the natural waterways, which acts as intermediate storage areas. Figure 5-3 shows cross sections of a polder during dry and wet periods.

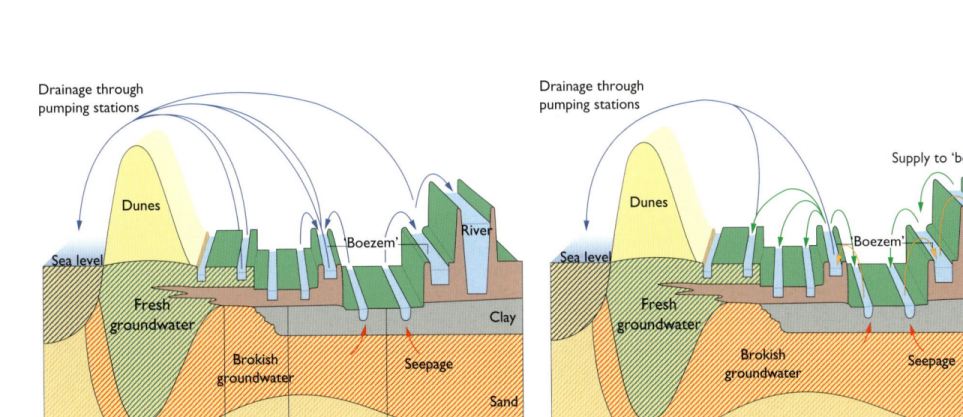

FIGURE 5-3
Cross sections of polders in the Netherlands in wet (left) and dry (right) periods

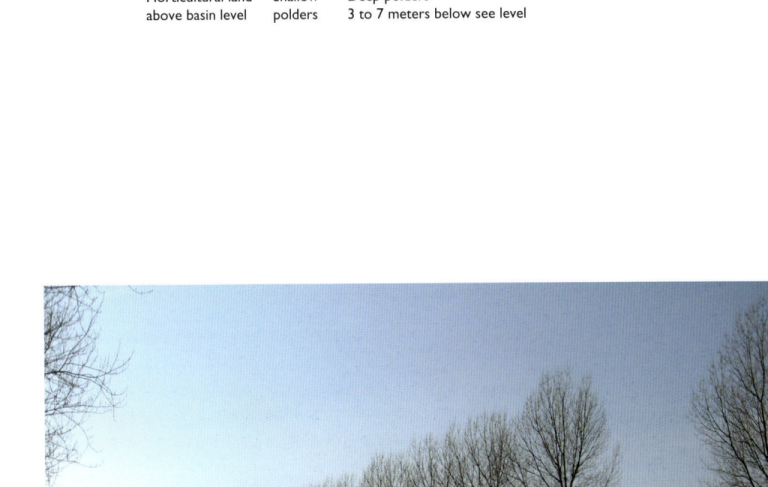

FIGURE 5-4
Overbroekse Polder, The Netherlands

Source: Chris Zevenbergen, 2007.

5.1.3 Water quality

Urbanisation often results in the pollution of urban streams and other receiving waters. The quality of runoff is influenced by many factors, including land use, waste disposal and sanitation practices. There are a number of pollutants of principal concern, see Table 5-2. These pollutants may interact to varying degrees in an antagonistic, additive or synergistic fashion, affecting organisms in receiving waters.

Significant amounts of pollutants ranging from gross pollutants to particulate and soluble toxins are generated from urban catchments. Roads and other transport-related impervious surfaces contribute a higher proportion of stormwater pollutants than other impervious surfaces (e.g. roof areas, pedestrian, pathways). Runoff from

TABLE 5-2

Environmental impacts of pollutants from urban stormwater runoff in receiving watercourses

Pollutant	Source	Environmental impact
Oxygen demanding materials	Vegetation, excreta and other organic matter	Depletion of dissolved oxygen concentration, which kills aquatic fauna (fish and macro-invertebrates) and changes the composition of the species in the aquatic system. Odours and toxic gases form in anaerobic conditions.
Nutrient enrichment (eutrophication) caused by inorganic compounds of nitrogen and phosphorus	Fertilisers and vegetative litter, animal wastes, sewer overflows and leaks, septic tank seepage, detergents.	In high concentrations, ammonia and nitrate are toxic. Nitrification of ammonia micro-organisms consumes dissolved oxygen. Excessive weed and algae growth block sunlight which affects photosynthesis and causes oxygen depletion.
Oils, greases and gasoline	Roads and parking areas (spillages and leakages of engine oil). Vegetable oils from food processing and restaurants and food preparation on the street	Carcinogenic, tumorigenic and mutagenic in certain species of fish. Pollution of drinking water supplies and affects on recreational use of waters.
Toxic pollutants - heavy metals, pesticides/ herbicides, and hydrocarbons	Industrial and commercial areas, leachate from landfill sites and improper disposal of household chemicals	Toxic to aquatic organisms and can accumulate in the food chain. In addition, they also impair drinking water sources and human health. Many of these toxicants accumulate in the sediments of streams and lakes.
Suspended solids, sediments, dissolved solids, turbidity	Erosion from construction sites, exposed soils, street runoff and stream bank erosion also drainage system decay (where one exists)	Sediment particles transport other pollutants that are attached to their surfaces. Interference with photosynthesis, respiration, growth and reproduction. Reduced hydraulic capacity of drainage channels, reservoirs and lakes etc. Sediments deposited also reduce the transfer of oxygen into underlying surfaces.
Higher water temperatures	Runoff flows over impervious surfaces (asphalt, concrete, etc.) and increased temperature from industrial wastewater.	Reduced capacity of water to store dissolved oxygen. Impact on aquatic species that are sensitive to temperature.

Source: Urban Stormwater Management in Developing Countries. Jonathan Parkinson and Ole Mark. ISBN: 1 84339 057 4. Published by IWA Publishing, London, UK.

transport-related surfaces consistently shows elevated concentrations of suspended solids and associated contaminants such as lead, zinc and copper, as well as other pollutants, such as hydrocarbons. In addition, problems related to microbiological pollution are mainly caused by the flooding of sanitation systems. The discharge of pathogenic bacteria (faecal coliforms, faecal streptococci, enterocci) and other micro-organisms (viruses, protozoa, etc.) can cause intestinal infections.

5.2 LAND USE AND RUNOFF

Increasing rates of urbanisation and land-use changes raise questions as to how this can affect rainfall–runoff processes and how this in turn can influence the occurrence of flood events and flood risk overall. The runoff process comprises three main components—surface runoff and subsurface runoff consisting of lateral flow and base flow. What can be influenced is either each component separately or the distribution of the total. At the initial stage, part of the rainfall is removed by evapo-transpiration and interception—which influences the amount of water remaining to form the flood.

The distribution of runoff is crucial for the formation of a flood and this is different with different types of floods (e.g. pluvial or fluvial floods). Since the runoff process is very complex, see Figure 5-5, it is beyond the scope of this book to precisely describe the influence of land use and land-use changes on this process. Therefore, in this chapter only the most important aspects will be described.

5.2.1 Amount of water forming runoff

Textbox 5-2 illustrates the fundamental relationship defining the components of water landing as rainfall (P, mm) on a surface; each term is time dependent. The first consideration is the amount of water which is available for runoff. Considering the precipitation as an input to the system, the amount of water available for runoff is the volume of precipitation reduced by evapo-transpiration and other losses. This reduction is significantly affected by land use in many ways. The main aspect is the vegetation cover which determines the rate of evapo-transpiration and the amount of interception. Therefore, areas with dense vegetation cover produce less runoff than areas with less or no vegetation. The evaporation process thus differs from surface to surface. Other important factors include temperature differentials, different volumes of areas available for surface storage (e.g. puddles) and different absorption of solar radiation.

The volume of space in depressions on the surface available for temporary water storage differs for different land use types and this may be formal or informal. Formal storage

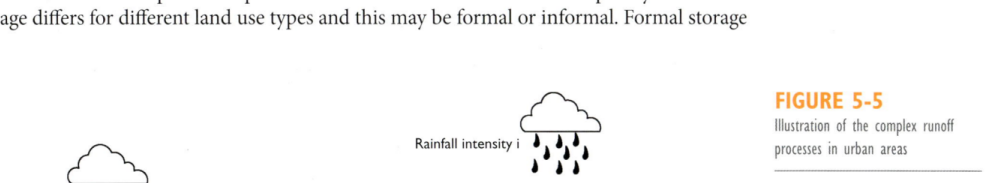

FIGURE 5-5
Illustration of the complex runoff processes in urban areas

Source: Slobodan Djordjević.

TEXTBOX 5-2

Expression of mass conservation law for water balance on surfaces

$$P = R + \Sigma L + \Delta S \tag{5-1}$$

Where

- P is precipitation (mm),
- R is runoff (mm),
- ΣL is sum of losses (evatranspiration, interception etc.) (mm),
- ΔS is change of storage (mm).

is intentional, whereas informal is what occurs in local depression that are part of the general landscape. This can significantly affect runoff processes in two ways. The first is that water trapped and stored on the surface can be evaporated. This effect can be added to the group of effects which reduces the total amount of water available for runoff. The second effect is the retention of water which then has the opportunity to infiltrate into the soil.

The factors so far described are those with the most significant influence on the reduction of water available for runoff. All other factors affect the distribution of the runoff into the three types outlined above and also the velocity of flow during the runoff process.

5.2.2 Water distribution into different types of flow

Another effect of the depression storage mentioned in the previous section can be observed in a change in the runoff distribution between surface and subsurface forms. In principle, the water which is stored in depressions can infiltrate and is runoff through lateral flow and/or base flows that are much slower than surface runoff.

The biggest influence on the distribution of flow is in the infiltration process itself. This process is mostly determined by surface cover and type and soil properties and thus the influence of land use is significant. The key parameter for infiltration is hydraulic conductivity. This parameter is connected mainly with soil properties and can be affected by land use (e.g. porosity, structure, soil moisture content). Furthermore, the hydraulic conductivity is time variable, as it is dependent on soil moisture content which varies with time.

Vegetation cover is an important land use factor. The vegetation (especially the root mass and rhizomes) influences infiltration mainly by affecting porosity, soil structure and regime of soil moisture content; which is also affected by exposure of the soil surface to sun. The exposure is again dependent on density of vegetation cover and its condition. The vegetation cover can also influence the occurrence of preferential pathways in the subsurface flow.

Another land use factor is the extent of urban/paved/sealed areas which are usually characterised by mostly impermeable surfaces causing rapid surface runoff as illustrated in textbox 5-3, which shows typical pre- and post-development runoff flows (hydrographs).

5.2.3 Affect on flow velocity

Considering surface flows as one component of the runoff process, the velocity of flow is also affected by land use. What is most important for the overland flow is the

FIGURE 5-6
Flood flow in the Służew Creek (May 28, 2002)

Source: Kazimierz Banasik, 2002.

FIGURE 5-7
The same place as in Figure 5-6 but six years later (May 25, 2008)

Source: Kazimierz Banasik, 2005.

roughness of the surface, which can vary considerably among different land use types. From another point of view, the whole process (mainly the flow velocity) is also affected by artificial drainage systems—whether they are surface or subsurface. These systems usually make the runoff process faster which can lead in higher peaks of runoff hydrograph.

TEXTBOX 5-3

Hydrological extremes in Służew Creek Catchment in 2007; case study (Warsaw)

The usefulness of the Instantaneous Unit Hydrographs (IUH) method estimated with Rao equations for predicting the flood flows in a small urban catchment.

The Służew Creek catchment (located in Warsaw, Poland) has been analysed by the Department of Water Engineering and Environmental Restoration (Warsaw University Life Sciences—SGGW) since the middle of the 1980s. Since then, the catchment has been changed from minimal urbanisation to strongly urbanised. As a consequence of land-use change, there are observed flash floods every year that lead to flood damage in the catchment.

A research team has conducted an analysis of comparison of two methods of Instantaneous Unit Hydrograph (IUH) estimation—on the basis of recorded data and on the basis of Rao equations, with the use of catchment characteristics—and assessed how the second one could be useful in predicting the flood flow in situations of limited information about the catchment. Both of the estimated IUH were based on the Nash model, in which the catchment is depicted as a cascade of N linear reservoirs with retention parameter k for each reservoir; and the CN-SCS method was used to evaluate effective rainfall. In the first method, based on recorded data, calculations used measured rainfall data and direct runoff data. Analysis was conducted for one hydrological year (2007), for a few chosen events with the highest water level. In the Rao method, the catchment was divided into two sub-catchments with different levels of urbanisation. The

total IUH was calculated for the catchment as a whole, taking differences of urbanisation for each sub-catchment into consideration. The estimations used the following equations, (Rao and Hamed, 2000):

$$LAG = 1.28 \cdot A^{0.46} \cdot (1 + U)^{-1.66} \cdot H^{-0.27} \cdot D^{0.37}$$
(5-2)

$$k = 0.56 \cdot A^{0.39} \cdot (1 + U)^{-0.62} \cdot H^{-0.11} \cdot D^{0.22}$$
(5-3)

Where:

LAG – lag time [h]

k – Nash model parameter [h^{-1}],

A – area of the catchment [km^{-2}],

U – rate of impervious area to pervious area in the catchment [-],

H – effective rainfall [mm]

D – rainfall duration [h]

The analysis has shown that IUH estimated with the use of Rao formulas based on catchment characteristics was similar to the measured IUH, so the Rao formulas could be useful to predict flood flows in urban catchments in situation of limited information. Estimated IUH are used to evaluate the maximum flood flows and the time of peak. The method allows the user to change parameters of land use, as the watershed is changing, which is important as far as sprawling cities are concerned.

FIGURE 5-8
Map of the Służew Creek catchment

Source: Kazimierz Banasik, 2005.

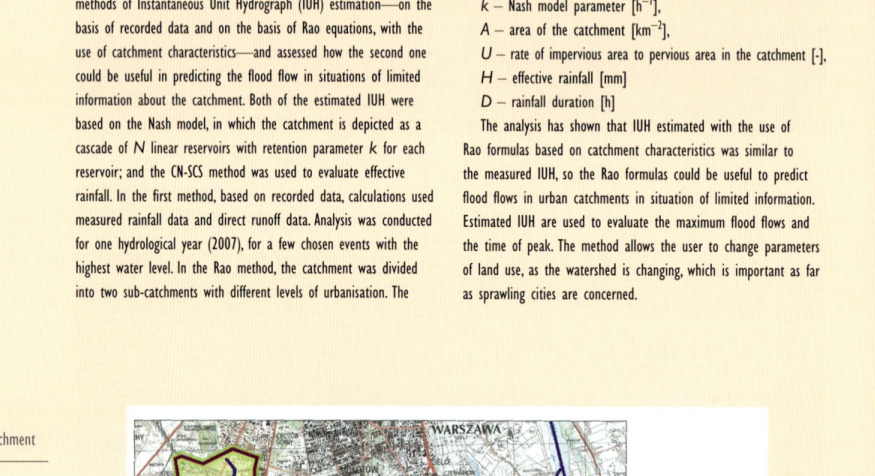

5.2.4 Influence of land-use change on runoff in a small urban catchment

The development of cities is inevitably connected with taking over new places, usually agricultural areas or open spaces, and changing their previous use into urban land. According to the European Environment Agency, some 48 per cent of the agricultural areas taken by growing cities, consist of arable land and permanent crops, followed by some 36 per cent that includes pastures and mixed farmland. As a consequence, the areas within watersheds, which produce a low value of runoff, are replaced by areas with much higher runoff. Therefore, less water is detained inside the watershed and more water generates sheet flow, which becomes flood flow after reaching a channel.

The increasing proportion of impervious areas to pervious areas in watersheds has an impact on the volume of runoff generated from watersheds and the time of concentration (the time from deposition of rainfall on the ground to formation of the runoff at the water gauge closing the catchment).

The most commonly visible consequences of the land-use change are flash floods observed in urban areas after high precipitation events. It is known that the larger urban rivers are reasonably well characterised because of the need to produce hazard ratings. However, the small urban rivers and watercourses are more problematic due to the limited information on them, their limited floodplain and a faster reaction to precipitation events. Limited floodplains in cities limit the ability of hydrologists to predict and therefore mitigate flood flows. Although there are available mathematical models that can be used for modelling the rainfall–runoff process their ability is dependent on the watershed data available. For catchments with limited information, the prediction of flash floods, although desirable, is problematic and subject to a high degree of error. However, the hydrological conceptual approach, based on SCS-CN (Soil Conservation Service—Curve Number) and IUH (Instantaneous Unit Hydrograph) method, has been successfully used.

The method requires parameters that are simple to estimate—like land-use, soil group and hydrological condition. As important as the value of runoff formed from the watershed, is the process of reconstruction of the shape of the runoff hydrograph, especially the estimation of the time and value of the peak flow. For catchments with limited information, the Rao method is a simple prognostic method, where the ratio of impervious area to pervious area is included.

Given that most cities are growing, the ratio of impervious area to pervious area in urban watersheds can be expected to rise. This raises the question about the possible usefulness of the application. The method could only become useful, if it takes the changing land use into account and allows for the change in input parameters.

5.3 MODELLING SURFACE RUNOFF

5.3.1 Introduction

Urban flooding is usually investigated and modelled by taking into account only the runoff generated in the urban area that is of interest, despite the fact that dependence of this event on the runoff from upstream parts of the relevant river catchment is well known. Nevertheless, separate modelling of urban areas and river basins is still a common practice.

The runoff amount and intensity on the catchment scale is strongly related to the impervious area cover, which is the predominant cover type in urbanised area. Small, densely urbanised catchments are more strongly affected by the urban runoff than large river basins. Urban floods can have both area-wide and local origin, and can be accompanied by water pollution problems. Generally, flooding in cities originates from extremely high flows and stages in major neighbouring rivers as a result of severe regional meteorological disturbances, or from local high intensity thunderstorms occurring over parts of the urban area. Modelling of runoff requires both the understanding of urban hydro-meteorological issues and knowledge of the physical characteristics of specific flood events.

5.3.2 Runoff modelling concepts

Many models have been developed for river catchments throughout the world, with various degree of complexity, from simple empirical formulae or correlations to the complex mathematical models, representing all phases of the water balance of a river basin. The choice of a specific model depends on the type of information required and how the results are to be applied. The number and types of assumptions within a model, the kinds of data needed and the level of complexity are important factors when deciding which model to use. Simulated runoff as a result of a hydrological model can be used further as input into hydraulic models to simulate floodplains, which is particularly important for urban areas.

It is important to understand the mechanisms that produce flooding in an urban catchment. The stormwater runoff process is generally split into two different subprocesses: *precipitation losses* and *retardance* of the flow on its way to the stormwater inlets and in the pipes of the sewer system. The precipitation loss is the amount of water that is not running off via the sewerage system.

The precipitation loss is expressed in the *discharge coefficient*, which is the ratio between the discharged quantity and the amount of precipitation that has fallen on the directly connected paved surface.

When dealing with precipitation loss, losses on paved and unpaved terrain need to be distinguished. Often it is assumed that unpaved terrain does not deliver runoff to the sewerage system. In a number of cases this is, however, incorrect. Aboveground runoff from unpaved surfaces occurs when the interception is filled, the infiltration capacity of the soil is exceeded and the surface storage in pools etc. has filled up. Retardance of the runoff is the conversion, delay and flattening of the inflow-curve due to the delay of the water that is flowing on its way to sewer inlets and in sewer pipes. Losses and retardance will be discussed more extensively in the subsequent sections.

Precipitation losses on unpaved areas consist of interception, infiltration depression storage and evaporation. On paved areas, the following loss processes can play a role:

- moisturising losses;
- depressions storage;
- infiltration losses;
- evaporation.

As a result of these loss processes, the runoff coefficient strongly depends on the characteristics of the urban basin as well as on the characteristics of the rainstorm. Observed runoff coefficients in the Lelystad basins mentioned above range between 20 and 100 per cent, with median values of 50–60 per cent for all rainstorms that produced over 1 mm of runoff. For storms producing over 5 mm of runoff, these median coefficients are 60–70 per cent.

The relationship between the flood behaviour of the entire river basin is often complex. The downstream characteristics of the flood differ from the upstream characteristics because of lag, routing and scale effects, and changes in geology, physiography and climate from headwaters to the outlet. Flood forecasting is much easier in large basins where the build-up of a flood and the transmission of the flood wave downstream can take days or even weeks and there are more chances for real-time forecasting and mitigation measures such as evacuation and flood protection. In small catchments, with short response times, real-time flood forecasting is much more difficult. The most extreme discharges in such catchments tend to occur as a result of localised convective rainfalls or high-intensity cells within larger synoptic weather systems.

In large catchments, both rainfall–runoff modelling and hydraulic modelling of the channels may be involved: the former to determine how much water will contribute to the flood wave; the latter to allow predictions of inundation of the floodplain and flooding of property during the event. In small catchments and dispersed settlements, the simulations using hydrological models are usually suitable for flood prediction.

Recently, an integrated approach of various models (surface runoff, sewer network, groundwater, water quality, etc.) has been commonly employed among the projects and researches. Additionally, the development of GIS technology provides increasingly detailed and accurate information. This has set a new standard in the next generation of urban flood modelling e.g. when modelling extreme events in which large surfaces are flooded, clearly the only way to obtain reasonably accurate results is by using GIS analytical tools. Therefore, advanced urban flood modelling requires an integration of GIS-based analysis of the surface terrain in order to reliably estimate surface flooding when the sewer network surcharged.

5.3.2.1 Simple runoff modelling methods

Precipitation is a common cause of floods, and types of precipitation determine the scale of modelling. At a very basic level, a simple rainfall–runoff model could just compute the amount of rain running off a hard surface. Since no bare ground areas are present in the case of parking lots covered with concrete or asphalt, infiltration would not take place. The amount of runoff from the parking lots would be about equal to the amount of rain that fell. Stated another way, the input would simply be rainfall and the output would be the runoff.

One of the simplest hydrologic situations to model is the peak runoff from a small area of a few square kilometres at most where the coverage of impermeable surfaces is important. The rational method estimates only the peak flow using rainfall intensity, area, and a land use factor (Equation 5-4):

$$Q_{\text{runoff}} = r_{D,n} \times \Psi_m \times A_{\text{total}} \times 10^{-4} \tag{5-4}$$

where Q_{runoff} is runoff of peak flow in l/s, Ψ_m is the dimensionless runoff coëfficient, see Table 5-3, $r_{D,n}$ is the rainfall intensity in l/(s.ha) and A_{total} is the catchment are in m^2. Using this method, there is no timing associated with the peak flow.

Although simple classical models, such as the rational formula, or time–area and unit hydrograph methods, are still in use, most of the advanced systems use more complex deterministic approaches.

5.3.2.2 Complex hydrologic models

While simple hydrologic models are limited by assumptions and calculations, more complex models allow us to better approximate different parts of the hydrologic cycle

TABLE 5-3

Runoff coefficients for land use types (after Chow, 1964)

Land use types	Runoff coefficients Ψ_m
Downtown business	0.70–0.95
Neighbourhood business	0.50–0.75
Heavy industrial	0.60–0.90
Light industrial	0.50–0.80
Multi-residential units, attached	0.60–0.75
Multi-residential units, detached	0.40–0.60
Single-family residential	0.30–0.50
Suburban residential	0.25–0.40
Playgrounds	0.20–0.35
Railroad yard	0.20–0.40
Unimproved areas	0.10–0.30
Parks and cemeteries	0.10–0.25
Asphalt and concrete	0.70–0.95
Brick	0.70–0.85
Roofs	0.75–0.95
Cultivated lands with loamy soils	0.40–0.45
Woodlands with sandy soils	0.10–0.15

and may take into account a complex array of natural and human-influenced factors (Figure 5-9).

The most basic approach for complex modelling of runoff from a basin is the lumped approach. This type of hydrologic model views a particular drainage area as a single unit and uses basin-averaged hydrologic and meteorological inputs. Some processes in the lumped models are described by differential equations based on simplified hydraulics laws and other processes by empirical algebraic equations. The output from a lumped model is usually a hydrograph at the basin outlet. Lumped models require less data input and less computational power than distributed models. In the past, lumped modelling methods were required due to data collection methods and software limitations. However, as new geo-spatial technologies become available, lumped models are being replaced by methods using more detailed spatial information to examine the basin on a finer scale. We draw a distinction between the semi-distributed and the distributed modelling approaches. In the semi-distributed approach, a basin is broken down into smaller sub-basins. Runoff amounts from methods, such as unit hydrograph, are used to estimate stream flow from each of these sub-basins. These runoff volumes are then routed downstream to estimate the stream flow output for the larger basin at the outlet.

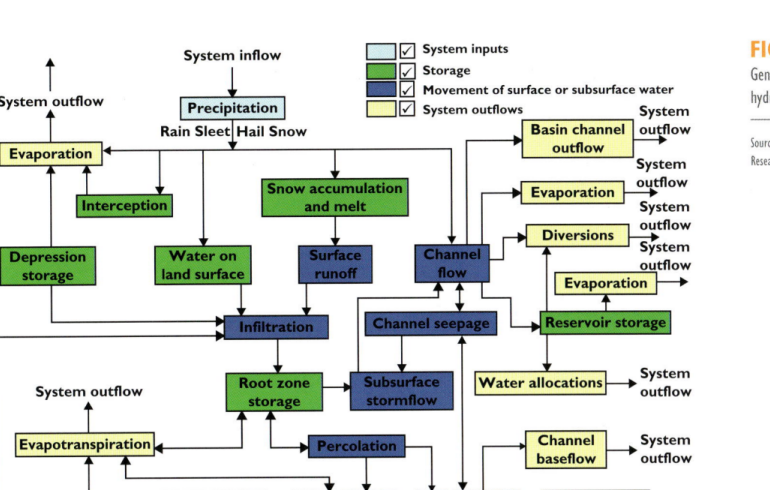

FIGURE 5-9
General components of a complex hydrologic model system

Source: University Corporation for Atmospheric Research (UCAR), Boulder, USA, 2006.

Distributed models take an explicit account of the spatial variability of processes, input, boundary conditions and catchment characteristics. This approach is one that represents processes as a grid and also provides for detailed predictions at the grid-cell level. Distributed models require more complex and more high-resolution data. Each cell has parameters allowing for its own stream flow estimates. The flow at any grid point can be estimated mathematically. Distributed modelling of hydrological processes has its limitations. One drawback to distributed modelling is the additional input data required for each cell. If these data are not available, they must somehow be estimated, introducing an uncertainty factor. Physics-based parameters and equations are used to route flows from cell to cell to the outlet. Distributed modelling methods allow modelling of progressively smaller areas. They also allow computation of more realistic runoff within a basin. Flash flood forecasts and urban flood simulations, for example, have improved with the implementation of distributed models.

Many distributed hydrological models have been developed, ranging from simple empirical equations, to complex systems of partial differential equations, which can incorporate the spatial distribution of various inputs and boundary conditions, such as topography, vegetation, land use, soil characteristics, rainfall, and evaporation, and produce spatially detailed outputs such as soil moisture, water table, groundwater fluxes, and surface saturation patterns. New data sources for observing hydrological processes can alleviate some of the problems facing the validation and operational use of hydrological models. Satellite, airborne and ground-based remote sensing has begun to realise some of its potential for hydrological applications, allowing monitoring and measurement of rainfall, snow, soil moisture, vegetation, surface temperature, energy fluxes and land cover over large areas. The main reason is that remote sensing data can provide large-scale, systematic land surface observations consistently over the large scale.

5.3.2.3 GIS in hydrologic modelling

In recent years, advances in Geographic Information Systems (GIS) have opened many opportunities for enhancing the hydrologic modelling of watershed systems.

5.3 MODELLING SURFACE RUNOFF

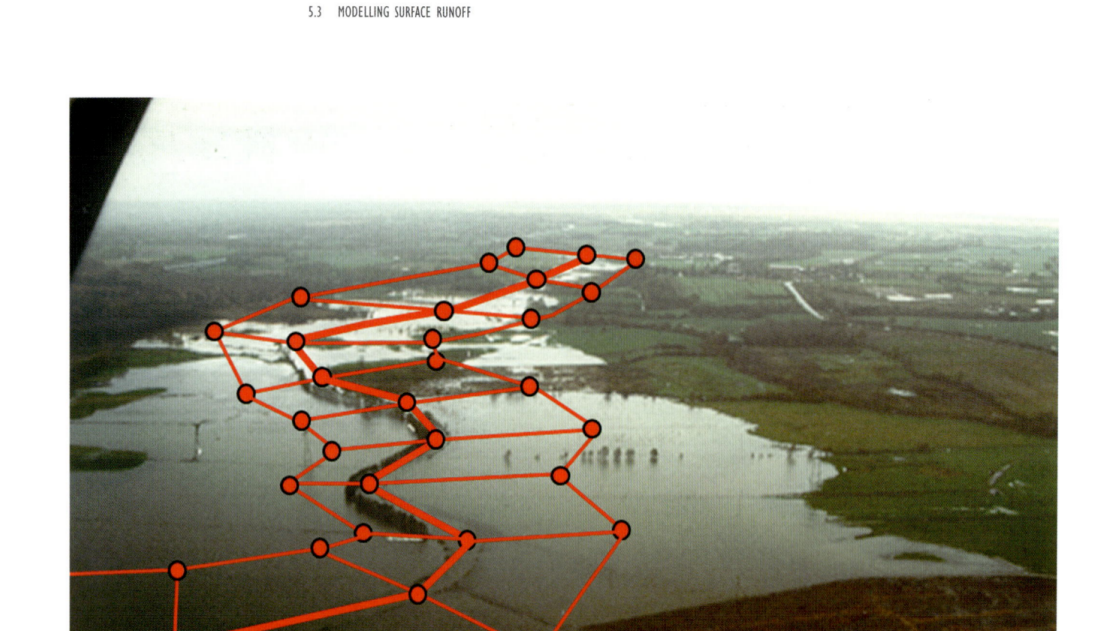

FIGURE 5-10
Floodplain and riverbed modelled by profiles in 1D-models

Source: Slobodan Djorjevic.

GIS technology provides a flexible environment for entering, analysing and displaying digital data from the various sources necessary for urban feature identification, change detection and database development. In hydrological and catchment modelling, remotely sensed data are valuable for providing cost-effective data input and for estimating model parameters. In addition, hydrological modelling has evolved to include radar rainfall and advanced techniques for modelling the catchment on a grid level. Rainfall and infiltration can be computed cell by cell providing greater detail than traditional lumped methods. These advanced modelling techniques have become feasible because the data manipulations can now be generated efficiently with GIS spatial operations. Weng (2008) presented the methodology which was developed to relate urban growth to distributed hydrological modelling using an integrated approach of remote sensing and GIS.

One of the main problems in urban hydrology is the uncertainty in the degree of imperviousness so that the spatial distribution of the impervious area cover is of great importance. This can only be taken effectively into account when a fully distributed hydrological model is used. The investigation of the assessment of imperviousness derived by a multi-resolution remote sensing technique may strongly improve these, as well as the need to apply fully distributed, grid-based hydrological models for urban runoff simulations.

Physically based modelling at the scale of an entire river basin requires large input databases. A GIS is very useful for linking to or integrating with environment models because of its advantages of data storage, display and maintenance. Therefore, there was a need for a GIS modelling environment that was capable of physically based modelling.

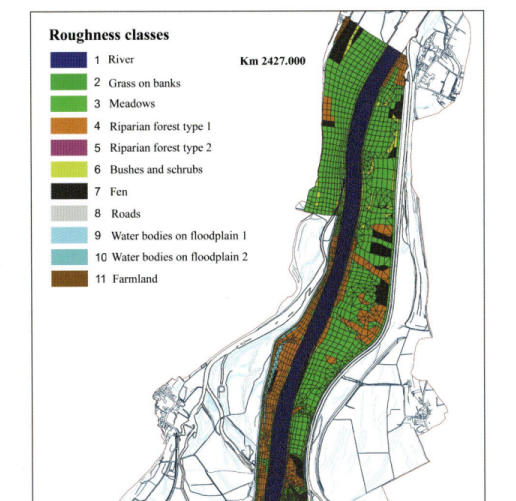

FIGURE 5-11
Floodplain and riverbed modelled by finite elements in 2D-models

Source: Slobodan Djorjevic.

As a result, a number of physically based and spatially distributed river flooding models have been developed for the simulation of runoff.

One such model, the LISFLOOD model, is used for the rivers Odra, Meuse, Elbe and Danube. It simulates river discharge in a drainage basin as a function of spatial data on topography, soils and land cover, precipitation amounts and intensities, and antecedent soil moisture content. The output of the LISFLOOD model consists of hydrographs at user-defined locations within the catchment area, time series of model parameters like evapo-transpiration, soil moisture content or snow depth, and can produce maps of any simulated variable, such as water source areas, discharge coefficient, total precipitation, total evapo-transpiration, total groundwater recharge or soil moisture. Since the physical process descriptions are universal, little or no additional calibration is needed if applied in a new catchment.

5.4 MODELLING PLUVIAL FLOODING

5.4.1 Causes

Pluvial flooding occurs when the local drainage system is not capable of collecting and conveying surface runoff. This may be caused by one or more of the following:

- the lack of a properly designed and built storm drainage and sewer system;
- heavy rainfall in excess of the 'design storm';

- catchment conditions worse than those assumed when the drainage system was designed (e.g. snow melting, imperviousness increased due to urban creep);
- partial or complete blockage of inlets and/or sewers pipes due to bad maintenance;
- failure of pumping stations, collapse of trunk sewers etc.

5.4.2 Data

Modelling of pluvial flooding requires high quality input data, which should include rainfall, a digital terrain model, land use and sewer system data. Rainfall data should ideally be provided by a dense network of rain gauges and/or the weather radar. Both sources are important, the former is generally considered as more accurate ('ground truth'), whilst the latter typically has higher spatial resolution, which enables advanced applications such as nowcasting (quick precipitation forecasting). Reliable modelling of urban pluvial flooding requires rainfall intensities given at 1 to 2 minute time steps.

Accurate digital terrain model (DTM) is essential for the simulation of pluvial flooding. It is used for sub-catchment delineation, creation of surface flow paths and ponds and as a basis for 2D modelling. Terrain data that contains information about buildings, walls, kerbs and other surface features is called Digital Elevation Model (DEM). DTM/DEM data for modelling of pluvial flooding should be provided in a 1 m or 0.5 m grid resolution. These days, data can even be obtained at 0.25 m resolution, which can be useful indeed, providing it does not cause computational memory problems.

Land-use data is used to automatically parameterise variables such as roofed and other impervious areas, type of connection to the sewer system, population equivalent (used to define waste water inputs), surface roughness, etc. Land-use images can be provided by remote sensing. Where this technology is not available, key features such as streets, car parks, housing, green areas, etc. can be distinguished from the existing maps and the corresponding imperviousness and roughness can be assigned to each surface type.

Finally, comprehensive and up-to-date sewer system data is crucial to enable the computation of the capacity of the drainage system, the exceedance of which leads to pluvial flooding.

5.4.3 Modelling approaches

Surface runoff, overland flow and, ultimately, pluvial flooding can be modelled by two general approaches, both of which have their advantages and disadvantages.

The standard approach—common in urban drainage modelling—is to split all processes into two phases, whereby rainfall is firstly transformed into effective rainfall by accounting for infiltration and other hydrological losses and routed along urban areas, to generate runoff hydrographs ('hydrological phase' of simulation). These are introduced as boundary conditions at the sewer network nodes and/or at the surface flow domain ('hydraulic phase' of simulation).

The alternative approach is to merge the two phases into one, i.e. to introduce rainfall directly onto the surface and simulate urban runoff (including hydrological losses), overland flow and flooding simultaneously.

The standard two-stage approach benefits from the option to use *calibrated* sewer models (even though they are most often calibrated via the observed results after events with little or no flooding). The downside of this approach lies in underestimating the speed and often also the volume of direct runoff from flooded areas, as well as in the common assumption that sewer network nodes can receive surface runoff hydrographs even during the surcharging.

Simultaneous modelling of all surface flow processes may appear to be more realistic because it has the potential to capture interactions between rainfall and flooding. Although more physically sound than the standard approach that involves a separate 'hydrological phase', simultaneous modelling of flooding and everything else that leads to it is more complex, and less knowledge exists about this emerging technology. Also, calibration and verification is a particular problem because field data from urban flooding events is rarely available at the desired spatial and temporal resolution, if at all.

5.4.4 Coupled models

Regardless of which runoff concept is implemented, modelling of pluvial flooding is specific insofar as it essentially involves the simulation of flows in two distinctly different computational domains, namely the urban surface (major system) and the buried pipe network (minor system), as well as complex interactions between them. Synchronised simulation of all of these types of flow can only be done by *coupled* hydraulic models, which solve flow equations in the two systems, whilst regularly computing water exchange between them. This interaction takes place via the links between flows above and below ground—inlets and manholes. The parameters of these links are a source of uncertainty in the modelling of pluvial flooding, because their geometry is complicated and the flow regime in them involves transitions—from free overflow when inlets collect relatively small surface inflow, via partially submerged flow in some of the inlet grating openings, to water flowing out from the manhole to the surface, and vice versa. The latter requires consideration of manhole covers—if they are sealed, hinged, loosely attached or freely resting—and consequently if they can be partially or completely removed during surcharging, which affects the area (hence the capacity) of a surface/sub-surface link. Figure 5-12 shows a photograph of a manhole taken during a flood event.

Coupled surface/sub-surface urban flood models are sometimes referred to as *dual drainage* models. Flood flow can be modelled either as a network of 1D (one-dimensional) open channels (surface flow paths) and ponds, or as a 2D computational flow domain. Corresponding coupled models are referred to as the 1D/1D (1D flow in the pipe

FIGURE 5-12
Manhole cover removal

Source: Marie-Claire ten Veldhuis, 2008.

network and 1D flow in the surface network) and $1D/2D$ models. Generally speaking, the former description is reasonably realistic as long as water remains within the street profile, though the uncertainties related to this approach greatly depend on the manner in which the surface network is defined.

5.4.4.1 1D modelling

A 1D surface network can be generated 'manually', assuming that surface flow paths follow street routes and have cross-sections the same as road profiles (with one or more gutters). The problem with this approach is that it is likely to generate a number of surface channels that will not see any significant surface flow, whilst some relevant flood routes—e.g. alleyways between buildings—may remain unaccounted for. The manual approach can also not identify and delineate the ponds in which stagnant water may accumulate. Therefore, a more intelligent way of defining a 1D surface network is by using a set of automated procedures, using GIS-based analytic functions that make use of DTM and land-use data. This involves five principal steps.

Firstly, the DEM may need to be enhanced. This could include data correction, identification of pit cells and flat areas, filtering of LiDAR data and detection of bridges. The second step is identification of spots prone to flooding, i.e. locating depressions or ponds (including ponds within ponds) and creating their stage–area curves. Thirdly, flow paths are generated between ponds using the rolling ball (or bouncing ball) method and the connectivity analysis is conducted to eliminate overlapping paths. The fourth step involves creating cross-sections of the flow paths by intersecting the DEM with vertical planes perpendicular to the paths and defining their roughness from the land-use data. The final step is visual checking the results, possibly followed by manual editing and exporting to the simulation model.

Thus, the created network of pathways and ponds contains maximum information that can be extracted from the DEM and land-use data and provides the best possible setup of an urban flood model within the $1D/1D$ framework. The resolution (hence the size) of the surface network can be set by the threshold defining minimum pond size (expressed as the minimum volume or horizontal area).

5.4.4.2 2D modelling

Treating the surface flow as two-dimensional is appealing since there is no need for predefining surface flow paths—floodwater finds its way on to the surface as the simulation evolves. In addition, 2D flow equations are clearly a more realistic description of flooding than the 1D model. Graphical representation of results—flood depths and velocities—is easier in 2D, since no post-processing is required (as in 1D).

A 2D flow simulation can be done on a structured or unstructured mesh—regular or irregular grid. The former directly uses LiDAR DTM data, and thus it has a simpler (matrix) data structure and is consequently easier to code. The latter, however, captures surface features much more efficiently, using a Triangular Irregular Network (TIN) that adapts the computational mesh to match the boundaries of buildings, kerbs, etc. With this type of mesh, it is important that the breaklines are included in the model. Figure 5-13 shows maximum pluvial flood depths calculated by two $1D/2D$ models, one based on the regular grid and the other using an unstructured mesh.

5.4.4.3 1D models versus 2D models

The $1D/1D$ approach is suitable for *designing for exceedance*, whereby the major system is envisaged to convey the overland flow in a controlled manner, minimisingw the consequences of pluvial flooding. In optioneering and solution development, overland

FIGURE 5-13
Flooding caused by extreme rainfall in the Mandaluyon Center, Philippines.

links can be edited, removed or added in 1D models with relative ease and numerical assessments can be directly implemented to determine the appropriate size of the channels. This is more complicated in 2D modelling, where the process is more 'try it and see' and the DTM is manipulated by 'burning' the new preferential pathways.

1D modelling is appropriate with *conveyance flooding* i.e. in areas with steeper topography where the flooding is usually shallow, with a relatively wide path (e.g. a road width). 2D modelling has advantages with *ponding flooding* when the topography flattens or where buildings have been constructed across valleys; in these cases, the floodwater accumulates in ponds and significant depths of flooding can occur.

2D models are generally more computationally demanding than 1D models, though to some extent this depends on the resolution. 1D models are therefore more suitable for real-time applications.

There is no doubt that 1D/1D and 1D/2D approaches are both needed, they will retain their important roles, regardless of other technological developments. The best results can be achieved by combining these two approaches—by applying them in accordance with the purpose of the project and the data availability.

5.4.5 Calibration

The main objectives in calibrating urban flooding models are to match the flood extent and the flood depth. Due to a lack of comprehensive sets of observed data, it is usually difficult to properly calibrate, verify and validate a pluvial flooding model (and urban flooding models in general). Observed flood extents, video records or photos can be invaluable for calibration. If high-tech equipment is not available, the areas affected by flooding and the highest flood levels can be cheaply recorded by tools such as resident gauges and chalk gauges.

Where no field data exists at all, some model parameters can be calibrated via the results of higher dimensionality models—relationships applicable to surface/sub-surface links can be determined from CFD (3D) models of flow through gullies and manholes, and 1D/1D models can be calibrated using the 1D/2D model results.

5.5 MODELLING COASTAL FLOODING

Coastal areas are subject to several different types of hydrodynamic processes each having different causes. Therefore, the modelling of flooding events on coastal areas has to consider not only the mainly marine agents but also the continental agents of flooding.

For these processes, the Navier–Stokes equations are the governing equations. They are integrated over the vertical axis and describe the flow and water level variations. This simulates the unsteady two-dimensional flows in single layer (vertically homogeneous) fluids. This kind of numerical model (flow model) is based on the same set of equations and it is only the application domain that changes. In fluvial channels, the 1D models have been used in the past due to the nature of the flow. In coastal areas, the 2D models are applied because the modelling domain is a basin rather than a channel.

Elevations are obtained from the solution of the depth-integrated continuity equation and the velocity is obtained from the solution of the momentum equations. Both formulations, conservation of mass and momentum, describe the physical processes using the following expressions:

Continuity equation:

$$\frac{\partial \zeta}{\partial t} + \frac{1}{R \cos \phi} \left(\frac{\partial UH}{\partial \lambda} + \frac{\partial UH \cos \phi}{\partial \phi} \right) = 0 \tag{5-5}$$

Momentum equations (written using traditional hydrostatic pressure and Boussinesq approximations):

$$\frac{\partial UH}{\partial t} + \frac{1}{R \cos \phi} \left(\frac{\partial UUH}{\partial \lambda} + \frac{\partial UVH \cos \phi}{\partial \phi} \right) - \left(\frac{U \tan \phi}{T} + f \right) VH \tag{5-6a}$$

$$= -\frac{H}{R\cos\phi} \frac{\partial}{\partial\lambda} \left[\frac{p_s}{\rho_0} + g(\zeta - \alpha\eta) \right] + M_\lambda + D_\lambda + \frac{\tau_{s\lambda}}{\rho_0} - \frac{\tau_{b\lambda}}{\rho_0} \tag{5-6b}$$

$$\frac{\partial VH}{\partial t} + \frac{1}{R\cos\phi} \left(\frac{\partial VUH}{\partial \lambda} + \frac{\partial VVH\cos\phi}{\partial \phi} \right) - \left(\frac{U\tan\phi}{R} + f \right) UH \tag{5-6c}$$

$$= -\frac{H}{R} \frac{\partial}{\partial \phi} \left[\frac{p_s}{\rho_0} + g(\zeta - \alpha \eta) \right] + M_\phi + D_\phi + \frac{\tau_{s\phi}}{\rho_0} - \frac{\tau_{b\phi}}{\rho_0} \tag{5-6d}$$

These equations can be solved using two different numerical schemes: one being the finite difference method (see Figure 5-14) and the other being the finite volume method.

After the definition of the governing equations and the modelling domain, it is necessary to describe the boundary conditions and the within domain conditions that establish the difference between the coastal models and the fluvial ones and that take into account the active agents of the marine environment.

With respect to coastal events, the main influence is the reduction of drainage capacity due to the rise of the sea level during intense or extreme precipitation events as well as fluvial discharges. Therefore, the modelling must address the time progression of this

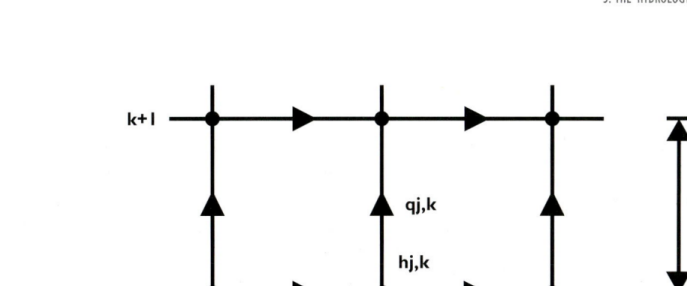

FIGURE 5-14
Example of finite difference scheme

Source: Slobodan Djorjevic.

rise of the sea level. In addition, the modelling must be capable of defining the extension of flooding level and its reach in land.

As a consequence, the boundary and domain conditions are characterised by the following forces: wind, wave and tides. Also, they will be characterised by the physical attributes of the water. Each one has to be considered as an input to the model. Some of them will be boundary conditions, others will be conditions within the domain and others will be a combination of the two.

The wind action has to be considered within the model, and can be simplified either as a constant action in the entire basin or introduced as a variable action in space and time. Wind action determines the movement of the superficial layer of the water and the rise of the sea level due to this phenomenon. Similarly, waves have to be to take into account in order to determine the currents as well as rise of the sea level due to the breaking waves.

In the case of the tides, it is necessary to distinguish between the astronomical tide and the meteorological one. The astronomical tide description has to be introduced with the sea level or with the water flows. This condition can be either a boundary condition or a domain condition. The meteorological tide is caused by the wind action and by the pressure variation that has to be introduced in the model.

The Coriolis effect is also taken into account, as this can have a greater or lesser influence according to the size of the modelling area. In addition, factors such as evaporation, precipitation, the friction with the bottom, and the sources and sinks of water flows also need to be incorporated into the modelling. This last component, sources and sinks is the nexus with the fluvial models. And then there are the salinity and temperature conditions that have an influence on the movement of the water masses.

Once these factors have been accounted for it is possible to simulate other events that lead coastal flooding like tsunamis.

And in order to incorporate terrestrial flooding terrain conditions: soil characteristics, buildings, etc. are also necessary.

TEXTBOX 5-4

Case study: Cancun–Nizuc coastal barrier

After Hurricane Gilbert (1988), the Cancun–Nizuc coastal barrier was severely eroded, a situation that was studied by the Autonomous National University of Mexico (UNAM) between 1989 and 1991. Following this, several recovery projects were carried out, but all of them only had limited results. At the turn of the twentieth century, many of the stretches of the barrier beaches had effectively disappeared, resulting in many of the hotels along the beach being subject to frequent wave attacks. Local authorities promoted new studies (2000–2001, 2003), in order to understand as accurately as possible the performance of the whole barrier system, and to come forward with proposals for the recovery of the beaches and the permanent protection of the barrier. The process of erosion on the beaches became even more serious after hurricane events Ivan (2004) and Wilma (2005). During 2004, additional studies were undertaken. Hurricane Wilma (2005) had a great impact on the barrier installations, and this probably had a determining influence on the decision to nourish the beaches along the barrier, which was made at the beginning of 2006.

The UNAM studies and the first ones by CFE established the true effect of the Gilbert hurricane (wind speed of 278 km/h, and instantaneous wind speed of 324 km/h) on the Cancun–Nizuc barrier. For that, the oceanographic conditions (wave and storm surge), under normal conditions and during the occurrence of Gilbert, were estimated from different databases, and with the use of numerical models. According to these studies, some of the average data during the occurrence of Gilbert were: $Hs = 11$ m, $Ts = 13$ s, wave direction $= 71.5°$ NE and storm surge $= 3.6$ m. Also, the sediment transport rates in ordinary and extreme conditions were calculated.

Although the problem of beach erosion in Cancun manifested after Gilbert apparently for the first time, these studies show that the hurricane only magnified an ongoing problem. Its origin must be attributed to the development and occupation of the barrier with building and hotel facilities.

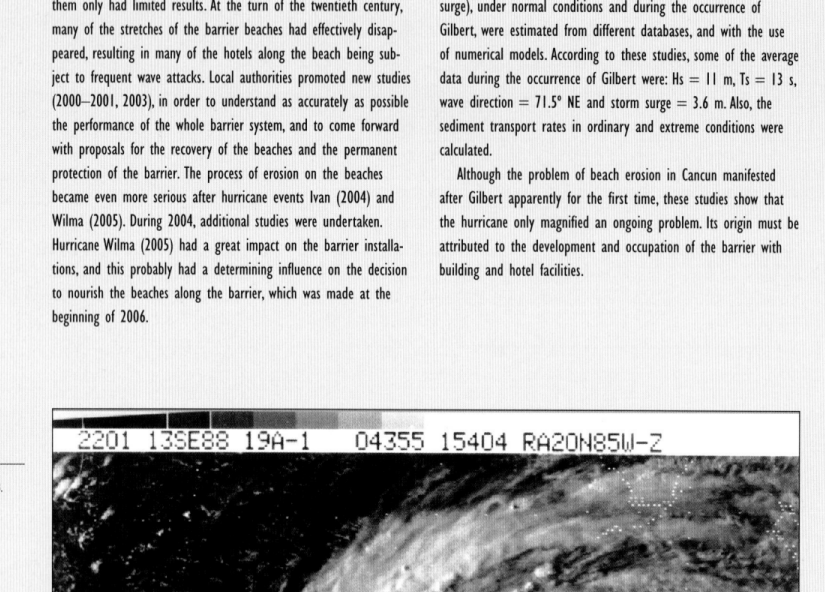

FIGURE 5-15
Huricane Gilbert

Source: National Hurricane Center (NHC), 1988.

Key Questions

The hydrological cycle

1. In what ways does the hydrological cycle in urban areas differ from the global hydrological cycle?
2. What aspects make urban hydrology different from conventional hydrology?
3. The urban water cycle, what does this mean and how might you go about 'closing' it?

Land use and runoff

1. How does the land-use change in rural areas influence runoff processes?
2. Can you describe some of the ways in which runoff processes can be affected by land use?
3. Is it possible to influence the runoff process through land-use changes that will lower flood risk, and is it possible for all types of floods?

Modelling surface runoff

1. What are the main differences between simple and complex hydrological models?
2. Describe how a distributed model works and the potential advantages as well as limitations?
3. Why is Geographic Information Systems becoming more important in hydrological modelling?

Modelling pluvial flooding

1. Pluvial flooding is caused by local heavy rainfall on the urban catchment. What other conditions may lead to pluvial flooding and why?
2. Modelling of pluvial flooding requires different sets of data. Can you explain the required resolution and the relevance of various data to modelling pluvial events?
3. Coupled hydraulic models link flow in the sewer network with the surface flood flow, whereby the latter can be treated either as the network of open channels and ponds, or as a two-dimensional flow domain. What are the areas of applicability, advantages and downsides of these two approaches?

Modelling coastal flooding

1. Why are two-dimensional flow models used for simulating coastal floods?
2. Describe the main parameters and their influences in setting up a numerical model for coastal flood studies?
3. What forces affect sea level rise that can lead to coastal flooding?

FURTHER READING

Chow, V.T. (1964). Runoff in *Handbook of Applied Hydrology*. Mc-Graw-Hill Book Company, New York.

FlOODSite (2009). *Development of Framework for the Influence and Impact of Uncertainty*. Report No T20-07-03. Available from: http://www.floodsite.net.

McLean, J. (2004). *Aurora—Delivering a Sustainable Urban Water System for a New Suburb*. Proceedings of the 4th International Conference on Water Sensitive Urban Design, 21–25 November, 2004, Adelaide: 22–23.

Rao, A.R. & Hamed, K.H. (2000). *Flood Frequency Analysis*. CRC Press, Boca Raton, Florida, USA.

Tucci, C.E.M. (2008). *Urban Flood Management*. WMO Cap-Net publication. APFM.

Van de Ven, F.H.M. (1990). Water balances of urban areas. *Hydrological Processes and Water Management in Urban Areas*. IAHS Publ. No. 198.

Van de Ven, F.H.M. & Voortman, B.R. (1990). The water balance of an urban area; Experiences of two monitoring sites in Lelystad, *The Netherlands (in Dutch)*, H_2O, *18(8)*, p. 170–176.

Weng, Q. (2008). *Remote sensing of impervious surfaces*. CRC Press, ISBN: 1420043749.

Urban flood-risk assessment

6

Learning outcomes

In the previous chapter you have been introduced to urban hydrology, how weather and climate affect urban areas, the tools that have been developed to model aspects such as surface water runoff, flooding that results from rainfall and runoff and flooding related to urban watercourses. In this chapter we will be considering how those tools and techniques are used and more particularly how to go about assessing the various risks associated with urban floods. The following topics are covered:

- the quantification of flood probabilities;
- what is meant by risk;
- the different types of damage that can arise from flooding and how they might be assessed;
- the estimation of flood risk and how this is depicted and mapped.

6.1 INTRODUCTION TO THE THEORY OF RISK

The philosophy of risk analysis finds its origin in the nuclear, aeronautic and chemical process sectors, where great hazards and financial losses are incurred as a result of the occurrence of unwanted events. Since the 1980s, literature on the subject of risk has grown rapidly. The word 'risk' is used and interpreted in many ways. In 1997, Stan $Kaplan^1$ in his address to the Society for Risk Analysis recalls how a committee established by this society laboured for four years to define the word risk and then gave up. The committee concluded it would be better to let each author define it in his own way explaining clearly what way that is.

Nevertheless, a number of common concepts are generally agreed upon in risk theory. The basic concept is that risk incorporates some probability of unwanted events and consequences following that event. Kaplan and Garrick in their paper in the first issue of the journal of the Society of Risk Analysis argued that the question 'What is risk' really includes three questions:

- What can go wrong?
- How likely is it to happen?
- If it does happen, what are the consequences?

¹ Stan Kaplan is one of the early practitioners of the discipline now known as Quantitative Risk Assessment (QRA), and a major contributor to its theory, language, philosophy and methodology.

6.1 INTRODUCTION TO THE THEORY OF RISK

The answer to the first question can be considered as a scenario; e.g. overtopping of a river dyke and collapse of a river dyke are different scenarios for flooding. Several qualitative and quantitative scenarios are available to find scenarios for unwanted events, see for example the fault tree handbook issued by NASA (NASA (2002) *Fault Tree Handbook with Aerospace Applications*. Version 1.1, *www.hq.nasa.gov/office/codeq/doctree/fthb.pdf*). Figure 6-1 shows an example of a first level fault tree for urban flooding.

The answer to the second question addresses uncertainty about the occurrence of hazardous events or scenarios. The answer takes the form of a frequency or probability of occurrence. The word probability is subject to even more semantic confusion than risk (see chapter 12). First, there is the statistician's meaning that refers to the outcome of a repetitive experiment and is close to the concept of frequency. Such a number is also called objective probability because it is, in principle, measurable in the real world. The Bayesian meaning of probability on the other hand expresses a degree of confidence or certainty which does not necessarily exist in the real world, but can exist inside our heads. This is typically the case of a 'one-shot' situation, like a satellite launch of a new type, for which we want to quantify our degree of confidence that it will succeed. It is important to note here that in everyday language the term risk is often used where really probability is meant as in: '*There's a high risk of another accident happening in this fog.*'

The answer to the third question, 'If it does happen, what are the consequences?' refers to a damage index, resulting from an unwanted event. Figure 6-2 illustrates aspects of damage that should be taken into account in answering this question. The damage index could be a vector or a multi-component quantity, which could be time-dependent and uncertain. In the latter case, damage can be expressed by a probability curve against the possible magnitudes of damage.

Kaplan and Garrick introduce risk curves which depict increasing severity on the horizontal axis and associated probabilities on the vertical axis, an example of which is shown in Figure 6-2. Risk curves draw a picture of risk for a range of possible scenarios with associated probabilities and severities or consequences. They give a complete representation of risk as opposed to a summarised risk value like average risk or expected value of risk, where risks related to events with low probabilities and high consequences are treated in the same way as those related to events with high probabilities and low consequences.

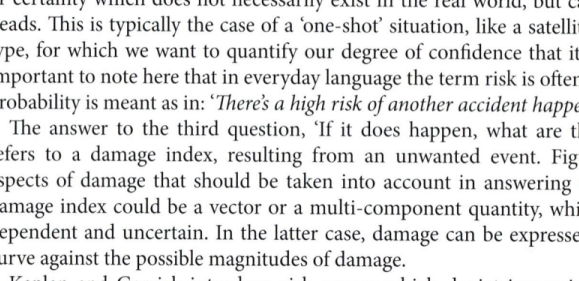

FIGURE 6-1
Example of a first level fault tree for urban flooding, depicting seven possible scenarios or failure mechanisms that can cause flooding. The scenarios in the squares are related to urban drainage systems, the other four are related to other water systems. Underlying causes of the depicted failure mechanisms appear in deeper levels of the tree

Source: Marie-Claire ten Veldhuis.

In reality, these different types of events are incomparable in terms of societal consequences and must in most cases be addressed with different prevention measures.

The approach describe here is referred to as quantitative or probabilistic risk analysis (QRA or PRA) as opposed to concern-driven risk management that is based on expert opinion, public or political consensus or combinations of those. Cox poses the question of whether concern-driven risk management can provide a viable alternative to QRA. QRA is typically based on a clear separation of facts or beliefs about the consequences

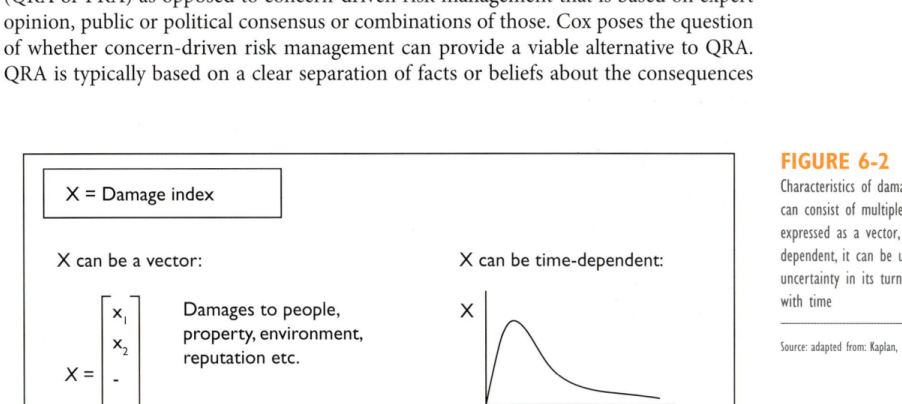

FIGURE 6-2
Characteristics of damage: damage can consist of multiple components expressed as a vector, it can be time-dependent, it can be uncertain and uncertainty in its turn can increase with time

Source: adapted from: Kaplan, 1997.

TEXTBOX 6-1
Example from a risk analysis: water defence system of a polder

Quantitative flood-risk analysis for a polder area aims to identify potential failure mechanisms, assess the corresponding probabilities of failure and to judge the acceptability of failure in view of the consequences. Initially, the entire water defence system of the polder is studied. Typically, this system contains sea dykes, dunes, river levees, sluices, pumping stations and high hills (Figure 6-3). In principle, the failure and breach of any of these elements leads to flooding of the polder. Thus, the probability of failure results from the probabilities of failure of all these elements. Within a longer element, e.g. a 2 km long dyke, several independent sections can be discerned.

Each section may fail due to various failure mechanisms such as overtopping, sliding, piping, erosion of the protected outer slope, ship collision and bursting pipeline. This approach has been tested on four polders. The analysis for the most important polder showed that the probability of human failure to close a sluice structure dominated the system with probability of 1/600 year.

Source: Vrijling J.K. (2001). Probabilistic design of flood defence systems in the Netherlands, Reliability Engineering and System Safety Vol. 74 Issue 3 pp. 337–344

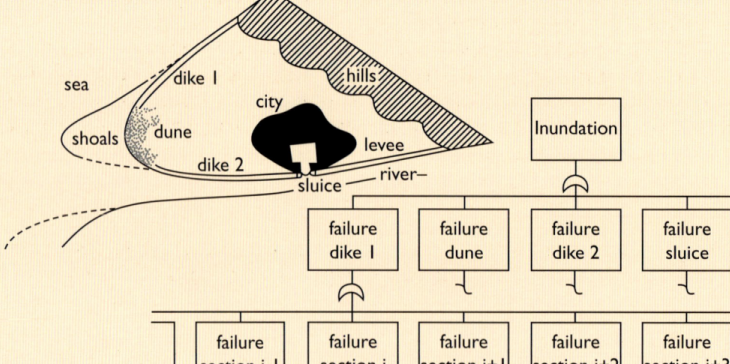

FIGURE 6-3
Flood defence system and its elements presented in a fault tree

Source: Hans Vrijling, 2001.

of actions. Each component is analysed separately, with beliefs about and preferences for consequences ultimately determining implied preferences for actions (e.g. via expected utility calculations).

Critics of QRA often encourage community members and other stakeholders to express holistic preferences directly for actions (e.g. banning animal antibiotics, GMOs, industrial emissions and use of DDT). From this perspective, a key role of the public is seen as being to express concerns and demand specific actions, rather than expressing clear preferences for consequences. Concern-driven risk management has several important potential political and psychological advantages over QRA, yet it is not clear that it performs better than (or as well as) QRA in identifying risk management interventions that successfully protect human health or achieve other desired consequences. Much of the explanation for the relatively poor performance of non-quantitative and judgement-based methods can attributed to the following: individual judgements are sensitive to logically irrelevant details of how information is presented, are often insufficiently sensitive to relevant information and information is often combined and used ineffectively. As a result, 'there are strong empirical and theoretical reasons to expect that judgment-based risk management in response to concerns, conducted without formal QRA, may lead to worse decisions and outcomes than would more quantitative models and methods'.

6.2 QUANTIFYING FLOOD PROBABILITY

6.2.1 Why probabilities?

6.2.1.1 Extreme variability over a wide range of space-time scales

A fundamental problem of hydrology is that the rainfall, which drives runoff, is extremely variable over a wide range of scales. Although this variability appears more obvious at smaller scales, which are particularly important in urban hydrology, it is also apparent at larger scales, e.g. yearly and decadal scales. Indeed, longer cycle periods are revealed when larger samples are analysed.

This multi-scale variability explains why it has become quite usual to consider the series of rainfall rates R_i measured at times t_i ($i = 1, n$), during a given duration d and at a given location, to behave randomly and therefore to consider R_i as a series of random variables. As much as possible the observation times, t_i, are regularly distributed with a fixed time increment Δt, i.e.:

$$t_{i+1} = t_i + \Delta t$$

Unfortunately, there are many practical reasons which prevent the realisation of this regularity; one example for instance is the ubiquity of missing data in rainfall series.

6.2.1.2 A general framework

Since the random behaviour rainfall is assumed to hold over a wide range of time increments Δt (especially for smaller and smaller Δt) and locations x (e.g. in a given basin or along a given river), one addresses in fact a stochastic space-time process $r(x, t)$, i.e. a continuous set of random variables. While it is worth bearing in mind that there is an underlying continuous stochastic process that is more general than the observed random discrete time series, it is usual to deal only with the statistics of the latter because, unfortunately, the corresponding concepts for the former are somewhat more involved.

The same observations hold for the resulting runoff, i.e. it can be seen either as a stochastic process $q(x, t)$ or a series of random discharges Q_i. Indeed, with the help of a mass balance equation the runoff corresponds to a nonlinear integration of the rain rate $r(x, t)$ over the catchment area. This integration yields a somewhat smoother variability, but does not eliminate it.

It is worthwhile noting that the previous remarks apply to urban and rural basins, as well as to the in-between case of peri-urban basins, therefore to both pluvial and fluvial (river) floods. Indeed, the interactions with the drainage network and the soil (infiltration and evapo-transpiration) are highly nonlinear and occur over a wide range of scales.

For simplicity sake, in this chapter we merely consider the basic property of a stochastic process $r(x, t)$ that yields, for given location x, duration d, and sequence of (increasing) times t_i, a series of random variables $R_i(d)$ corresponding to the integration of $r(x, t)$ over a time window $w_i(d)$ of duration d and including the time t_i:

$$R_i(d) = \frac{1}{d} \int_{w_i(d)} r(x,t)dt \tag{6-1}$$

the corresponding rainfall accumulation (or water depth) is defined by:

$$A_i(d) = \int_{w_i(d)} r(x,t)dt = R_i(d).d \tag{6-2}$$

There are two classical types of windows:

- fixed time windows:
  the duration d corresponds to Δt:

$$\Delta t = d; \ w_i = [t_{i-1}, t_i] \tag{6-3}$$

and $R_i(d)$ corresponds to the average of $r(x, t)$ between consecutive measurements.

- moving time window:
  d is no longer related to Δt:

$$w_i = [t_i - d, t_i] \tag{6-4}$$

and $R_i(d)$ corresponds to the moving average of $r(x, t)$ over the duration d.

For instance, daily rainfall rates correspond to a fixed window defined by a given beginning of the day (often defined at 6am), whereas 24-hour rainfall rates are obtained with the help of a moving window of a duration of 24 hours. Although both have the same duration (24 hours), they are not identical and furthermore do not have identical statistics. For instance, the daily rainfall extremes are known to be lower than the 24-hour ones.

It can be mentioned that somewhat more sophisticated windows correspond to wavelets, i.e. windows that are both compact in the physical and Fourier spaces, but we will not discuss them in this book.

6.2.2 Probabilities and frequencies

6.2.2.1 The frequentist approach

Due to the fact that some concepts of the probability theory are somewhat involved, hydrologists have tended to extensively use a frequentist approach that has led to the framework of the Flood Frequency Analysis, which is rather ubiquitous in hydrology.

The frequency approach is based on the interpretation of probability that defines the probability $\Pr(A)$ of an event A, e.g. the event where the rain rate exceeds a given threshold:

$$A = \{R=s\} \tag{6-5}$$

as the limit of its relative frequency of A (whose number of occurrences is N_A) in a large number N of trials:

$$\Pr(A) \approx \frac{N_A}{N} \tag{6-6}$$

and more precisely:

$$\Pr(A) = \lim_{N \to \infty} \frac{N_A}{N} \tag{6-7}$$

This indeed yields an intuitive understanding of the probability and furthermore the inverse of this frequency is naturally understood as a return period, although it should not be understood in a periodic manner. The return period concept is particularly used for the annual maxima, then it is expressed in years. For instance, an annual maxima with a probability of 10^{-3} is said to have a return period of a thousand years.

This frequency interpretation is well founded if a few theoretical requirements are not forgotten: this is an asymptotic result and trials have to be independent. When applied to an empirical series R_i ($i = 1, n$) it corresponds to a strong hypothesis of stationarity: the R_i should be identically distributed, i.e. they should have the same probability, which is therefore independent of the time. The process is therefore said to be stationary. Furthermore, the R_i should be independent. Both requirements can be at best understood as approximations that we discuss below.

6.2.3 The simplistic framework

6.2.3.1 Statistical moments

Let us first consider, in spite of its limitations, the simplest framework where the R_i fully satisfies the requirements of the frequency approach, i.e. the R_i are identically independently distributed (i.i.d.) random variables; i.e. they correspond to independent trials of the same random variable R. The goal is now to determine the attributes of the common probability distribution of these variables, not forgetting that it should be achieved not only for a unique duration, but also for various durations. Each duration defines a random variable R with a given probability distribution. One possibility is to use estimates of the statistical moments of various orders q, i.e. estimates of the mathematical expectations (denoted by $<.>$) of $R(d)^q$. These estimates are obtained by attributing a uniform probability to all the outcomes $R_i(d)$:

$$\left\langle R(d)^q \right\rangle \approx \frac{1}{N} \sum_{i=1,N} R_i(d)^q \tag{6-8}$$

The law of large numbers ensures that these estimates converge in the large N number limit to the theoretical statistical moment value (right-hand side of Equation 6-8) if (and only if) the later is finite (e.g. Feller, 1971).

For an illustration, Figure 6-4, from (Lovejoy *et al.*, 2008), displays the statistical moments of the rain radar reflectivity measured by the satellite TRMM for various space increments $d = \Delta x$ along the satellite path².

6.2.3.2 Rank analysis and quantiles

Although statistical moment analysis is a classical tool, it became usual, especially for an extremely fluctuating random variable, to proceed to a rank analysis of the empirical quantiles. The later are defined with the help of the inverse function of the probability distribution, i.e. instead of computing the probability $p(s)$ to exceed a given threshold s, a quantile is the threshold $s(p)$ such that the exceedance probability equals a given probability p:

$$\Pr(R > s) = p(s) \Leftrightarrow \Pr(R > s(p)) = p \tag{6-9}$$

As for any inverse function, graphically it corresponds to exchanging the two axes (see Figure 6-5 and 6-6), and the inversion exists due to the fact that $p(s)$ is obviously monotonous, more precisely a non-increasing function. Indeed, the probability to exceed a threshold cannot increase with the threshold. Given a sample R_i ($i = 1, n$), the values R_i are considered as empirical quantiles, which are ranked in descending order (i.e. from the largest value with the rank $m = 1$, second largest with $m = 2$ and so on to $m = N$). In agreement with Equation 6-6, the corresponding probability is defined with the help of a 'plotting position' formula:

$$p_m = (m - a)/(N + b) \tag{6-10}$$

where the couple of constants (a, b) is usually not taken as (0, 0), as expected in the large N limit (Equation 6-7), but (0, 1) for the Weibull plotting position, (0.5, 0) for the

² Obviously, the above definitions also apply for space increments, not only for time increments.

6.2 QUANTIFYING FLOOD PROBABILITY

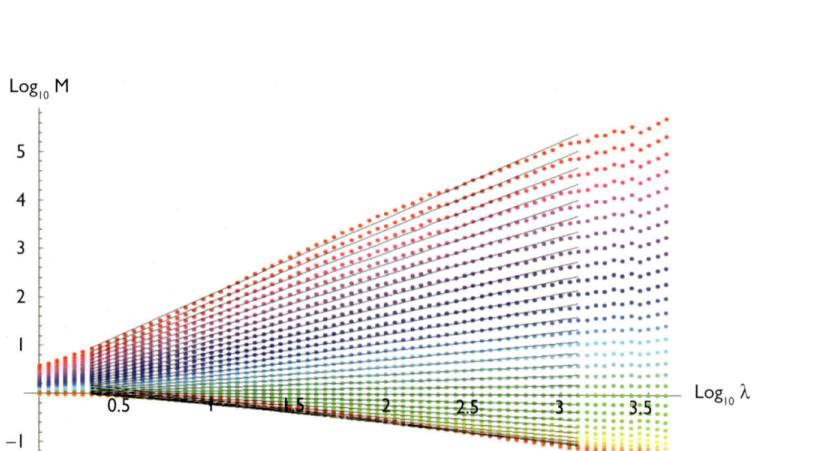

FIGURE 6-4

Log of the statistical moments of radar reflectivity $< Z_{\lambda} >$ observed by the radar of the satellite TRMM, normalised by the mean over all the orbits $<Z_i>$, of order $q = 0–3$, $\Delta q = 0.1$, with the spatial resolutions $\lambda = L_{earth}/\Delta x$ ($L_{earth} =$ 20,000 km). Black lines are linear regressions and therefore correspond to power laws for the statistical moments w.r.t. to the resolution, λ.

Source: Daniel Schertzer.

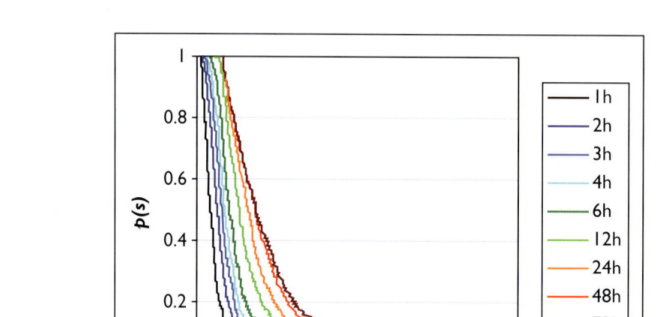

FIGURE 6-5

Empirical probability distributions $p(s)$ of exceeding a given threshold s

Source: Daniel Schertzer.

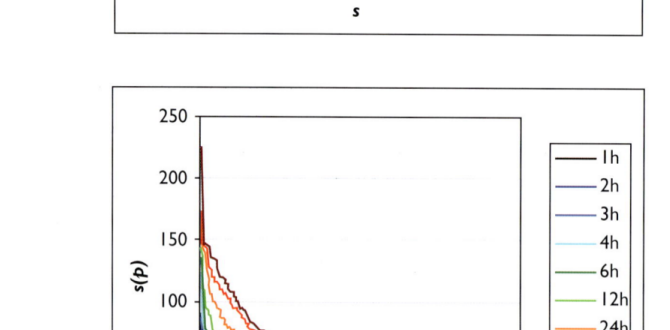

FIGURE 6-6

Empirical quantiles $s(p)$ of having an excess probability p, for the maxima of rainfall accumulation for durations from 1 to 72 hours during the 273 episodes at Nîmes (1972–1990) that are further discussed below.

Source: Daniel Schertzer.

Hazel plotting position and (0.4, 0.2) for the Cunnane plotting position, to take into account the finiteness of N.

6.2.3.3 Intensity duration frequency (IDF) curves: a closer look

If we repeat the same operation for various durations, d, we can now draw a set of curves $s(p, d)$ called Intensity Duration Frequency (IDF) curves (see also Chapter 2.3) because each of their points are defined by these three quantities, where the intensity (e.g. the rainfall rate) corresponds to an empirical quantile. Usually (see Figure 6-7), they are projected on the two-dimensional Intensity-Frequency plane and their projections are therefore labelled by the durations. Furthermore, they are usually log-linear plots with the logarithm of the inverse of the frequency (hence the logarithm of the return period) along the horizontal axis and the intensity along the vertical. In other words, the IDF curves are a convenient log-linear plot of a set of quantile functions (or exceedance probability distributions, when inverting the two axes) for various durations. This explains why they have been used so much in hydrology and other disciplines (e.g. economy) to assess risks. To respect the independence condition, the IDF curves are usually not drawn from the original time series, but from the extrema extracted from the latter.

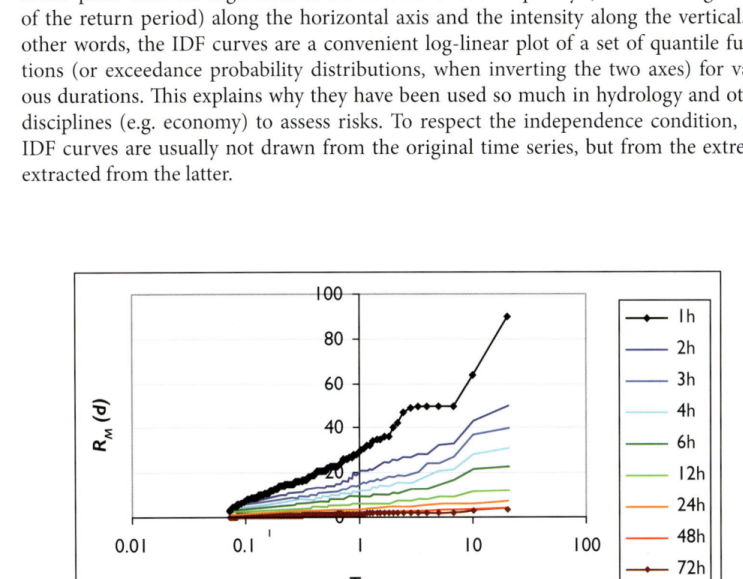

FIGURE 6-7
Empirical Intensity Duration Frequency (IDF) curves obtained on maxima of 273 rainfall episodes at Nîmes (1972–1990) for durations from 1 to 72 hours (moving time window). The left figure (a) corresponds to rainfall intensities, decreasing with an increase of durations. The right figure (b) corresponds to rainfall accumulations, increasing with increasing durations. Hence, the last figure is usually referred as the Depth Duration Frequency (DDF) curves. In both cases, the return period T corresponds to the inverse of the frequency being expressed in years

Source: Daniel Schertzer.

6.2.4 How to adapt hydrological data to this framework?

Obviously, river discharges Q_i at consecutive times t_i are not independent. On the contrary, they exhibit a strong persistence, which might even correspond to a long dependence over the past, contrary to Markoffian processes that have a finite time memory. Therefore, various strategies have been designed to obtain derived time series with more or less independent components. The classical strategies correspond to a particular sampling of the extremes (either maxima or minima) because they are the main quantities of interest (e.g. floods or droughts), whereas most recent strategies try to uncover an underlying white-noise process (a continuous generalisation of a discrete i.i.d. time series).

6.2.4.1 Sampling of the extremes

6.2.4.1.1 *Extremes over a given period*

Taking the extremes over a given period (e.g. the year) of a time series with a given duration (e.g. daily data) is the classical sampling of the extremes. The underlying argument being that extremes are much less correlated than average events. At first glance, this is rather convincing, but the 'much less' should be assessed, which is rarely the case. Furthermore, there is the opposite phenomenon of the clustering of the extremes, i.e. they are far from being regularly distributed, but are rather clustered. This widely observed phenomenon brings into question this sort of sampling, because it will miss secondary events of a given cluster that are more extremes than some primary events of other years. Besides, the hydrological time series data are rather short. Nevertheless, this sampling is straightforward to implement, noting that time windows for the period and the duration can be chosen as either moving or fixed windows, independently of each other.

6.2.4.1.2 *Event analysis: peak over threshold (POT) and variants*

The limitations of the annual extreme sampling lead to consider a first generalisation called 'Peak Over Thresholds (POT)' sampling. Initially, only events which are continuously above (below for droughts) a given threshold are kept. The threshold value (e.g. 20 mm for daily rainfall accumulation) is chosen to get a given average number

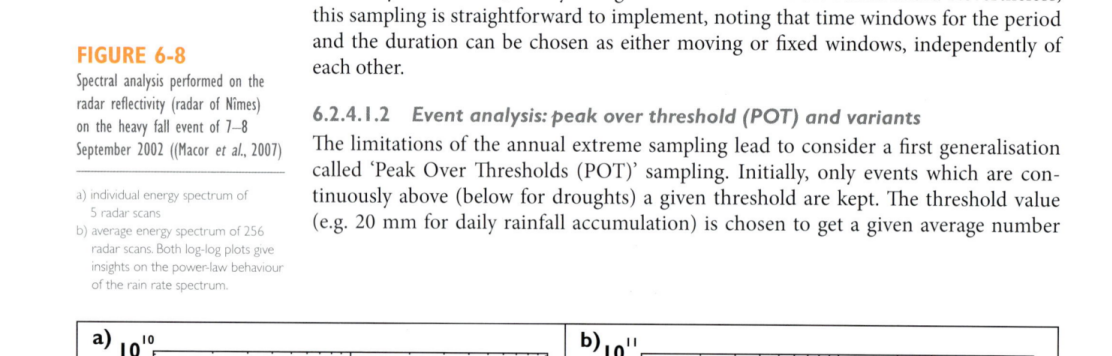

FIGURE 6-8
Spectral analysis performed on the radar reflectivity (radar of Nîmes) on the heavy fall event of 7–8 September 2002 ((Macor *et al.*, 2007)

a) individual energy spectrum of 5 radar scans
b) average energy spectrum of 256 radar scans. Both log-log plots give insights on the power-law behaviour of the rain rate spectrum.

of such events per a given period (e.g. a year). The final derived series is that of the extreme values of these events. This procedure also has the advantage of increasing the number of analysed extrema. Unfortunately, the introduced flexibility can be questioned: decreasing/increasing the threshold will decrease the independence of the selected peaks/lows.

Extra criteria were therefore introduced to define the rainfall/flooding events: two consecutive events above a given threshold (e.g. 20 mm for daily rainfall accumulation) will be considered as distinct events only if they satisfy these criteria, otherwise they are fused into a unique event. A Student test on their relative independence can be used, but very often the criteria corresponds to a long enough (e.g. 72 hours) 'no rain' period. The latter is generally defined by a lower threshold (e.g. 2 mm vs. 20 mm, both for daily accumulation). Furthermore, as longer events than those of the POT arise, instead of only extracting the unique global extremum of each event (as done for POT), it is natural for each given duration to extract the extrema over moving time windows of this duration all over this event (and over the no rain periods on both sides over this event). Overall, this extraction of maxima (corresponding to a well-defined duration) is obtained by combining the two types of time windows. Figure 6-9 displays the resulting series of maxima for 273 rainfall episodes at Nîmes (1972–1990), which on average corresponds to 14.4 episodes per year that exceed the fixed threshold of 20 mm for daily rainfall accumulation. It might be important to note that compared to Figure 6-7, these data exhibit a rather different type of scale dependency. Hence, this example underscores the importance of the type of windowing and sampling in use.

6.2.4.1.3 Derivative series

Because the strong dependence of a data time series can result from a time integration, it can be eliminated, or at least greatly diminished by considering the corresponding series of increments, i.e.:

$$\Delta Q_i = Q_i - Q_{i-1} \tag{6-11}$$

It can also possibly yield a stationary time series (ΔQ_i) from a non-stationary one (Q_i). This is precisely the case for a stochastic process with stationary increments, where the increments ΔQ_i are i.i.d., but not the Q_i themselves.

This approach has been broadly generalised with the help of fractional derivatives, i.e. not integer order derivatives. Although the later concept seems at first glance rather awkward, it allows us to define in a rather straightforward manner a broad family of processes with non-classical scale dependency (e.g. fractional Brownian motions). For instance, it yields:

$$\Delta Q_i \propto \Delta t^H \tag{6-12}$$

with H being not necessarily an integer, contrary to the classical deterministic case ($H = 1$). Such a non-classical behaviour is easily observable with the help of spectral techniques (Fast Fourier Transforms) since the corresponding energy spectrum will follow a power-law:

$$E(\varpi) \propto \varpi^{-(2H+1)} \tag{6-13}$$

Figure 6-8 gives an illustration of this behaviour for the rainfall rate estimated by the (ground) radar of Nîmes (Macor *et al.*, 2007).

6.2.4.2 Scaling and Montana laws

An important property of the IDF curves (see Chapter 2.3) in a log-log plot i.e. Log $R_M(d,T)$ vs. Log T for various durations, d, where $R_M(d, T)$ is the rainfall maxima for duration d and period T- is that they look rather parallel over a wide range of return periods and therefore often well superpose after a simple normalisation (Figure 6-9). If this normalisation only depends on the ratio of the respective durations it will be necessarily a power law:

$$R_M(d_1,T) = R_M(d_2,T) \cdot (d_1/d_2)^b \tag{6-13}$$

which is known in hydrology as a Montana law and a scaling law in many other domains. The Montana parameter b corresponds to the slope of the straight line behaviour of Log $R_M(d, T)$ vs. Log d for given periods, T. Montana laws are often invoked for various parameter related to hydrological extrema. Figure 6-9 gives an example of a Montana law for the mean of the rainfall maxima, which gives support to the idea of independency on durations for the rainfall maxima when normalised by their mean as illustrated by Figure 6-8b. Due to the quality of the straight line fitting (in a log-log plot), Figure 6-10 could be used to justify a unique scaling model over the full range

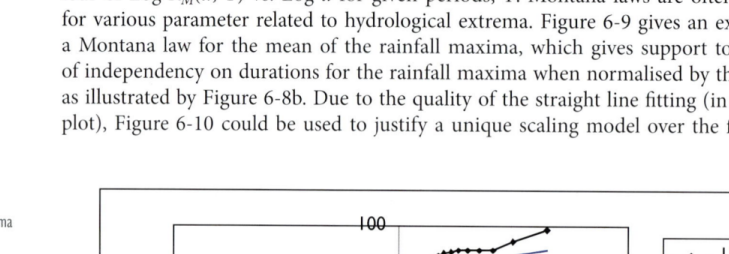

FIGURE 6-9

The IDF curves obtained on maxima of 273 rainfall episodes at Nimes (Figure 6-7) represented in log-log coordinates

The curves for durations from 1 to 72 hours, look almost parallel for a wide range of return periods.

Source: Daniel Schertzer.

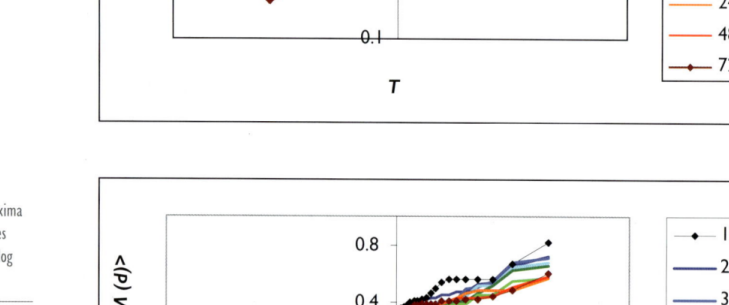

FIGURE 6-10

The IDF curves obtained on maxima of 273 rainfall episodes at Nimes (Figure 6-7) represented in log-log coordinates

Being normalised by the corresponding mean, these curves superpose well over the return periods ranging from at least 3 weeks to about 2 years. For longer return periods, the increase of dispersion among the curves for different durations could be attributed either to a limited sample size or to two different behaviours for durations—shorter and longer than 6 hours.

Source: Daniel Schertzer.

of durations (Rosso and Burlando, 1990), however the relevance of the latter will be questioned below.

6.2.5 Extreme value theory (EVT)

Because the long hydrological series are unfortunately rather short (their length rarely reaches a century but is typically of 20 years), and, furthermore, far from being continuous, prior statistical laws are needed to interpolate data points when missing data are numerous and to extrapolate them for larger return periods. The choice of this prior distribution usually takes hold in the classical theory of extreme values (see also Textbox 2.5), which deals with the asymptotic ($n \to \infty$) behaviour (if any) of the maximal values:

$$M_n = \max\{R_1, R_2, .., R_n\} \tag{6-14}$$

of a sequence of random variables $\{R_1\}$ and more generally the supremum of a stochastic process. For i.i.d. random sequences (Fischer and Tippet, 1928; Frechet, 1927; Gnedenko, 1943; Gumbel, 1958), with some extensions to the case of short-range correlations (Leadbetter and Rootzen, 1988; Loynes, 1965), it was demonstrated that there are only three universality classes, called 'types', for the distribution of the suitably renormalised maximal value M_n (rescaling $a_n > 0$ and a recentering b_n):

$$n >> 1: \Pr(a_n(M_n - b_n) < x) \approx G(x) \tag{6-15}$$

In agreement with common sense, these types are determined by the probability tails (i.e. large quantiles) of the original series not by other details. Indeed, one obtains a Gumbel law (often called EV1 for Extreme Value of type 1), a Fréchet law (EV2) and a Weibull law (EV3) for a probability tail having an exponential fall-off (often called a 'thin tail'), a power-law tail fall-off (a 'fat tail') and a bounded tail, respectively. Ironically, the Fréchet law was historically the first law to be derived, whereas its type is only the second one and it was too often disregarded, in particular in hydrology. In spite of their profound differences, these three types of distribution can be put into a unique asymptotic distribution of the maxima, often called Generalised Extreme Value (GEV) distribution:

$$G(x) = e^{-(a(x-b))^{-1/\kappa}} \tag{6-16}$$

where EV1 corresponds to $\kappa = 0$, EV2 to $\kappa > 0$, $a > 0$, $s > b$, and EV3 to $\kappa < 0$, $a < 0$, $s < b$.

It is therefore a current practice to use GEV, and, in particular, EV1, to interpolate and extrapolate empirical IDF curves. However, the general limitation of the (classical) Extreme Value Theory to short-range correlated data, calls for its extension in a more general framework (Schertzer *et al.*, 2009b).

6.2.6 Stochastic simulations of the extremes

Instead of relying on pure statistical extrapolations of the IDF with the help of more or less ad hoc statistical laws, long runs of stochastic models whose parameters are a fit to the (short) recorded time series can be used. Usually the stochastic core corresponds to a rainfall simulator whose results are input into a deterministic rainfall–runoff model. This is, for instance, the structure of the SHYPRE model (Arnaud and Lavabre, 1999) which has been validated by comparing its IDF outputs to empirical ones and has been used to extrapolate IDF curves up to return periods reaching 10,000 years. Figure 6-11 illustrates an example of numerical simulations obtained with the help of another stochastic model (Schertzer and Lovejoy, 1987). Although it is defined by only three parameters that undertake the scaling behaviour of rainfall series, it may reproduce sufficiently well the empirical IDF curves, as shown by Figure 6-12. To convert this unique dimensionless curve to (physical) IDF curves for various durations, a Montana law has been used.

The SHYPRE model can also be used to test the duration range on which the Montana law holds. Figure 6-13 displays a map of the Montana parameter b estimated with the help of the SHYPRE model with the assumption of a unique Montana law for each 1 km^2 pixel of France (Arnaud *et al.*, 2007). By contrast, Figure 6-14 displays the map of the ratio of the Montana parameters estimated on durations smaller and larger than four hours. This figure clearly suggests that a unique Montana law remains an acceptable approximation over a rather limited domain, which nevertheless includes Nîmes. Therefore, the parameters estimated on hourly data cannot be generally used to estimate quantiles for durations of a few minutes, which are the most pertinent durations for urban hydrology. This again highlights the necessity of further development of innovative and more physically based frameworks.

FIGURE 6-11

Example of a Montana law: empirical relation (in log-log plot) between duration and the corresponding mean of the series of the maxima extracted from 273 rainfall episodes at Nîmes (Figure 6-7). These means rain rates were used to normalise these series as well as the corresponding IDF curves so that they collapse into a unique curve

Source: Daniel Schertzer.

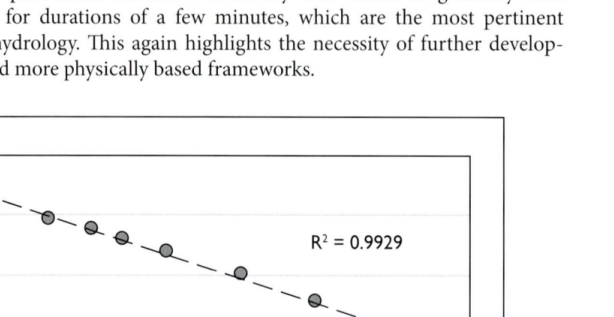

FIGURE 6-12

Intensity–frequency curve obtained on the long stochastic simulation from Figure 6-11

Source: Daniel Schertzer.

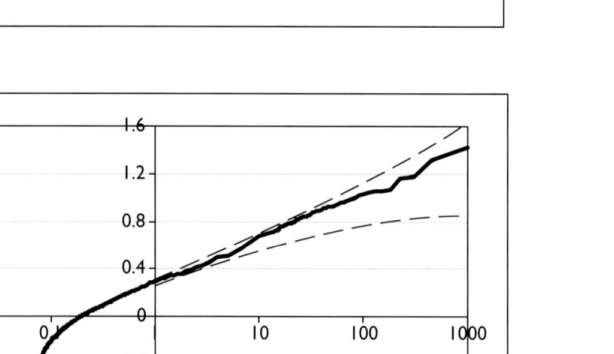

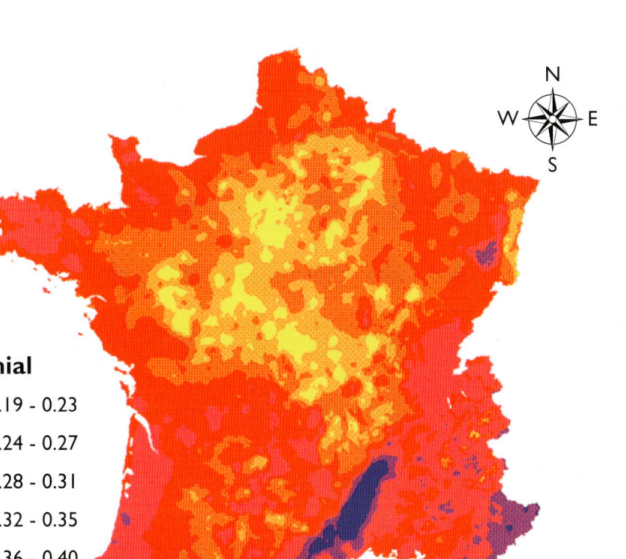

FIGURE 6-13

Example of a map of the Montana parameter b estimated with the help of the SHYPRE model with the assumption of a unique Montana law for each 1 km^2 pixel of France for the return period of ten years

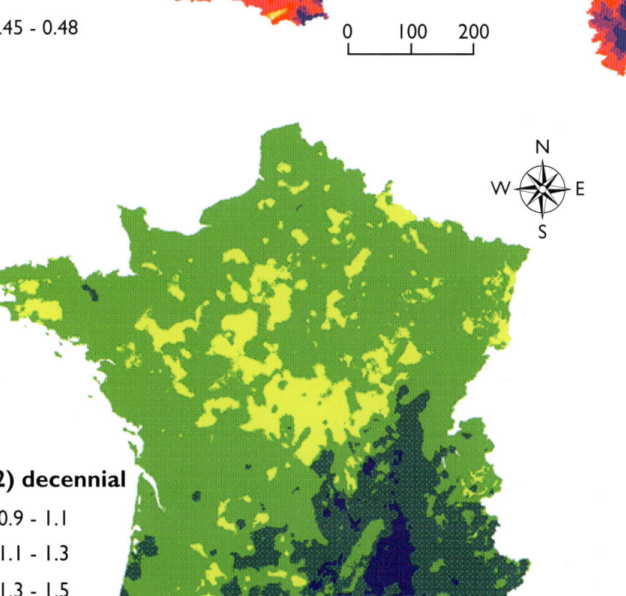

FIGURE 6-14

Map of the ratio of the Montana parameters estimated on durations smaller and larger than 4 hours

FIGURE 6-15
Relationship between hazard identification and assessment, vulnerability identification and assessment and risk characterisation.

6.2.7 Beyond the present limitations

We started from the observed extreme variability of hydrological fields over wide ranges of scale as a motivation to introduce probabilities. We are back to it when examining the limitations met by the present statistical and stochastic approaches in hydrology.

While IDF curves are very convenient to represent the quantile distribution in a multi-scale manner, they unfortunately rely on two hypotheses that are questionable: stationarity and independency of the components of the time series. We pointed out that beyond the classical sampling of the extremes and its limitations, there is the possibility of eliminating long-range dependency by uncovering a white-noise process whose (fractional) integration would generate the observed long-range dependent process. Stationarity has been at best a questionable approximation that has become even more questionable due to climate change. To escape from this hypothesis, the parameters of the prior distribution should no longer be estimated over the full sample, but the time evolution of their estimates obtained over sub-samples of smaller sizes should be analysed. Since estimate uncertainty becomes larger with smaller sub-samples, this requires a physically based prior distribution, rather than a mathematically convenient one. This is a current challenging field of research.

6.3 TANGIBLE AND INTANGIBLE DAMAGES

6.3.1 Introduction

The economic value of damages arising from natural hazards is, among other factors, a function of the affluence and wealth of the society impacted by hazards such as floods. As societies become relatively wealthier a greater proportion of income is invested in 'goods' rather than spent on day to day requirements such as food. A further factor to consider is how the wealth in a society is distributed and how this is reflected in its geographic distribution.

It is customary to categorise flood damage and losses into direct or indirect, and by whether or not they are tangible or intangible. Direct damages result from the physical contact of floodwater with damageable property. The magnitude of the damage may be taken as the cost of restoration of the property to its condition before the flood event, or its loss in market value if restoration is not worthwhile. Indirect flood losses are losses caused by the disruption of physical and economic linkages of the economy, and the extra costs of emergency and other actions taken to prevent flood damage and other losses. These indirect effects of floods can be suffered by productive activities and emergency services both within and beyond the area of immediate direct physical flood impact. Even though the division into direct and indirect, tangible and intangible damage is commonplace, the interpretations and delineations of what is considered direct and indirect damage differ. Table 6-1 gives examples of damages in the four categories.

Much of the magnitude of flood damage loss is a function of the nature and extent of the flooding, including its duration, velocity and the contamination of the floodwaters by sewage and other contaminants. All these affect damages and losses, and the location of the flood will affect the economic networks and social activities causing indirect losses.

6.3.2 Measures for direct and indirect tangible damage assessment

Because of the nature of different assessments, techniques are necessary to accurately gauge the magnitude and extent of flood damages. Many methods and models have been developed but there is little agreement among experts and academics which should be preferred. This is partly due to the difficulty in establishing the relevant 'boundaries' for any analysis. Variation is also found in the temporal and spatial scale of analysis; spatial scales varying from local to urban, regional or national level and temporal scales varying from minutes to hours, days, weeks or years.

Quantification of economic damage comprises three main elements (Figure 6-16):

1. determination of flood characteristics;
2. assembling information on land use data and maximum damage amounts;
3. application of stage-damage functions.

TABLE 6.1
Distinction between categories of flood damage

Type of damage	Tangible	Intangible
Direct	Damage to buildings and contents	Fatalities and injuries
	Business interruption inside flooded area	Loss of memorabilia
	Agricultural losses and cattle	Damage to physical and mental
	Road, utility and communication infra structure	health, including inconvenience
	Clean-up costs	Historical and cultural losses
		Environmental losses
Indirect	Road traffic disruption	Inconvenience of post-flood
	Rail communications disruption	recovery, psychological traumas
	Damage for companies outside flooded area	Societal disruption
	Evacuation	Undermined trust in public
		authorities

FIGURE 6-16
Schematisation of the process of assessment of direct damage based on land use mapping, flood modelling and stage-damage functions.

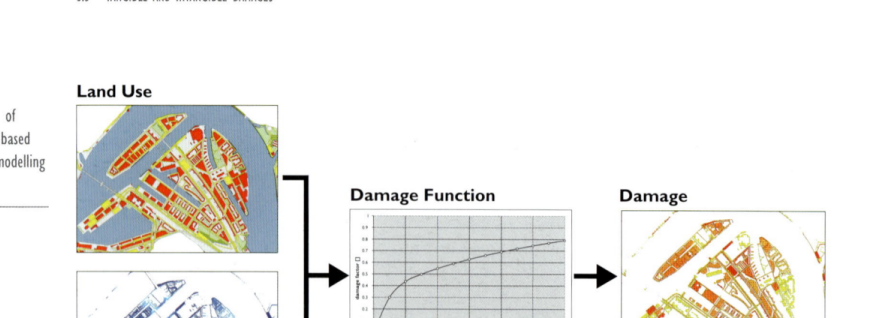

Flood characteristics are usually derived from hydrodynamic models, which generate output information about the development of the flood flow over time and provide insight into flood characteristics, such as water depth, flow velocity and the rate at which water levels rise. All these characteristics can be depicted on a map.

Indirect flood damage includes disruption of physical and economic linkages of the economy, and the extra costs of emergency and other actions taken to prevent flood damage and other losses. These indirect effects of floods can be suffered by productive activities and emergency services both within and beyond the area of immediate direct physical flood impact. Quantifying indirect flood effects has always been problematic, yet a recognition of the increasing interconnectedness of modern industry leads to suggestions that losses in one location could spread to other economic activities well away from the flood-affected area. To estimate indirect economic damage connected to interruptions in business flows, a model is needed that can capture and analyse the interconnectedness between economic agents in an economic system on the one hand, and the impact of stochastic shocks on this system of interconnected activities on the other hand.

Indirect economic damage becomes important in cases of severe flood damage over large affected areas. When insufficient data are available to estimate indirect damage, indirect damage is sometimes estimated as a percentage of the direct damage.

6.3.3 Intangible flood damage

Intangible flood damages include loss of lives and injury, health-related issues, disruption and inconvenience, preparedness such as warning plans and community education, isolation and evacuation. It can also include loss of aesthetic and cultural valuables e.g. historical buildings or art lost during the flooding of a museum. Often it can be problematic to determine and/or express all consequences from flooding directly in monetary value. In practice, it is extremely difficult to rate the impact of different intangible flood damages into a framework; e.g. what's the value of social disruption? Although attempts have been made to define some way of quantifying intangible damages, generally they are excluded from damage assessment and are only dealt with qualitatively. Note that intangible flood losses are not always congruent with tangible damages.

6.3.4 Trauma, stress and anxiety

Research into the intangible consequences of severe flood events has shown that physical and mental health effects on people whose properties were flooded can be serious. People who suffered flooding even regard the intangible effects of flooding to be higher than their direct material damage. Intangible effects include loss of memorabilia, psychological stress during the flood and during recovery, where discussions with insurance companies are specifically mentioned. The UK Department for Environment Food and Rural Affairs (Defra) conducted a survey involving 1,510 face-to-face interviews (983 flooded and 527 at risk respondents) in 30 locations across England and Wales in autumn 2002. All 30 locations had suffered fluvial or surface water flooding to varying degrees since January 1998. The results of their survey demonstrate that flooding causes short-term physical effects and, more significantly, short- and long-term psychological effects.

It is clear that after floods that post-traumatic stress is a serious though often unrecognised problem which can remain for some time after the actual event. The extent of the potential health impacts is shaped by the general health of the affected population, socio-economic conditions and cultural as well as personal factors. The recognition of its importance suggests that dealing with post-traumatic stress should be incorporated into post-event recovery planning. This is not to underestimate the extent to which people are able to adapt.

The degree of health impact was associated with a wide range of factors including socio-demographic factors (especially prior health and age), flood characteristics (especially flood depth) and post flood events (especially problems with insurers in settling claims for flood damage which emerged as the most important factor). More than 60 per cent of flooded and at-risk respondents expressed a willingness-to-pay (WTP) to avoid the health impacts associated with flooding. Of those that did not provide a value, some provided genuine zero value bids (for example, on the grounds of not being able to afford to pay extra amounts). When these were accounted for, the overall mean WTP values for flooded and at risk respondents were about £200 and £150 per household per year, respectively. As for the health impacts, the WTP values provided by flooded respondents were associated with a wide range of factors but income and extent of long-term psychological effects (i.e. stress) emerged among the most important influencing factors. However, the most important factor was age, with people in their 50s having the highest WTP values. It was also this age group which suffered the greatest short- and long-term psychological effects.

6.4 LOSS OF LIFE ESTIMATION IN FLOOD-RISK ASSESSMENT

6.4.1 Introduction and background

The loss of human life is one of the most important consequences of flooding. In particular, floods of low-lying areas protected by flood defences can be catastrophic and such situations are generally found in river deltas. In the Netherlands, a storm surge in the North Sea led to flooding of the Southwestern part of the country and 1,836 fatalities. The flooding caused by hurricane Katrina in 2005, caused more than 1,100 fatalities in the state of Louisiana and hundreds of fatalities occurred in the flooded parts of the city

FIGURE 6-17
Mortality function for the zone with rapidly rising water (Jonkman, 2007)

Source: Bas Jonkman, 2007.

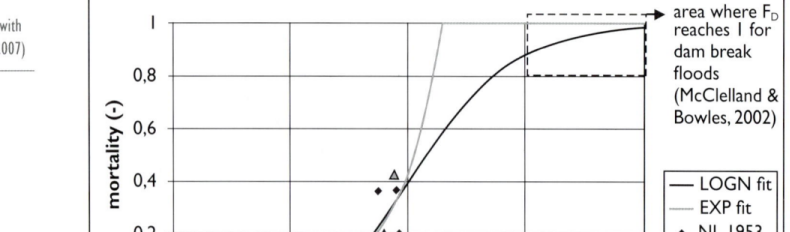

of New Orleans. Globally, deaths from flood events have shown a tendency to decline as flood warnings have improved.

6.4.2 Method for loss of life estimation

Jonkman has proposed a method for the estimation of loss of life due to floods (the general approach is described in the following formula) (see Jonkman *et al.*, 2007):

$$N = F_D(1 - F_E)N_{PAR} \tag{6-17}$$

where: N = number of fatalities; F_D = mortality amongst the population exposed; F_E = fraction of the population that can be evacuated; and N_{PAR} = number of people at risk in the area.

Depending on the level of detail of the analysis, the above factors can be quantified by general (expert) estimates or more detailed calculations. A more detailed assessment requires insight in the flooded area and flood characteristics (e.g. depth, velocity and rise rate). The number evacuated can be estimated by using evacuation and traffic models and considering the effects of warning. The general mortality fraction depends on the severity of the flood and it is around 1 per cent for coastal floods and lower for river floods. The mortality amongst the population exposed can be estimated by so-called mortality functions. These relate the mortality amongst the exposed population to the flood characteristics for different zones in the flooded area. By analysing empirical information from historical floods, such as the floods in the Netherlands in 1953, mortality functions have been developed. An example of the mortality function for a zone with rapidly rising waters is shown in Figure 6-17.

6.5 CROSS-SCALE FACTORS AND INDIRECT DAMAGES

6.5.1 Introduction

The financial consequences of flooding can be substantial. Buildings, roads and other objects need repairing or sometimes reconstruction. Areas need to be cleaned or

TEXTBOX 6-2

Case study: Loss of estimation for South Holland, The Netherlands

The method presented above has been used to estimate the loss of life for South Holland. This is one of the largest flood-prone areas in the Netherlands. The area has 3.6 million inhabitants, is the most densely populated area in the country and includes major cities. As an example, the output for a more severe coastal flood scenario with breaches at two locations (Den Haag and Ter Heijde) is considered. In this case, an area of approximately 230 km^2 could be flooded with more than 700,000 inhabitants. It is expected that the possibilities for evacuation of this area are limited because the time available for evacuation (approximately one day) is insufficient for a large-scale evacuation of this densely populated area. It is calculated that this flood scenario could lead to more than 3,000 fatalities. Figure 6-18 shows the flooded area and the spatial distribution of the number of fatalities estimated with the method described above. This approach can also be used in the context of flood-risk assessment by evaluating different flood scenarios and their probabilities and consequences (Jonkman *et al.*, 2008b).

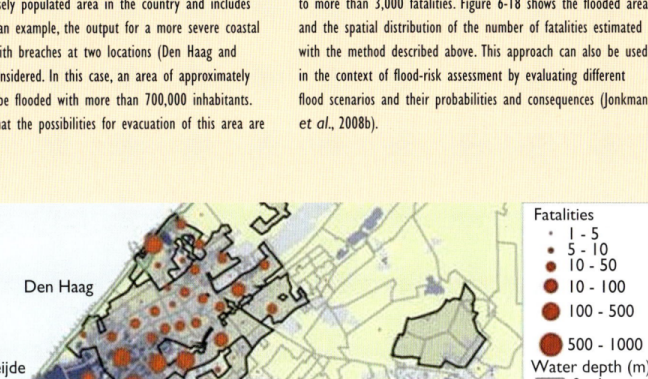

FIGURE 6-18
Fatalities by neighbourhood and flooded area for the scenario with breaches at Den Haag and Ter Heijde (Jonkman, 2007)

Source: Bas Jonkman, 2007.

decontaminated and flood defence works need to be restored. Yet, these costs only reflect part of the equation. During and after a flood, industries and other sectors might suffer from business interruptions. Destroyed highways and railroads might prevent transportation of goods and people. Interruption of water supply, electricity or gas might prevent operation and production. Businesses outside the flooded area that depend on the delivery of materials from the area might in turn stop or interrupt production. All these indirect consequences of flood impacts are called indirect damages. Indirect damages have been defined by Parker *et al.* as: Indirect flood damages are those caused through interruption and disruption of economic and social activities as a consequence of direct flood damages.

The notion of indirect damages assumes that we look at economies as networks of activities through which goods, labour and capital flow. Thus, disruption of individual

6.5 CROSS-SCALE FACTORS AND INDIRECT DAMAGES

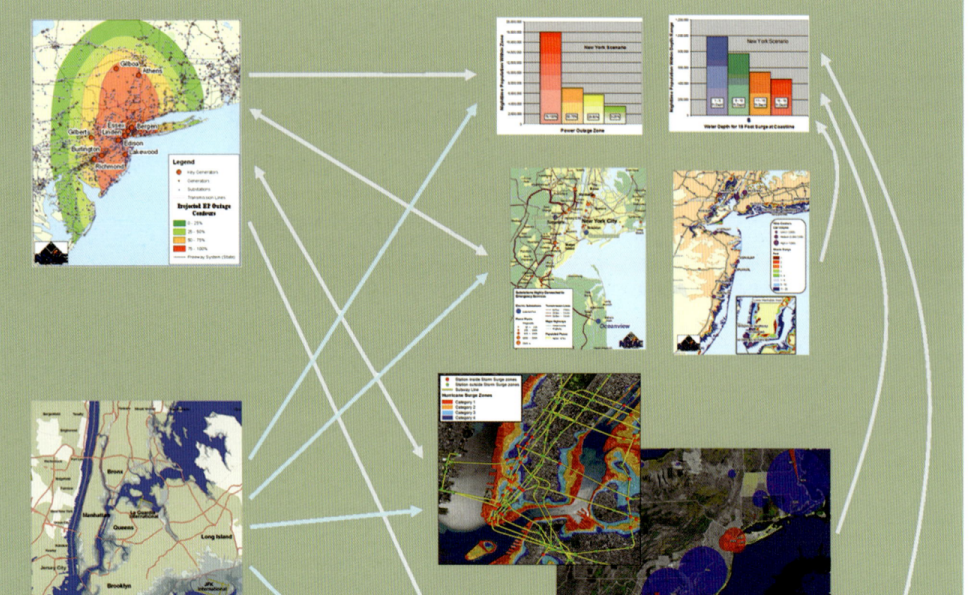

FIGURE 6-19
Indirect damages resulting from electric power outage (from: National Infrastructure Simulation and Analysis Center, 2003, Electric Power Consequence Analysis

Source: NISAC, 2003.

businesses can propagate through the network and cause *ripple effects* in which many other businesses are affected.

Since small business interruptions of one company can lead to disproportionately large consequences for other businesses, these are also referred to as *higher order effects* (e.g. think of how the disruption of yeast production in a single production facility can affect the production of bread somewhere else). Since business disruption can have regional, national or even international consequences (think of the costs caused by interruption of a stock market), the dependencies between companies on a spatial scale determine the spatial extent of the ripple effect: *cross-scale factors*. Ripple effects are often only temporary phenomena; companies that suffer supply shortages due to business interruption in the flooded area choose other suppliers outside the affected region. This process is referred to as *substitution*.

6.5.2 Measuring indirect damages

Measuring or estimating indirect damages is not at all straightforward, since it involves a clear understanding of all the dependencies and responses within the economic system towards interruptions initiated on a local scale. This covers both the spatial and temporal dimension of the indirect consequences. In fact, indirect damage assessment is more related to urban and regional economics than to flood impact assessment, since the cause of business interruptions is hardly relevant for how the consequences propagate through the economic network; for indirect damage modelling it is hardly relevant if the cause of business interruption is a flood or an earthquake. Modelling indirect damages is therefore often performed using methods derived from urban and regional economy: regional input–output models, Computation General Equilibrium (CGE) models or econometric models. Indirect damage modelling is a highly volatile field. The current state-of-the-art leaves a lot of room for improvement; current assessment is imprecise and model predictions differ substantially between different methods. In particular, further refinement of CGE models is promising, since these incorporate sophisticated behavioural response schemes that simulate loss-reduction behaviour (e.g. through substitution) of economic agents and thus are able to mimic the resilience found in the dynamic market-driven contemporary economy.

6.5.3 Comparing direct to indirect damages

When compared to direct flood damages (or damages resulting from natural disasters in general), the extent of indirect damages is often regarded as limited. Due to substitution

TABLE 6.2
Secondary indirect effects of floods: the case of the lower Thames area (from: FHRC, 2003, Multicoloured Manual, chapter 3)

Loss category		Damage and losses caused by a major flood (0.5 annual probability)		Size of economy: Input linkages (£m) (purchases from floodplain businesses)	Secondary indirect losses as percentage of input linkages
		Potential losses*	**As percentage of direct damages**		
Direct flood damage		£85,404,000	100.00%		
Indirect flood losses		£2,866,041	3.36%		
Secondary indirect effect:	Locality	£171,962	0.20%	86.73	0.20%
	Subregion	£278,006	0.33%	140.25	0.20%
Loss of income from wages loss:	Subregion	£68,000	0.08%	140.25	0.05%

* It is recognised that these figures are given in a form that is too precise; this is done for illustrative purposes only.

and other compensation methods, ripple effects often last for only short periods of time. In other words, the network of interdependencies between businesses is resilient. Nevertheless, the indirect damages on a local or regional level can be substantial. In the aftermath of Hurricane Katrina, the economic consequences due to business interruption, restoration of utility lifeline systems (water supply, electricity and gas) and demographic and social changes were estimated significantly larger than the direct damages. This was to some extend due to the long recovery period. Note that the estimation of indirect damages is to a large degree dependent on the scale of perspective. Because of substitution, businesses outside the affected area may profit from disaster impact since suppliers within the region are knocked-out. Consequently, when taking a regional perspective the indirect damages might appear minimal since the economic impact suffered within a local area is compensated by an increased turnover outside the flooded area. In this case, the indirect damages only reflect those suffered during the reconfiguration of the regional economy in the first days following the disaster.

The spatial distribution and sectoral variation between industries is a good indicator of the expected indirect damages. Concentration of vital industrial sectors in flood-prone areas is a recipe for disaster, and thus a high level of indirect damages, because the consequences of business interruption are widespread. This also holds for major infrastructure (one major highway for accessing the area) and utility lifelines.

6.6 FLOOD-RISK MAPPING

6.6.1 Use of flood hazard mapping

Although flood hazard maps can serve a variety of purposes, two are perhaps of greatest importance; the ability to identify zones that are at risk and following from this their use as a communication tool. In both cases, the ultimate aim is to seek to reduce human losses due to floods. Flood hazard maps are designed to convey crucial flood-related information and prepare people for future floods. The information contained on such maps can include the extent of past disastrous events, simulated inundation areas, depth of flooding, evacuation routes and centres, disaster management centres, danger spots, communication channels and systems. Flood hazard maps are becoming increasingly important, especially in developing countries, as a non-structural response as the implementation of structural measures often requires not only the devotion of considerable resources but also takes considerable time to implement.

In developed countries, flood hazard maps are a basic tool for preparedness and mitigation activities, including flood insurance programmes. In countries such as the United States and some European countries e.g. the United Kingdom, flood hazard maps are critical in determining the insurability of properties, though their use for such purposes depends on the national approaches to the provision of flood insurance. At the international level, over the past decade there has been increasing interest in the importance of developing meaningful financial risk-sharing arrangements such as insurance and reinsurance against disasters, through imaginative public/private partnerships. Small developing island states have raised the question of access to insurance for damage caused by natural catastrophes because of the progressive pull-back by insurance companies or, at the very least, a sharp rise in costs.

The financing of schemes to compensate for damage caused by natural catastrophes should be a component of systems to combat these catastrophes and should constitute part of proper governance. It applies to EU member countries, even though some of them have no national insurance system despite being highly exposed. A basic necessity in any insurance scheme is the availability of information on which to assess hazards and risks, which is where flood hazard mapping plays a pivotal role.

The effectiveness of flood hazard maps depends not just on the accuracy of the information contained within them but also, more importantly, on the level of understanding and awareness within the communities that make use of them. Experience has shown that when communities have a better understanding of the potential disasters in their area they are better able to prepare and respond. Flood hazard information also assists authorities and communities in making decisions about future building and land-use development proposals and the potential risks associated with flooding.

It is for this reason that, internationally, increasing attention is being paid to the development of flood hazard maps. Exploratory analysis through the UNEP and the University of Grenoble has demonstrated that global approaches to flood hazard mapping is possible with the increasing availability of basin descriptors providing regional flood frequency forecasts, which, combined with globally available elevation data, can be used to map inundation patterns associated with 100-year floods. Studies have already been carried out for North and South America and Mozambique. Increasingly, processed satellite imagery is being combined with historical events and GIS spatial databases to indicate areas prone to flooding from catastrophic events such as storm surges, tsunamis and intense rainfall events within river basins. These are of particular importance to urban areas, for example in Bangladesh, these approaches have assisted in the planning of river works for flood countermeasures to protect urban and industrial areas.

In the Caribbean since 2001, efforts have been made to improve the identification of flood hazards. Caribbean agencies together with the Japanese and Canadians have been working to develop multi-hazard maps and vulnerability assessments that show inundation as a result of differing return events. These are used by planners to assist with the development of disaster preparedness plans as well as to inform insurance decisions. In Nicaragua, flood hazard maps developed in conjunction with international bodies have been developed for use by central government agencies and municipal authorities to assist them in their planning.

Europe, as a result of rising concerns over the social and economic impacts of flood, has developed the Flood Directive which sets out specific requirements for flood mapping and zoning. Competent authorities use these zones to identify avoidance measures. The probabilities of flooding associated with these zones are also used as a starting

TEXTBOX 6-3
Flood mapping in Lagos, Nigeria

Lagos is the biggest city in West Africa and by 2015 it is projected to be one of the world's five largest cities. It is the industrial and commercial hub of Nigeria and forms a continuous urbanised area stretching along the Atlantic coast and inland with an average population density of some 8,000 people per square kilometre. Some 54 per cent of the urbanised area is located on what were previously floodplains. The combined effects of heavy rainfall and storm surges together with poor planning and the coastal location makes Lagos particularly prone to flooding with severe human, environmental and socio-economic effects. This has led to the need for a flood disaster management plan. Using data from LANDSAT, Google Earth and the Surveys Department as well as rainfall data, flood-risk analysis was carried out in one area within the North-eastern region of Lagos megacity. This demonstrated that it would be possible to develop information systems that could be used to reduce urban sprawl in flood-prone areas.

point for the appraisal of measures to alleviate flooding. The zones for flood hazard maps are:

- floods with a low probability, or extreme event scenarios;
- floods with a medium probability (likely return period $\geq$ 100 years);
- floods with a high probability, where appropriate.

The flood hazard maps should be enhanced to flood-risk maps by including the following information:

- the indicative number of inhabitants potentially affected;
- type of economic activity of the area potentially affected;
- installations which might cause accidental pollution in case of flooding and potentially affected protected areas;
- other information which the member state considers useful, such as the indication of areas where floods with a high content of transported sediments and debris floods can occur and information on other significant sources of pollution.

In the case of coastal and groundwater flooding, the risk maps may be limited to defining the area of extreme flooding.

6.6.2 SAFER moves forward the flood hazard mapping

The hazard mapping in the south of Germany was part of the European funded INTERREG IIIB project SAFER. Due to the international cooperation with relevant stakeholders and authorities as well as the successful exchange with other European partners the project's experiences in the mapping process can be taken as a best practice example (Haimes, 1998) .

In general, SAFER's flood hazard maps use scenarios based on the characteristics of an area and describe the specific flooding situations of different types of inland water bodies and coastal zones. The maps indicate for various recurrence intervals where overflows occur and to what extent.

6.6.3 Flood hazard maps in the federal state of Baden-Württemberg

It is essential to any effort of flood-risk reduction, that the hazard is known. In this regard, the flood hazard maps contain extremely valuable information. There is a huge variety of end-users that profit and build on this kind of information. For instance, planning of the emergency services is based on the flooding conditions illustrated in the maps.

The flood hazard maps of Baden-Württemberg indicate the flood extend and the water depth (see Figures 6-20 and 6-21) based on today's conditions. The maps provide information for several recurrence intervals, e.g. in Germany 10, 50 and 100 year and the extreme event.

The flood hazard mapping approach in Baden-Württemberg is innovative as it is multilateral, cooperative and interactive. The approach ensures close cooperation with local communities and represents a regular exchange with end-users such as water management authorities, spatial and regional planning institutions, fire brigade and disaster management (Haimes, 1998). Overall, this broad approach ensures that the public are more aware of flooding.

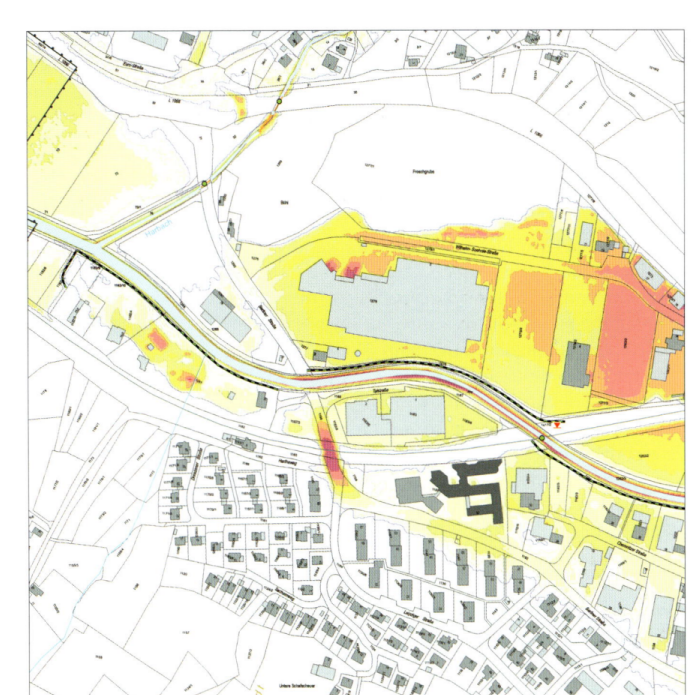

FIGURE 6-20 AND 6-21

Hazard maps of Baden-Württemberg. Type 1 (water depth) and Type 2 (water extent)

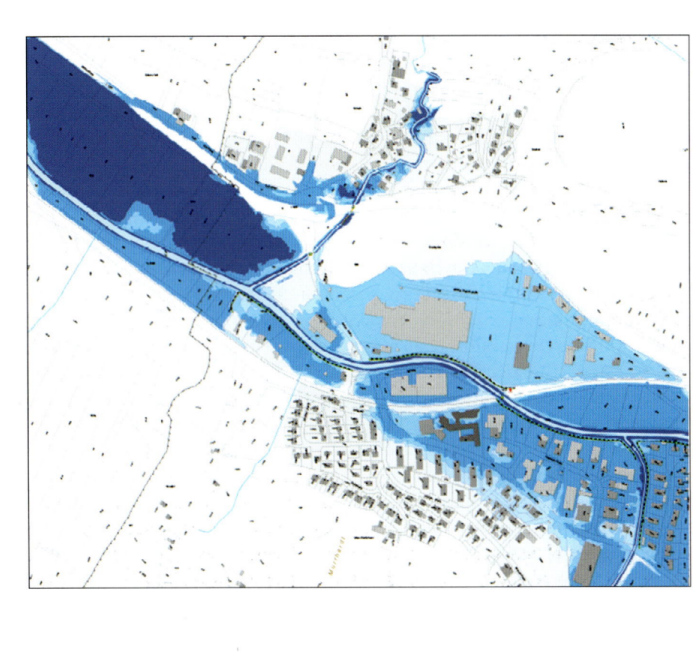

The processing of the flood hazard maps embraces the following working steps: definition of water bodies and demand analysis; terrestrial watercourse survey and development of digital terrain model; hydraulic computations; combination of results; layout; and publishing.

6.6.4 Strong and continuously local involvement

Since the beginning, SAFER's flood mapping was done in a very close cooperation with stakeholders, regional governments and local communities. Their in-depth involvement was scheduled for all crucial working packages. This is especially advantageous as firstly, the quality of the maps is increased by the on-site knowledge of the local stakeholders and secondly, the acceptance of the maps is raised.

During the start-up phase, water management authorities, spatial and regional planning institutions, the fire brigade and disaster management have been asked about what kind of content would be useful to include in the hazard maps. For instance, the fire brigade and the disaster management mentioned a certain grading of the water depth. According to this specific information, adequate vehicles and instruments can be organised and allocated for evacuation measures (Haimes, 1998).

Also in the beginning, the definition of relevant watercourses was discussed with local communities. Generally, the relevant watercourses are defined by the regulation catchment size bigger than 10 km^2—following the Water Framework Directive. However, there are exceptions. Nonetheless, certain watercourses that have a smaller catchment size but represent an especially high flood danger are included in the process.

6.6.5 Different ways of publishing flood hazard information

The publication of flood information is a very crucial step in the hazard mapping. Today, the most effective way to achieve this is via the publication on the Internet. In the SAFER project, however, all ordinary ways serve to achieve the raising of public awareness and to provide flood information for the different user groups. Printed copies as well as different programme files support the practical application. Additionally, several interfaces have been developed to promote a sophisticated appliance.

Generally, Internet publication has practical advantages. The flood information is presented and available to everyone. If necessary, the digital information can be further processed. At any location, the user can request online the water depth, water level and topography. The hazard maps can be kept up to date at source; hence, the latest available information is always accessible. During a flood event, the information can be embedded into flood warnings.

For the administration, additional features are included in the online viewer, for instance, an extended search function for administration districts and communities as well as streets and house numbers. The online viewer provides an excellent opportunity for the citizens to be prepared and to refresh the knowledge about flooding in their home area. Baden-Württemberg's hazard maps are provided on the Internet for free (see Figure 6-22).

For municipalities, an additional feature has been developed, the print-on-demand viewer. The viewer allows users to choose the type, scale and location of interest according to their needs and to send an order to receive the selection as pdf file. The print-on-demand viewer allows a very flexible and user-orientated way of providing the hazard maps.

FIGURES 6-22
Online viewer for the hazard maps. The viewer provides both types (extent and depth) to a scale of 1:5,000 (http://www.hochwasser.baden-wuerttemberg.de/servlet/is/15783/).

Key Questions

Flood risk

1. What is a common definition of risk?
2. Draw a simple fault tree for sewer flooding.
3. What is involved in a quantitative risk analysis (QRA)?

Flood probability

1. In the light of expected climate change effects, a lot of attention is paid to rare flood events with severe consequences; how do these events relate to the more frequent flood events with smaller consequences?
2. How might you expect climate change and urban development to affect the different types of events?
3. Urban flood incidents are inherently unpredictable which makes data collection for flood-risk assessment and management a difficult task. How can we monitor occurrence, extension and damage of urban flood incidents in consistent way and what new monitoring techniques can support this type of data collection?

Tangible and intangible damages

1. What are the different problems associated with measuring tangible and intangible damages?
2. How can intangible consequences related to physical and psychological health be measured?
3. How can intangible consequences of flooding fully be incorporated in flood-risk assessment and evaluation of flood alleviation schemes?

Loss of life estimation in flood-risk assessment

1. Explain the concept of a mortality function in the context of loss of life estimation.

Cross-scale factors and indirect damages

1. What is meant by indirect damages? Can you give an example?
2. What are the problems associated with measuring indirect damages?

Flood-risk mapping

1. What are the benefits of online accessible flood hazard maps? In your opinion, what are the risks of providing information of and for hazard circumstances by means of an electronic system?
2. There are many end-user groups that profit from flood hazard information. How and in what ways can flood hazard maps improve technical flood protection at either public interest or private property?
3. The flood hazard mapping is very complex. Nevertheless, there are key players that increase the quality of the final product and strengthen the on-site acceptance. Who are these key players and what kind of input can they provide?

FURTHER READING

FlOODSite (2009). *Development of Framework for the Influence and Impact of Uncertainty*. Report No T20-07-03. Available from: http://www.floodsite.net.

Jonkman, S.N. (2007). *Loss of Life Estimation in Flood Risk Assessment, Theory and Applications*. Dissertation, Delft University of Technology, Delft.

Maksimovic, C. & Tejada-Guibert, J.A. (2001). *Frontiers in Urban Water Management: Deadlock or Hope?* London: IWA Publishing.

Tucci, C.E.M. (2008). *Urban Flood Management*. WMO Cap-Net publication. APFM.

Schertzer, D. and Lovejoy, S. (1987). Physical modelling and analysis of rain and clouds by anisotropic scaling and multiplicative processes. *J. Geophys. Res.* 92(D8): 9693–9714.

Vrijling J.K. (2001). Probabilistic design of flood defence systems in the Netherlands, *Reliability Engineering and System Safety* Vol. 74 Issue 3 pp. 337–344.

IV

Responses

When the probability and specifics of a future external driver are difficult to define, the enhancement of system resiliency is a rational strategy to deal with risk. This also holds true for flood risk. Part IV elaborates on responses to flood risk and introduces you to a resilience-focused approach. This approach is based on two pillars: anticipation and precaution. We will argue that the shift from traditional approaches to a resilience-focused approach will require cultural changes in regimes, institutions, decision-makers and professional actors. It will need changes in the planning and risk assessment procedures for the flooding system and requires also that planning and building processes are effectively integrated. You will be made familiar with a wide range of available responses to reduce flood risk, including both structural and non-structural measures ranging from urban drainage systems and technology to flood proofing the urban fabric to the management of land use and urban planning and recovery. In Part IV you will learn that urban flood management is all about anticipating conditions that do not yet exist, to be able to cope with uncertainties and be prepared to reconsider.

A row of amphibious houses lines the waterfront at Maasbommel, panelled in blue, yellow and green. They have a hollow concrete cube at the base to give them buoyancy

Source: Dura Vermeer, The Netherlands, 2007.

Responding to flood risk

Learning outcomes

In this chapter, we discuss the different ways in which flood risks are and can be responded to. In reading the chapter, you will learn:

- how responses to floods can be grouped together and form the basis for the adoption of a suite of responses;
- about examples of how different levels of risk can be coped with;
- about the range of institutional and individual responsibilities for taking action to manage flood risk, emphasising that it is not just government agencies who have a responsibility to act;
- how the concepts of resilience, vulnerability, robustness and sustainability have been used and applied by various agencies to inform their flood-risk management plans and actions;
- about the critical connection between land-use planning and flood-risk management;
- about the way in which buildings and other forms of infrastructure affect the nature of the risks associated with flooding in urban areas.

At the end of the chapter you will find a number of questions that you might like to think about.

7.1 RESPONSES

Responses to flood risk are defined here as 'changes to the flooding system that are implemented to reduce flood risk'. Thus, they are usually active measures introduced as part of the adaptation process. Of course, it is possible to 'do nothing' and accept the risk, but even with this perspective, if a flood occurs it will be necessary to deal with it by responding in some way.

Responding to climate change may entail *mitigation*—reducing the causes (a main focus for policymakers); whereas *adaptation* is the adjustment in natural or human systems to actual or expected climatic stimuli or their effects, which moderates harm or exploits beneficial opportunities.

In general, responses are introduced in combination as a *portfolio* of response measures. Responses can be grouped into locational sectors, e.g. coastal or inland flooding; into types of measure, such as structural or non-structural; and into physical, regulatory

TABLE 7-1

UK's foresight future flooding response themes and groups—updated 2008

Response theme		Response group
Managing the rural landscape	1	Rural infiltration
	2	Catchment-wide storage
	3	Rural conveyance
Managing runoff from the urban fabric	4	Urban storage
	5	Urban infiltration
	6	Urban conveyance
Managing the urban fabric	U1	Building development, operation and form
	U2	Urban area development, operation and form
	U3	Urban source control and above-ground pathways
	U4	Urban groundwater control
	U5	Urban storage above and below ground
	U6	Main drainage form, maintenance and operation
Managing flood events	7	Pre-event measures
	8	Forecasting and warning
	9	Flood fighting
	10	Collective damage avoidance
	11	Individual damage avoidance
Managing flood losses	12	Land-use management
	13	Flood-proofing
	14	Land-use planning
	15	Building codes
	16	Insurance, shared risk and compensation
	17	Health and social measures and policies
River engineering and maintenance	18	River conveyance
	19	Engineered flood storage
	20	Floodwater transfer
	21	River defences
Coastal engineering and management	22	Coastal defences
	23	Coastal defence realignment
	24	Coastal defence abandonment
	25	Reduce coastal energy
	26	Morphological coastal protection

TABLE 7-2

Combined response groups used in the UK's foresight future flooding update (Evans *et al.*, 2008)

Response theme	Associated groups (Table 7-1)		Combined response group
Managing runoff from the urban fabric	4	Urban storage	Managing urban runoff
	5	Urban infiltration	
	6	Urban conveyance	
Managing flood events	8	Forecasting and warning	Real time event management
	9	Flood fighting	
	10	Collective damage avoidance	
Managing flood losses	12	Land-use management	Land-use planning and management
	14	Land-use planning	
	13	Flood-proofing	Flood-proofing buildings
	15	Building codes	

or behavioural. Different publications and guidance approach responses specifically in relation to their particular vision or context. Here the responses are considered in terms of performance, vulnerability, resilience, robustness, sustainability and adaptability. The application of other decision systems to the selection of responses is also reviewed.

The UK Foresight Future Flooding study in 2004 used a number of response themes and groups that were updated in a further review in 2007 as part of the Pitt Review as shown in Table 7-1.

These were used in a framework that included the scenarios outlined in Chapter 7.3 and as part of the 'Source-Pathway-Receptor' model (see Textbox 7-1), where responses are targeted at each of these three stages in the flood-risk model. In practice, and as shown in Table 7-2, portfolios of measures were used rather than the isolated, individual responses labelled 1–26 and U1-U6 in Table 7-1.

It was concluded that engineering (infrastructure) responses are indispensable to future flood-risk management and that portfolios of integrated engineering and non-structural responses will be required to deal with future flood-risk management. However, the sustainability aspects of the way in which engineered responses are implemented in the UK, risk compromising the effectiveness of these measures (Section 7.2).

7.2 PERFORMANCE STANDARDS AND EXPECTATIONS

According to Tapsell *et al.* (2005):

flood management policy is becoming more 'people oriented', rather than 'engineering oriented' as in the past. It is people who should determine flood risk standards, and people who will have to take many of the actions that are required to minimise risk, by buying insurance, flood proofing their homes, and reacting and responding to flood warnings.

The capability of individuals to respond differs according to age, gender, disability and wealth. To avoid creating victims, special care should be given to the most vulnerable individuals: children, elderly people and those who cannot move or communicate without support. Such categories of individuals should receive special care from neighbourhood and civil protection services. In general, individuals with greater wealth are more ready to move from their property and do not care as much to leave what they have behind. This is one of the reasons why there are more flood victims in poor areas and less wealthy countries.

Responses need to provide the appropriate level of performance. Design performance standards against flooding risk vary internationally. They can relate to the probability occurrence of certain water levels or even flow rate or velocity of flow, normally expressed as a return period. Or they may attempt to deal with risk by defining combinations of probability and consequence in some form of matrix as illustrated in Table 7-3. These days, these performance standards need also to take climate change into account (see Chapter 4). Often public expectations do not match what can be provided, as a risk-free society is impossible to achieve within realistic resource availability. It is increasingly recognised that responses need to be based on adaptable approaches

TABLE 7-3

Example of a risk matrix for triggering action on flooding

Note: Red shading indicates high priority, green indicates no current driver for action, but long-term aspirations may not be met, orange indicates low priority which may be acted upon if other problems exist in the same area and cost savings can be made if combined actions are taken.

leading to greater resilience for the system as a whole and should, where feasible, be incremental, reversible and 'no-regret'. This requires a new way of looking at responses, especially those that entail 'hard' engineering and are seldom reversible. In future, these need to be able to accommodate changes (adaptations) in response to new knowledge, demands, expectations and in the assessment of performance, with attendant flexibility in standards and codes of practice.

The performance of a flood-risk management system depends on the capacity of each component and on how it interacts with other components. In some parts of the area, a system's capacity might appear to be performing well, but this may be because in other parts, perhaps upstream, performance is such that flows are not being passed on downstream.

Flood-risk management is one of the less well appreciated aspects of urban living for dwellers and property owners, although rural dwellers and those living at the coast may be more aware. There is an awareness of the need to maintain the structure of buildings and even external areas in the curtilage, but in general, there is very little awareness of the need to maintain buried assets, such as pipes and drains or even river frontages. This is often an 'out-of-sight' problem but is also linked to the willingness or interest to invest in the maintenance or replacement of poorly performing drainage or flood defence systems. The same applies to privately owned, major flood-risk management infrastructure.

The expectation that flood risk will be managed by 'others' has been one that communities have adopted across the oldest urbanised communities, with the belief for example, in the Netherlands that the external flood defences were effectively unbreachable. This perception needs to be changed. The European Floods Directive and other national policies now expect those at risk to contribute to improved flood protection by implementing personal preparatory measures through personal or community initiatives and at the individual's own expense.

There are major differences in performance of the various drainage systems: coastal, major, minor, building, local drainage and sewerage. The formal expectations of system performance of the major and minor drainage systems are based on 'aspirational' performance levels, i.e. the expected capability in terms of flood frequency of occurrence (probability) for major and minor drainage systems in urban areas based on the main sources and published standards used in a particular country (Table 7-4). These depend on the location, and may or may not distinguish between the various types of flooding, i.e. coastal, river and sewer. National standards may be seen as the starting point of the process of ensuring that adequate performance of the flood defence and drainage network is sustained. In most countries, these standards are implemented via the town and country planning systems and the standards laid down for buildings, given in e.g. Building Regulations.

For example, in Europe there are defined standards, ENs, that are interpreted for each member state. In England, EN12056 (BSI, 2000), for building drainage, gives four categories for estimating design rainfall rate. This includes the use of 'probable maximum rainfall', a dubious concept given the variability in spatial and temporal behaviour of rainfall. When climate change is also introduced, it makes the concept even more elusive. This EN makes no reference to climate change, nor does the UK standard used for Building Regulations in England (Part H), although Planning Policy Statement 25: Development and Flood Risk (PPS25) suggests there is a need to 'maintain the integrity of existing land drainage arrangements on development sites' (Building Regulations, Part C). This is clearly applicable to new buildings. However,

TABLE 7-4

Examples of aspirational performance of major and minor drainage systems (England and Wales)

Average annual event frequency (years)	Average annual probability	Performance measure	Source
10	0.1	Conveyance capacity of minor drainage system in rural areas	EN 752
20	0.05	Conveyance capacity of minor drainage system in residential areas (presumably at the periphery of the drainage system)	EN 752
		Typical conveyance capacity for rehabilitation of existing sewerage system	
30	0.33	Conveyance capacity of new minor drainage systems in residential areas	WRc (2006)
		Conveyance capacity of minor drainage systems in city centre/commercial areas	EN 752
30	0.33	Building drainage	EN 12056
50	0.02	Conveyance capacity of minor drainage system where railways and underpasses may otherwise be flooded	EN 752
75	0.013	General level of protection of buildings from flooding for insurance purposes	ABI (2004; 2005)
100	0.001	Minimum level of protection of buildings from flooding in new developments	CLG (2006)
200	0.005	Maximum level of protection of buildings from flooding in new developments	ABI (2004)
		Minimum level of protection for residential properties for flooding giving 'normal terms of cover'	
Beyond 200	residual	'a well designed drainage system should ensure that there is little or no residual risk of property flooding occurring during events well in excess of the return period for which the sewer system itself is designed'	CLG (2006)

the main stock of existing properties may be more than 100 years old, with a replacement turnover rate of up to 100 years. Currently there is no simple guidance available to inform property owners about changing flood risk, or to encourage their greater commitment to the maintenance of their existing drainage systems, although changing approaches by insurance companies may invalidate cover in future if adequate maintenance is not undertaken.

There is an assumption in many countries that building development should not take place in areas that are at risk of flooding. An example of this is the planning guidance for England and Wales which can lead to conflict between economic development needs in communities and safety. Even recently, despite a decade of investigation, organisations such as the Association of British Insurers (ABI) have unrealistic expectations about what is achievable:

> New developments, wherever possible, should be located in areas which are not at risk of flooding from any source, both now and in the future. However, in those exceptional cases where development must take place in areas where there is a risk of flooding, it is necessary to design the development to minimise the potential damage to each property and the overall damage to all properties, both new and old, that might be caused in all areas affected at the same time.

There is nowhere on the planet that can be presumed to be not at risk of flooding now and in the future especially as rainfall can occur anywhere, even now. With a changing climate, the risks of extreme rainfall occurring can only increase.

Performance standards may be allowed to reduce when the external drivers change the impacts where this is acceptable. For new developments in England, PPS25 indicates a need to 'maintain little or no residual risk to property' from development. Hence, it should be possible to presume that in the future, existing property will be safe from external sources of flooding from new developments and that 'a priori' unknown climate changes are properly factored in. In fact, adaptation potential (capacity) must also be provided in these new developments due to the uncertainties of climate change and responding thereto; requiring at the least, design for exceedance flows.

Even in the most developed countries, regulations and standards are not necessarily well defined in relation to the management of flood risk. For example, flood resistance or resilience for buildings is not currently a requirement of the national Building Regulations 2000 in England and Wales. Approved Document C of the Building Regulations provides no advice on particular measures to be taken for flood-prone areas or to alleviate flood risk. Part C does, however, provide guidance for different levels of exposure to wind-driven rain and to the impacts of ground contaminants. Part C now directs builders to particular flood issues and to the provision of further guidance on flood resilient construction. The UK Government Department of Communities and Local Government (CLG) produced guidance on the construction of new buildings to cope with flood risk in 2007. This guidance should be the first source of guidance for new buildings. The main drawback is that the guidance would not necessarily result in flood resilient construction in small urban catchments. The guidance does not specifically deal with sewer flooding and is concerned only with inland river and coastal flooding.

Stakeholders and indeed homeowners would expect building regulations to include flood resilience in the future. Building regulations are concerned primarily with health and safety, accessibility and energy efficiency (carbon reduction), and flood impact on buildings can certainly be considered as a health and safety issue. However, there are likely at least to be conflicts with accessibility and energy efficiency requirements of any building regulations.

Building regulations require a regulatory impact assessment to be carried out prior to being made mandatory. It is likely that this type of exercise would result in the economic case being made for flood resilient construction. However, flood-risk management and control is viewed as a planning consideration. In addition, it is unlikely that any development of building regulations would be applied to all newly constructed buildings. It is more likely to try to target specific flood-risk areas, similar to the approach to building design to manage driving rain levels in different parts of the country. In this case, small urban catchments are liable to be missed.

Building regulations generally apply to new construction, but they are also used for major refurbishment and for extensions. Some parts cover work to existing buildings (e.g. Part L deals with energy efficiency in England & Wales). In the UK, it might therefore be argued that the role of building regulations in flood management should be undertaken as follows:

- For new construction—where the flood risk is estimated as greater than 1 in 75 years (or 1.3 per cent annual probability of flooding) guidance on resilient construction would be as given in the existing CLG guidance (2007).
- For existing construction—for all flooded buildings there should be a requirement to consider flood resilient repair or retrofitting. This would include a flood-risk assessment to be undertaken and appropriate technical measures to be undertaken when there is a risk of a return flood, even when this occurred outside a floodplain or coastal flood area.

At present, repairs to buildings after a flood are not typically subject to building regulation control through the municipality (or other building control authority). Bringing repair within the aegis of building control may result in a delay to repairs occurring and extend the time involved in repair work and cost. There would therefore be a need to 'fast track' any flood resilient repair to buildings by the municipality, although not at the expense of other work.

7.3 RESILIENCE, VULNERABILITY, ROBUSTNESS AND SUSTAINABILITY

Responses to flood risk should aim to:

- increase the resilience (reducing vulnerability of the receptors);
- be robust and sustainable.

Resilience and vulnerability are discussed in Sections 1.3 – 1.5 and again in Sections 11.2 and 11.3 where robustness is also considered in the context of the ability or not to self-organise and restore functioning. In Section 7.1, the resilience of buildings is introduced in the context of regulations. This section is concerned with the wider concepts of sustainability and resilience (see also Textbox 13.4), related to the sustained functioning of the urban system as a whole, rather than just to one building or a neighbourhood.

The Scottish Government has devised a framework for dealing with flood risk more sustainably; where resilience of urban and rural areas in the face of flood risk and after a flood event, has been taken as the main measure of sustainability, as illustrated in Table 7-5: 'Sustainable flood management provides the maximum possible social and economic resilience against flooding, by protecting and working with the environment, in a way which is fair and affordable both now and in the future.' Where resilience is further defined as: [being] 'able to recover quickly and easily'. This broadly fits with the EU Floods Directive's 3Ps—Preparedness, Prevention and Protection—and the associated emergency response and recovery as shown in Table 7-5.

As well as examining the applicability, value and cost of responses, the assessment of the likely sustainability in terms of scenarios was also examined in the UK Foresight Future Flooding study. This included environmental impacts and benefits as well as social factors, especially the influence of governance. This is illustrated in Figure 7-1 for responses using measures related to main sewerage systems and showing the four socio-economic scenarios: World Markets, Global Responsibility, National Enterprise and Local Stewardship. In Figure 7-1, the particular response measure is

TABLE 7-5

Scottish Government's strategic 4 As approach to flood-risk management and the EU flood directive's 3Ps (Water Information System for Europe, WISE)

Four As	WISE
Awareness: enhancing the awareness and engagement in all aspects of flood risk and the means of managing it at the policy level (politicians/decision-makers), among the professionals (of the involved authorities and elsewhere) and at the public level (people, companies, developers, insurance companies).	**Preparedness**: informing the population about flood risks and what to do in the event of a flood.
Avoidance: limiting flood damage and easing recovery by planning and adapting buildings, infrastructure, surfaces and economic activities and building capacity in individuals and institutions to become more resilient.	**Prevention**: preventing damage caused by floods by avoiding construction of houses and industries in present and future flood-prone areas; by adapting future developments to the risk of flooding; and by promoting appropriate land-use, agricultural and forestry practices.
Alleviation: reducing flood risk by implementing physical, technical, non-structural and procedural measures for the management of water systems.	**Protection**: taking measures, both structural and non-structural, to reduce the likelihood of floods and/or the impact of floods in a specific location.
Assistance: providing support to recovery processes and engaging and building capacity in communities, and others prior to, during and after flood events.	**Emergency response**: developing emergency response plans in the case of a flood.
	Recovery and lessons learned: returning to normal conditions as soon as possible and mitigating both the social and economic impacts on the affected population.

more sustainable under the particular scenario the further the line is from the centre of the diagram.

In the Foresight 2004 study, the categories used were:

- cost effectiveness: the value for money of implementing the measure;
- social justice: the impact of the measure on different types of household;
- environmental quality: the impact on biodiversity, and the area and quality of habitats;
- the ability of the measure to cope with uncertainty relating to scenario differences in socio-economic factors and climate change (robustness);
- uncertainty relating to the ability of the measure to cope with extreme events and how they would operate (precaution).

Although no single response measure was simultaneously both robust and did not compromise sustainability criteria, the study indicated a wide range of measures that were attractive and did not entirely compromise the sustainability objectives, in particular land-use planning and management, building regulations and codes. Whereas, realignment of coastal defence, flood-proofing and engineered flood storage all provided flood-risk reduction benefits but were not good in sustainability terms, particularly as regards social justice and the differential impacts on the poorest members of society. Social justice was contravened more in terms of the way in which the response measures are traditionally implemented in the UK, rather than the response itself being inherently

7.3 RESILIENCE, VULNERABILITY, ROBUSTNESS AND SUSTAINABILITY

FIGURE 7-1
'Spider' sustainability assessment diagram from the UK Foresight Future Flooding study (adapted from Evans *et al.*, 2004a)

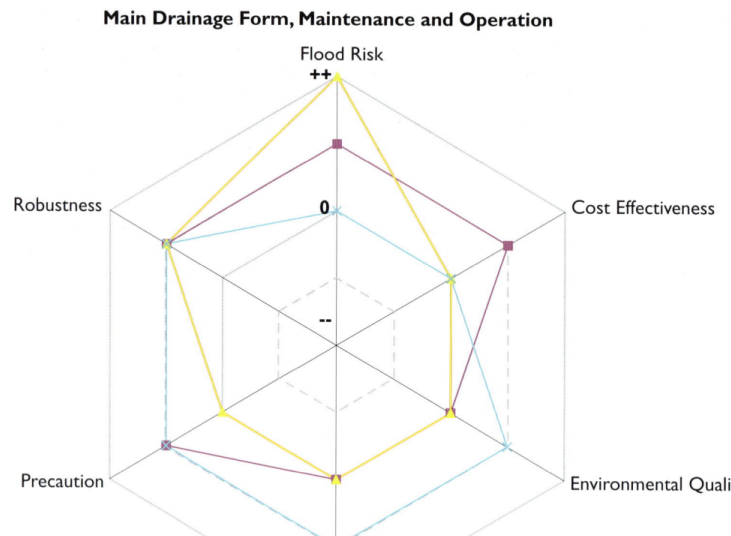

Main Drainage Form, Maintenance and Operation

unjust. In the subsequent analysis updating the Foresight study, catchment-wide storage was a new option that also appeared both robust and sustainable.

Recent guidance from the Communities and Local Government Department of the UK Government sets out the approach for flood resilience of new developments. The guidance sets out an approach that can be summarised as follows:

- For flood depths of less than 300 mm depth use dry proofing.
- For flood depth between 300 mm and 600 mm use dry proofing with some wet proofing measures.
- For flood depth greater than 600 mm use wet proofing measures.

For small urban catchments, especially for repair of existing buildings, the preferred approach would be dry proofing even up to 900 mm depth. In this way, the following main objectives are achieved:

- contaminated flood water is prevented from entering the building;
- the degree of recovery and repair to the inside of the building is limited or the need is eliminated.

A range of measures can be taken to dry proof homes (see also Section 9.2). These included the following categories:

- temporary dry proofing measures;
- permanent dry proofing measures.

The measures may be expensive. In the case of temporary measures, they will require installation quickly once the threat of a flood is realised. This latter point can be critical to the success of the measures undertaken. An obvious problem with small urban catchment floods is that many areas are not well covered by existing flood maps or warnings (Section 6.6). Floods from surface water runoff, sewers and small streams can occur with little warning (flash floods) and therefore the opportunity to install dry proofing equipment is limited. Repair should therefore focus on the use of permanent dry proof measures where possible and be supported by temporary measures. Additional wet proofing should be included where the nature of the repair required presents the opportunity, or where there is a specific risk that more than 900 mm of flood will be experienced.

7.4 PRECAUTIONARY AND ADAPTIVE RESPONSES

In the preceding sections of this chapter a variety of responses have been considered, many of which could be implemented as precautionary measures or as part of an adaptive strategy. Where external drivers are reasonably well defined within a narrow uncertainty boundary, such as expected sea-level rises, a 'precautionary' approach may be appropriate, which, as shown in Figure 7-2, is feasible.

This shows the approach used when the external drivers have been fairly predictable. The problem with uncertainty and precautionary responses is acknowledged in the Pitt review report into the 2007 flooding in England: 'Early action will also avoid lock-in to long-lived assets such as buildings and infrastructure which are not resilient to the changing climate.' However, this still allows major asset investors to claim that their large infrastructure (as invested in as shown in Figure 7-2) has been designed to be resilient—when this is clearly impossible with current limited knowledge.

Over time, the external drivers (like climate change) are now expected to change more rapidly than has previously been anticipated at the design stage (Figure 7-2)

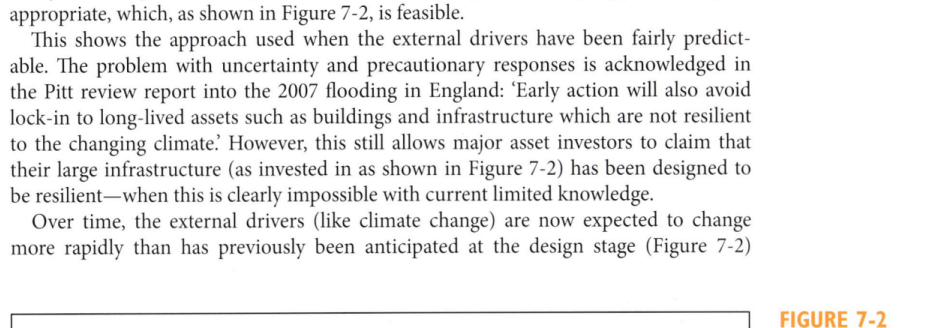

FIGURE 7-2
The traditional response to external driver changes using large asset investments

Source: Berry Gersonius, 2009.

7.4 PRECAUTIONARY AND ADAPTIVE RESPONSES

and the flood defence asset can no longer deliver the required service performance. Modification or abandonment in favour of an alternative is not possible or not acceptable due to the large investment at the outset; hence, another major asset investment becomes necessary to meet the changing driver. A better alternative, where there are significant uncertainties in the estimates of future drivers (Section 1.5), is an adaptable response which allows for current uncertainties to become better defined over time, as illustrated in Figure 7-3.

Figure 7-3 is a simplified version of the UK Government's guidance on climate change. The performance standard here is shown to be constant with time and adaptation allows for the system to respond by performing around the expected standard. Responses may include non-structural as well as structural measures and as these are effected gradually over time, they can use evolving contemporary knowledge about the drivers at the time of responding (e.g. about climate change, as illustrated). Deviation from the expected performance standard (risk) may need to be restricted to an acceptable level. A constant performance standard is unlikely and in general, society's expectations are that performance will improve with time, hence the performance line should have a positive slope and the adaptation measures would need to reflect this.

Such adaptations as illustrated in Figure 7-3 are seldom undertaken in response to climate change alone but embedded within broader sectoral initiatives such as coastal (erosion) defence. Many adaptations, such as non-structural measures can be implemented at low cost but comprehensive estimates of adaptation costs and benefits are currently lacking. Adaptive capacity is uneven across and within societies, although it is more and more assumed that by taking an adaptive approach to water resources management, overall costs can be reduced.

The capacity to adapt is dynamic and influenced by economic and natural resources; social and professional networks and cultures; entitlements; institutions and governance; human resources; and technology. However, there are substantial limits and barriers to adaptation e.g. the inability of natural systems to adapt to the rate of climate change; and financial, technological, cognitive and behavioural, social, institutional and cultural constraints.

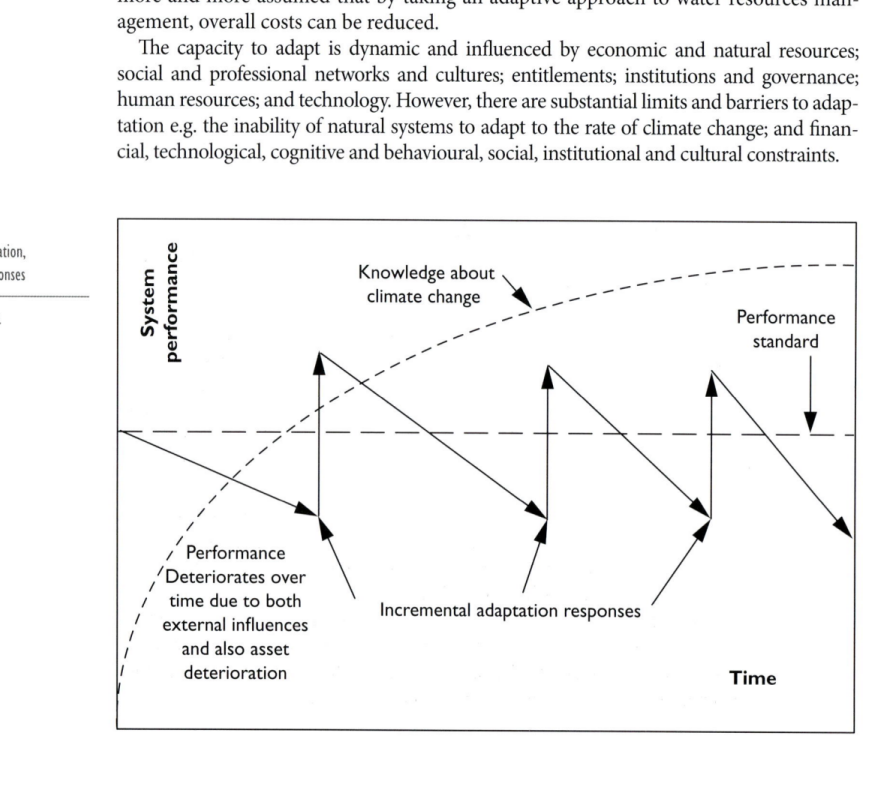

FIGURE 7-3
The concept of adaptation, with incremental responses

Source: Berry Gersonius, 2009.

There is conflicting evidence as to whether or not catastrophic events are effective at promoting adaptive responses. More often than not, policymakers and those charged with policy implementation rely on traditional 'we have always done it this way'—so-called 'common-sense' responses to crises. Ironically, it is often just these responses that have been inadequate to prevent the catastrophe occurring in the first place. There is no better example of this than the return to 'business as usual' by the banking sector following the public bailout after the crisis of 2007–09.

High adaptive capacity may not necessarily translate into successful adaptations. In Norway, research on adaptation to changing flood risk has shown that high adaptive capacity is countered by incentives that are too weak to encourage proactive flood management. Despite increased attention to potential adaptation options, there has been less understanding of their feasibility, costs and effectiveness, and the likely extent of their actual implementation.

7.4.1 Non-structural measures

At the community or neighbourhood scale, the capability to respond is most strongly related to social structure and social capital. Where there are more younger and middle-aged people, social structure usually has the greatest ability to respond, although enhanced social capital can provide a greater willingness to help or to deliver a common response. Tapsell *et al.* (2005) define:

> Social capital as made up of the networks and relationships between individuals and social groups that facilitate economic well-being and security, with disasters being socio-environmental by nature and their materialisation the result of the social construction of risk. We are all vulnerable in one way or another. This complexity makes it very difficult to plan for simple 'fixes' to flood risk. No single investigation into vulnerability indicators will provide a holistic and comprehensive answer, however, there are aspects of vulnerability and resilience that can be explored and represented through the development and application of quantitative indicators.

The ageing society in many countries, especially across the EU, may have more disabled people and others less able to respond personally to flood risk. Better communication and the spreading of information may also require institutional strengthening to develop the capacity to respond within communities.

Structural responses, as well as referring to physical infrastructure systems to prevent flooding, can also refer to physical actions such as moving vulnerable people out of hazard areas, from hospitals, schools, homes and facilities for the elderly. Other responses could also be in the construction of living facilities on a higher elevation than the likely flood level. *Non-structural solutions* in relation to capability are aimed mainly at institutional and social strengthening; providing the capacity to deal with the hazards in a more vulnerable society.

Communities and wider regions are very vulnerable to the potential damage to public infrastructure, such as transport, power supply, information transfer, and water supply and sewerage systems. Damage to such facilities disturbs supply in a wider region far beyond the directly affected hazardous area. For example, in the summer of 2007, more than 300,000 people were without piped water supplies for more than a week in England due to flooding of a water treatment plant. In such cases, responses may be structural with movement of the population out of the hazardous area or in increasing system robustness high enough that there will be no loss of service.

7.4 PRECAUTIONARY AND ADAPTIVE RESPONSES

These may be considered as a set of mitigation measures that do not make use of traditional structural flood-risk management measures. They are part of an integrative risk management approach that addresses all components of the Source-Pathway-Receptor model (see Textbox 7-1). This strategy encompasses:

- information (flood maps and other material such as brochures, public presentations, internet portals, etc.);
- education and communication (training, collaborative platforms);
- spatial planning (flood-risk adapted land use);
- building regulation;
- improvement of buildings;
- flood resistance (wet proofing and dry proofing);
- flood action plan at a local scale (infrastructure maintenance);
- financial preparedness (insurance of residual risk and reserve funds, emergency response (evacuation and rescue plans, forecasting and warning services, control emergency operations);
- emergency infrastructure (allocation of temporary containment structures, telecommunications network, transportation and evacuation facilities, recovery measures, disaster recovery plans, financial provisions of government).

TEXTBOX 7-1
The Source-Pathway-Receptor (S-P-R) model in brief

The flooding system is formally defined as the complex, dynamic risk producing-risk response system that adapts to changes in drivers for flood risk and other factors. The distinction between 'drivers' and 'responses' relates to the extent to which flood managers and policy makers have power to influence change and the policy level at which power is influenced.

The Source-Pathway-Receptor (S-P-R) model is a well-established framework for the analysis of both drivers and responses (DETR (2000)). Figure 7-4 shows how the 'drivers' and 'responses' (the flooding system state variables) act upon the flooding system resulting a change of the flood risk.

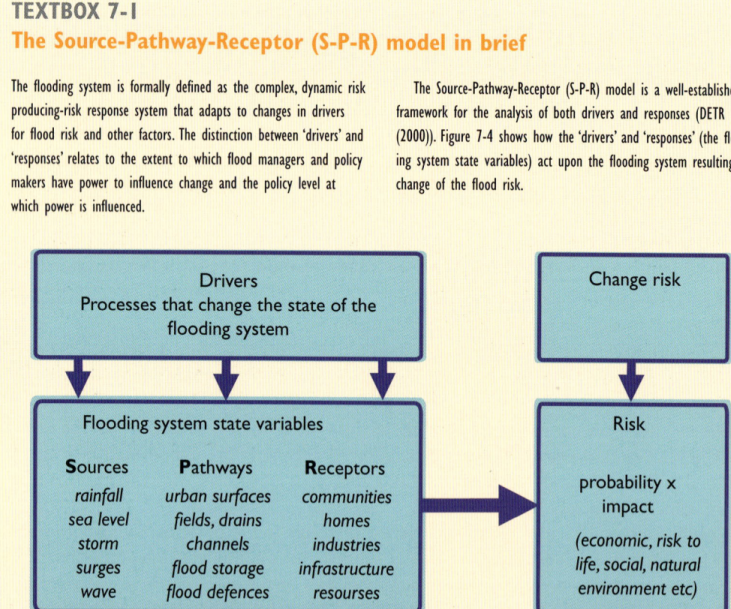

FIGURE 7-4
Conceptual framework of the Source-Pathway-Receptor (S-P-R) model

Source: DETR, 2000.

These measures can be divided into two broad categories that correspond to different types of opportunities and assessment problems: flood probability reduction measures and flood resilience measures. They can also be divided into measures that are already generally in use (traditional) and those that are beginning to be used (emergent).

7.4.1.1 Flood resilience measures

The shift towards integrated flood-risk management (IFRM) and the 'living with floods' concept has triggered a demand for active involvement and more interactivity among different stakeholders, such as experts in different fields, authorities or residents. Capacity building should support the effective participation of stakeholders within their role in IFRM. The key issue in building capacity is to increase flood-risk awareness among the stakeholders, for instance using flood maps. But it should go beyond this and aim to develop the capability of stakeholder groups to understand and engage in all facets of the challenge of flood-risk management.

For the public, information and communication (brochures, flood marks, etc.) should always allow the community to identify the problem and those who are expected to respond.

This rethinking and substantial change in the traditional approach to flood-risk management have to be developed within an appropriate regulatory framework, and necessary restrictions in the land-use planning in areas prone to flooding have to be included in development plans for all types of flooding (Section 7.2). Flood preparedness can be mainly influenced by increasing the resilience of the built environment (in the widest sense) (see also Section 10.3). In parallel, contingency planning encompasses all activities and resources in case of a hazard event that include: emergency response, emergency infrastructure, financial preparedness and recovery plans.

7.4.1.2 Flood probability reduction measures

Flood probability reduction measures control the source (such as surface runoff, sewerage overflow) and/or act on the pathway of flooding. Such measures have to be treated in an integrated manner and to be planned in a way that would not cause any increase in flood risk downstream. Only in this way, would the sustainability of flood management practice conform to the requirements of new EU legislation that postulates a catchment-based approach. In urban areas away from the coast, this can usually be delivered by a combination of sustainable drainage systems (SUDS or BMPs) or restoration of retention or storage areas through either natural elements or artificial structures in urban watercourses (Section 8.3).

There is a range of factors which can preclude or restrict to some extent the application of Best Management Practice (BMP) for urban stormwater control, these include: hydrological or geological (e.g. water table level, sediment input); with others being more technical (liability, maintenance); social (the social acceptability of multifunction devices); or economic (affordability).

7.4.2 Buildings

Research has shown that where flood insurance is available, homeowners look primarily to the insurance industry for redress. Furthermore, homeowners who have never experienced flooding and do not consider themselves to be in areas at risk of flooding feel that an insurance policy is their most effective flood-risk mitigation strategy.

In the UK, The Association of British Insurers encourages householder action on flood resilience through raising awareness, and they are actively discussing with government how best to roll out large-scale resilience programmes. However, homeowners are generally still poorly informed about how to make their properties more resilient to floods. Whilst insurance companies are currently unable to finance flood resilience measures, the Council of Mortgage Lenders have confirmed that their members are willing to provide loans for resilience measures.

There has been some considerable debate in recent years on the role of flood resilient construction in flood management at the local level. As a non-structural measure, flood resilience can involve either dry proofing and/or wet proofing measures (see also Section 9.2). In order to be effective, it is unlikely that one or other of these options will be successful unless other non-structural measures are also operating. For example, flood warning will be necessary in order to ensure that flood products can be fitted when flood warnings are given. If no system of flood warning is in operation, then there is a substantial risk that building owners will not be able to determine flood risk and to fit temporary defences or products in place.

Wet proofing strategies also need to work in combination with flood warning systems. There must also be places for people to seek safe refuge during a flood where a wet proofing strategy is adopted, as it assumes that the property will be allowed to flood thus giving the occupants no option but to abandon the property.

In new buildings, resilience can be designed in from the outset (Chapter 9). The location of individual buildings can be determined as well as their vulnerability through carrying out a risk assessment for a site. Actions such as raising the level of the ground or using certain types of building material can be considered. In existing buildings, the opportunities may be limited by the design, construction and materials used, as well as by the location and level of risk.

The delivery of flood resilient construction as a non-structural measure for flood management has not as yet achieved full policy support by government agencies and municipalities across Europe. There is typically a requirement to avoid flood risk wherever possible through the planning legislation, which makes the encouragement to use such measures less attractive to regulators. The value of resilient construction may be in protecting existing buildings as opposed to allowing substantial new developments in flood-prone areas. In a 'political' sense, this is more appealing to those in government than relocating buildings and communities. It does, however, raise the question of preventing new development in flood-risk areas where existing communities are well established. To prevent any new development will undoubtedly result in the stagnation of the community and quite possibly economic failure. Therefore, flood resilient construction could play a substantial role in balancing the conflicting pressures of development and planning.

Research for Era Net $Crue^1$ in 2008 showed that there is typically a reluctance by homeowners to invest in flood resilience for buildings and a preference for the professional engineer to remove the risk through management of flood defences, drainage and infrastructure. Homeowners typically did not indicate any level of trust in the performance of flood products. This displays both a lack of understanding of the use of flood products to protect a property, but also the lack of maturity of such products. There is a need for better demonstration of the function and performance of such products. The acceptance of the public generally as regards wet proofing approaches where the building is allowed to flood, but can more readily recover, is as yet uncertain. There is undoubtedly a preference to avoid flooding and resilience measures are only addressed after a flood has occurred.

¹CRUE ERA-NET action on flooding aims to develop strategic integration of research at the national funding and policy development levels within Europe for the sustainable management of flood risks by improving co-ordination between national programmes.

7.5 CONFRONTING FLOOD MANAGEMENT WITH LAND-USE PLANNING: LESSONS LEARNT

The way in which land is used and managed is possibly the most significant aspect of flood-risk management (see Chapter 3); hence, it is essential to ensure that this is an integrated process. For example, the key messages from the update of the United Kingdom's 2004 Foresight Future Flooding study were that:

- river and coastal defences have the greatest potential impact for reducing overall flood risk;
- better land-use planning and the flood-proofing of buildings still appear among the most important risk reducers;
- finding the space in urban areas to accommodate increased overland flows is one of the most important responses.

These messages are also reflected in the Delta Commission report on the future of flood-risk management in the Netherlands, where the way in which people live and the use of land are central to the plans for the next century and beyond.

Hence land use, land-use planning, together with planning the layout of urban areas (also a planning consideration) are seen as being amongst the most important of all responses for managing increasing flood risk. This is in terms of both the drivers and pathways of flood risk and also the vulnerability of the receptors (see also Chapter 3). Of considerable significance is the amount of paved or sealed, impervious surfaces, from which runoff can occur. Urban development can be responsible for large increases in the rate and volume of runoff and this is a problem that is by no means peculiar to Europe or developed countries.

When discussing an idealised approach to managing flood risk via urban form there have been very strong pleas for better urban planning:

The dominance of economic issues in the development of urban form has created a legacy of exposure and vulnerability to flood risk, and a growing recognition of the limitations of this methodology has led to a desire to manage flooding in a way more harmonious with nature.

There is a need to balance all societal needs not only for flood-risk management, and often there are seemingly conflicting needs regarding for example, building houses on floodplains, which appears to be self-evidently a bad idea. Hence, other considerations, such as the availability of alternative space for new homes and travel distances to work, may be more important. In any case, it *is* possible to build resilient properties in floodplains (Chapter 11).

In the context of urban flood risk, it is not only the land management in the urban area itself that is important, but also that of the surrounding rural and built-up areas within the river catchments draining into and through the urban area. At the coast, responding to changing flood risk may require loss of land in a managed retreat process and relocating buildings and communities. Tables 7-1 and 7-2 show the updated, 2008 UK flood response groups. Within these, Table 7-1 illustrates the potential rural land planning and management response processes adapted from the 2004 study on Future Flooding and Coastal Erosion Risks. Similar groups of responses are presented for tackling coastal, river and urban flood risk that entail responses that change land use and require significant planning, in Tables 7-7, 7-8 and 7-9, respectively.

TABLE 7-6

Potential flood-risk management responses related to rural land planning and management

Response group	Specific measure	Type of response
Water retention through management and infiltration into the catchment	Arable land-use practices	Land-use management
	Livestock management	Farm and land-use practices
	Tillage practices	Farm and land-use practices
	Field drainage (to increase storage)	Farm and land-use practices
	Buffer strips and buffering zones	Farm and land-use practices
	Afforestation	Land use management
Water retention through catchment–storage schemes	Ponds, bunds and ditches	Farm and land-use practices
	Wetlands and washlands (i.e. large-scale floodplain storage)	Planning and land-use management
	Impoundments	Planning and land-use management
Managing conveyance	Management of hillslope connectivity	Land-use management
	Channel maintenance	Farm and land-use practices
	Channel realignment	Farm and land-use practices

TABLE 7-7

Potential planning and land use flood-risk management responses related to coastal flooding and erosion (adapted from Nicholls *et al.*, 2007)

Response group	Specific measure	Type of response
Realignment of defence infrastructure: landward relocation of coastal defences, managed or unmanaged abandonment of defences	Change configuration of coastline: <15 years for managed realignment <100 years for unmanaged realignment	Land-use management
Morphological protection: allow or engineer desirable changes in the coastal morphology to develop self-sustaining forms which provide flood and coastal defence functions	Promote broad-scale formation of natural landforms to provide protection, including the above measures	Land and land-use management

There continues to be a conflict between the needs of the rural, especially agricultural community and the downstream urban areas, where rural land use is managed to reduce the downstream flood risk. This on-going conflict between countryside and farming (food production) and the adaptation of farming practices to the benefit of downstream urban areas (including flooding arable land) has a long history and waxes and wanes in accordance with society's perceived need for locally produced food.

TABLE 7-8

Potential planning and land use flood-risk management responses related to river engineering (adapted from von Lany and Palmer, 2007)

Response group	Specific measure	Type of response
Changing river conveyance and flood flow pathways on floodplains to reduce water levels	Canalisation of river reaches sometimes creating multi-stage channels	Planning and land use management
	Forming flood bypass channels	
Flood storage	Dam the river creating storage area	
	Create washlands	
	Enhance natural floodplain storage areas	
	Provide off-line storage adjacent to the river	
Flood defences	Ring dykes around vulnerable areas	

TABLE 7-9

Potential planning and land use flood-risk management responses related to urban flood-risk management (adapted from Ashley and Saul, 2007)

Response group	Specific measure	Type of response
Building development, operation and form	Form of building	Planning and land use
	Form of curtilage	Planning and land use
	Form of local area drainage system	Land use
Urban area development, operation and form	Urban form, density and layout	Town planning and land-use management
Source control and above-ground pathways	Source controls (e.g. local storage, pathways)	
Storage above and below ground	Above ground storage	
Main drainage	Where this is above ground in channels, culverts, etc.	

In terms of the effect of upland land management practices on downstream flood risk, there is still only limited evidence for a causal link other than for rural areas directly adjacent to urban areas, from which direct runoff from saturated soil can cause 'muddy flooding'; so-called because such flooding is accompanied by lots of soil and sediment which blocks drains and leaves deposits see Figure 7-5. Nonetheless, there are examples of flooding attributed to runoff from saturated soils, such as the 2007 English floods, where high river flows occurred as a result and the 2009 flash floods caused by a tropical storm in Manila, Philippines.

7.5 CONFRONTING FLOOD MANAGEMENT WITH LAND-USE PLANNING: LESSONS LEARNT

FIGURE 7-5
Sediment deposits from muddy flooding in Chamonix, France

Source: Ian Goodall, 2009.

Where significant risks do exist, responses to muddy flooding require changes in land use and land drainage to prevent or divert such flows. Otherwise, it is recommend that rural land management practices should be considered as part of a programme of flood management measures and not as stand-alone measures. This is because extreme rainfall events will lead to runoff despite attempts to control it by rural land-use management. There are, however, a range of potential multi-benefits from the better management of rural and agricultural land that result in improvements to diffuse pollution, amenity and enhancements to landscapes and wild life habitats.

In coastal areas, responses may require structural measures (Section 11.4) that in themselves will also require formal planning approval. However, in these areas an increasingly common approach to the protection of properties is known as 'roll-back'. This is where plans have been made that an eroding coastline will not be defended and hence any existing urban areas that will be affected are re-located away from the risk area. In England, the management of coastlines is dealt with in Integrated Coastal Zone Management plans, and municipalities are responsible for the associated town planning. In many areas there are formal planning policies for the roll-back of properties at risk that provide preferential planning permission for the relocation of those properties to areas that will not be affected for at least a century. Reluctance to relocate by property owners and dwellers despite these policies has been attributed to a lack of appreciation of the imminence of threats and the costs, which have to be borne personally.

In most urban areas and those that are urbanising, the planning of the urban space, area and form has the most significant implications for flood-risk management. Frequently, flood-risk management is not given a major place in this planning; being an important consideration, but subordinate to the more 'important' aspects, such as layout and density of buildings, roads and transport. Traditionally not only flood risk, but also all aspects of water management have been considered as something to be sorted out once all of these other more important aspects have been decided on.

White, in arguing for an 'absorbent City' that is more resilient to flood risk, suggests that a coherent and reflexive strategy is required. This is one that should not be seen purely from a structural standpoint, but should encompass the internal operation of urban institutions and their ability to influence resilience locally and also elsewhere. Spatial planning is proposed as a very effective mechanism to influence the market and instigate a more interventionist approach addressing the potential market failure brought on by flooding, as insurance is subsequently unobtainable, compelling state intervention. The reflexive strategy is one that learns by doing and is part of the need to inculcate an active learning momentum on the part of all stakeholders suggests in this context that:

> A long-term view would be gradually to adapt the urban form and function within any city to be more sensitive to its geography and move towards a more sustainable pattern of development; not solely determined by socio-economic factors, but also its local geographical, climatic and environmental constraints. Thus far, the focus has been on structural resilience measures to protect against flood risk. This, however, repeats the technocratic mistakes of historic approaches to flood defence and essentially commits future generations to potentially unsustainable defence measures. Resilience to flood risk should be systematically built into the planning process, with a break away from the ad hoc treatment of the past and a move towards sustainable, strategic city-wide solutions based on prevention rather than protection. Furthermore, the seeds of cultural change contained in the natural management approach argue that a paradigm shift away from defence towards proactive action and management should occur and therefore spatial planning needs to adapt to manage the environmental threats of the 21st century.

However, we must add a cautionary note. No matter how well planned a system might be or how well prepared the possibility of being overwhelmed by nature will always remain. Whilst human misbehaviour does compound the severity of the impacts of flooding there will be cases where it will come down to the management of the emergency. In the case of the 2009 Manila floods, it was argued that no amount of planning could have easily coped with a storm in which 400 mm of rain fell in 12 hours, the September average is 390 mm and the previous record for 24 hours was 330 mm (the Philippines chief weather forecaster blamed climate change for the mass downpours).

7.6 BUILDING TYPES, INFRASTRUCTURE AND PUBLIC OPEN SPACE

White proposes that a city that is adaptive to changing flood risk is comprised of:

- blue infrastructure;
- green infrastructure;
- built environment;
- wider catchment.

Blue infrastructure recognises the need for functioning space for water within the urban area and that there may be existing washlands or storage areas that should only have flood resilient or resistant buildings located therein. In some urban areas, as knowledge advances about the extent of potential inundation, such areas may need to be created that potentially require the relocation or removal of existing buildings, as has been done in London and Ontario, Canada. Ideally, such spaces in urban areas should be multi-functional and include green (planted) areas that may be used for recreation or other purposes. These will also help with reducing the urban heat island effect by promoting local evapo-transpiration as temperatures increase in the future and also, where sited on pervious soils, act as a sink for excess stormwater runoff. Although care should be taken when promoting the latter in order to ensure that groundwater problems (quantity or quality) do not occur elsewhere in the urban area.

These blue-green spaces may also serve as flood flow exceedance pathways designed to cope when the formal drainage systems are overloaded. Currently urban planning and design does not include such considerations in urban layouts; assuming that flood risk can be dealt with by 'tagging' such considerations on at the end of the other aspects of urban planning. In many countries there is a predominance of interest in using existing redevelopment 'brownfield' areas for further development of buildings and infrastructure, but some of these areas may contribute to biodiversity, reductions in heat island effects and flood storage/pathways, and should perhaps be better maintained without being built on. Where it is not possible to bring about the maintenance or restoration of blue-green corridors, it will be necessary to provide resistant and resilient construction or retrofit this to existing buildings.

Densification of the buildings is now something of a mantra in new urban development planning. This increases the rate and amount of runoff and provides less space internally for surface drainage features. This is illustrated in Figure 7-6 at Elvetham Heath in England. However, it may be possible to use land external to the development which may have green features and hence be available for multifunctional use as illustrated in Figure 7-7 from Baltimore, Maryland, USA.

In common with other aspects of developing resilience and adaptability to uncertain futures and flooding, fixed and inflexible regulations compelling all developments to comply with specified urban densities are foolish and should be avoided in

FIGURE 7-6
Dense urban development in Elvetham Heath, England

Source: Richard Ashley, 2009.

FIGURE 7-7
Stormwater storage basin in the area surrounding high-density developments in Phoenix, USA

Source: Chris Zevenbergen, 2004.

order to allow for the most appropriate building density to be defined based on local circumstances, rather than national or regional perspectives. Where they exist, such prescriptions should be used as guidelines only.

Advantage should be taken of non-critical infrastructure space to store or route floodwater. This may include certain roads and even railways within cuttings (lowered areas). However, this must be prepared for, as has been the dual flood-storage road tunnel in Kuala Lumpur in Malaysia that can be used to pass extreme flood events by closing the road.

Where there is existing green space, it may be appropriate to consider this for development. Embargoes on development on greenfield or green belt land are unrealistic as are embargoes on developments in floodplains. The approach should also be one of flexibility in the light of local circumstances, as one single type of approach will not be appropriate for all situations. As flood risk is usually one of the least important of the considerations in development planning; a proper, flexible and adaptable approach balancing the needs of all aspects of the development, including flood risk, needs to be taken, but with due regard and care for all of the challenges faced. Included in this should be the problem of 'urban creep'—where there is paving over of small areas for e.g. car parking in urban areas which needs to be controlled as part of the overall city planning process.

Urban areas cannot be seen in isolation and must also make use of land outside of the local administrative boundary. This requires information on flood source areas and flow pathways in the surrounding rural areas as well as in the urban area. Upstream flood storage or attenuation options should be exploited where appropriate, including using low-value green space or agricultural land.

The idealised urban form for a flood-resilient city should reflect its individual geographical constraints, but must also be in tune with other socio-economic, environmental and particular challenge specificities. For example, in Australia, flood-risk management in urban areas is now seen as simply one part of the integrated water cycle management approach that also entails using water in the city as a fundamental component of the water supply system. This is because the main challenges are not water excess but water shortages.

The maintenance of critical infrastructure against flood risk is a major challenge. This includes energy supplies, transport links, health services and education provision. Even without direct experience of flooding, victims may experience loss of key services and utilities. For example, following the devastating floods in New Orleans, returning residents, who had not been flooded, often found that key services like hospitals, transport and schools had not returned to effectively functioning levels.

Dealing with this requires good planning to protect and respond to potential flooding incidents. For example, in the summer of 2007, some 350,000 residents in England suffered loss of piped water supplies for several days due to the inundation of a Water Treatment Works. This, together with other flooding in England and Wales, prompted the national inquiry by the government, leading to the Pitt Report which contains a complete section on 'Maintaining power and water supplies and protecting essential services'.

Much of the response requires effective cooperation and coordination between the various agencies and the service or utility suppliers and hence is locally specific. As well as flood-risk management, critical infrastructure also needs to be managed to deal with all risks together in a coordinated and integrated approach. As an example, the US Department of Homeland Security has developed a National Infrastructure Protection Plan (NIPP), Figure 7-8.

The NIPP is supported by 18 sector-specific plans, one of which is for flooding. For flood risk, these rely on the Federal Emergency Management Agency (FEMA), which is part of the DHS, as FEMA coordinates production of detailed flood maps for the entire USA.

The NIPP framework calls for critical infrastructure and key resources (CIKR) partners to assess risk from any scenario as a function of consequence, vulnerability and threat (see also Section 6.1), as defined below.

$Risk = f(C, V, T)$, where

C = Consequence—the social, economic and environmental impacts of an event

V = Vulnerability—degree of susceptibility to disruption

T = Threat—something with the potential to cause harm (a hazard).

This framework is central to the Strategic Homeland Infrastructure Risk Assessment (SHIRA) process. The DHS maintains, updates annually and is enhancing a comprehensive catalogue that includes an inventory and descriptive information about the assets and systems that make up the nation's critical infrastructure.

Risk assessments are undertaken using the Risk Assessment Methodology for Critical Asset Protection (RAMCAP) as outlined in the NIPP. As part of the risk assessment,

FIGURE 7-8
NIPP Risk Management Framework—Continuous improvement to enhance the protection of critical infrastructure

Source: The National Infrastructure Protection Plan (NIPP).

utilities develop an inventory of asset components including physical, cyber, IT and personnel, and identify which components are most critical to their continued operation.

Responding to the risk and threats can be framed using a hierarchy as shown in Table 7-10.

Standards of protection for critical infrastructure will depend on the consequences of failure, perhaps more than the probability. As for example, inundation of one of

TABLE 7-10

Flood-risk management hierarchy for critical infrastructure (adapted from CIRIA, in press)

Measure	Description	Example
Avoid	Ensure that the critical components of the asset system are not unacceptably exposed to the flood hazard.	Critical infrastructure such as power stations should be located in areas where the risk of flooding is negligible.
Substitute	Where some facilities are found to be at unacceptable risk; ensure that the essential services they provide can be substituted with alternatives during the period of disruption.	In the regeneration of a riverside urban area, existing housing within floodplain areas could be partially or wholly substituted with amenity open space.
Control	Implement flood-risk management measures to reduce flood frequency making the asset flood resistant or flood resilient.	Construction of floodwalls and embankments to contain river or tidal waters.
Mitigate	Implement measures, such as flood forecasting, warning, incident management and emergency response procedures to mitigate residual risks.	Educate affected parties on the nature of the residual risk. Provide a flood warning service. Ensure emergency evacuation plans are in place.

TEXTBOX 7-2

How the influence of land use on the runoff process varies

Often after each flood, the view is expressed that if there were more forests in the basin area there would be fewer or no floods. This is mostly in the context of flash floods where the surface runoff is the most important contribution to the flow. The argument is always the same: forests have higher retention capacity than e.g. arable land or urban areas. This can be assumed as a truism according to many publications but it is important to assess how the whole runoff process and formation of floods can be affected by land use. The retention capacity derives from interception, surface storage and often also infiltration. However, the retention capacity is always limited and can be assumed more or less constant according to the degree of infiltration *irrespective of the rainfall.*

There are many theories describing the runoff process, e.g. infiltration excess overland flow (Horton), partial area infiltration excess overland flow (Betson), and saturation excess overland flow (Cappus and Dune). Most of these suggest that at the start of a storm some initial losses are applied which are represented by interception and surface storage and then the infiltration process starts. Surface runoff starts at the moment when the interception and surface storage capacities are full and when the infiltration rate is less than the rainfall intensity. However, the infiltration capacity is also more or less constant and can reach a point where the infiltration is very low. Therefore, from some point in the storm event all precipitation becomes runoff through the surface runoff.

Hence, with increasing amounts of precipitation the relative influence of initial losses and infiltration decreases. In effect, infiltration capacity and initial losses are more or less constant, irrespective of the magnitude of the rainfall. With increasing precipitation, more of the rainfall thus contributes to surface runoff. Therefore, the relative importance of land use is much higher for shorter return periods or lower precipitation. The demonstration of this relationship is presented in the figure below.

When investigating the influence of forests and meadows on flood reduction, it is instructive to use hydrologic models which can demonstrate and quantify the effect on the runoff process due to land-use changes. A number of studies, mostly on small catchment have shown that similar results for the variation in the coefficients of peak runoff for a range of precipitation totals and return periods. The variation coefficient is defined as the standard deviation of the peak runoff data for the catchment divided by the mean. This value expresses the relative dispersion of the data around the mean. Therefore, if the influence of land use is smaller, the variation coefficient will also be lower.

the 12 nuclear power stations situated adjacent to the Loire River in France, would be catastrophic, not only locally but also nationally and potentially, globally. Hence, standards of at least 1 in 1,000 years or the maximum probable event for probability are typically set for such installations.

Key Questions

Performance standards and expectations

1. Can you explain what a portfolio of response measures might be?
2. Put forward reasons why a uniform performance standard for urban drainage infrastructure is not feasible.
3. What do you see as the relative strengths and weaknesses of the Scottish response to flood-risk management?

Resilience, vulnerability, robustness and sustainability

1. Can you explain the difference between resilience and robustness in the context of flood-risk management?
2. How can scenarios be used to inform flood-risk management strategies?
3. Can you give examples of how adaptation measures can play a role in flood-risk management?

Precautionary and adaptive responses

1. What are the differences between precautionary and adaptive responses?
2. What role can non-structural measures play in mitigating/minimising flood risk?

3. Give examples of measures that have been adopted to reduce the probability of flooding in urban areas.

Confronting flood management with land-use planning

1. In what ways does land-use planning influence urban flood management?
2. Name some of the potential flood management responses and give examples for rural, urban and coastal areas.
3. How might land-use practices in upstream catchments affect the nature of flooding in downstream urban areas?

Building types, infrastructure and public spaces

1. In what ways might local circumstances influence the goals and approaches of flood management?
2. Under what circumstances do you think it is acceptable to develop floodplain areas, and what stipulations would you add to that development?

FURTHER READING

Andjelkovic, I. (2001). *Guidelines in non-structural measures in urban flood management.* IHP-V Technical Documents in Hydrology, no 50, UNESCO, Paris.

DETR (2000). Guidelines for Environmental Risk Assessment and Management, 2nd edition, The Stationary Office, London, Institute of Environmental Health.

Evans, E., Ashley, R., Hall, J., Penning-Rowsell, E., Sayers, P., Thorne, C. & Watkinson, A. (2004). *Future Flooding Scientific Summary: Volume II—Managing Future Risks.* London: Office of Science and Technology.

Evans, E.P., Ashley, R., Hall, J.W., Penning-Roswell, E.C., Saul, A., Sayers, P.B. Thorne, C.R. & Watkinson, A.R. (2004a). *Foresight Future Flooding, Scientific Summary: Volume 1: Future Risks and their Drivers.* Office of Science and Technology, London, http://www.foresight.gov.uk/OurWork/CompletedProjects/Flood/index.asp

Evans, E.P., Simm, J.D., Thorne, C.R., Arnell, N.W., Ashley, R., Hess, T.M., Lane, S.N., Morris, J., Nicholls, R.J., Penning-Rowsell, E.C., Reynard, N.S., Saul, A.J., Tapsell, S.M., Watkinson, A.R. & Wheather, H.S. (2008). *An update of the Foresight Future Flooding 2004 Qualitative Risk Analysis.* Cabinet Office, London, http://www.archive.cabinetoffice.gov.uk/pittreview/thepittreview/final report.html

Peter H. von Lany & John Palmer. (2007). Future Flooding and Coastal Erosion Risks , chapter 22, River Engineering Responses, ISBN: 9780727734495.

Tapsell, S., Burton, R., Oakes, S. & Parker, D.J. (2005). *The Social Performance of Flood Warning Communications Technologies* (No. TR W5C-016): Environment Agency.

The Pitt Review (2008). Learning Lessons from the 2007 floods, Cabinet Office, 22 Whitehall, London SW1 A 2 WH.

White, I. (2008). The absorbent city: urban form and flood risk management. *Proceedings of the Institution of Civil Engineers: Urban Design and Planning,* December (DP4), 151–161.

Urban drainage systems

8

Learning outcomes

In this chapter you will learn about the following key concepts:

- The scope of urban drainage and that it covers more than just stormwater.
- The objectives and functioning of Sustainable Urban Drainage Systems (SUDS) and Low Impact developments (LIDS).
- Recognition of the importance of stormwater quality and the need to include appropriate hydrological techniques for analysis of these issues.
- Elementary design principles and construction information.
- Relevance of catchment modelling systems for the design and analysis of drainage systems.
- The philosophy of Water Sensitive Urban Drainage and the practices and tools necessary for their implementation.

8.1 A HISTORICAL PERSPECTIVE

Urban drainage systems are typically considered to include all aspects of surface water collection, storage and conveyance that together manage stormwater in the urban area.

Conventional urban drainage systems aim to collect and transfer surface water runoff and foul sewage to a location remote from the point of origin. These systems have developed to deal with flooding and, in combination with sanitary drainage, public health and more recently to protect the natural environment.

Open channels were initially constructed to prevent street flooding, conveying surface water to the nearest watercourse. Unfortunately, the population of towns disposed of household wastes in the system, creating problems with blockages, filth and odours and so the open channels were covered. From about the mid-nineteenth century, the channels were replaced with pipes and culverts, carrying the waste to rivers by the shortest possible route. In the last quarter of the nineteenth century, in industrialised countries, intercepting sewers were laid to divert sewage away from the more central parts of cities to remote, downstream outfalls.

This early approach resulted in combined surface water and foul sewerage systems ('combined sewers') which are generally still in use, together with the 'relief valves' for excess flows, known as combined sewer overflows (CSOs). As urban areas developed and expanded, both surface water runoff and foul sewage volumes increased as permeable (natural) ground was replaced with impermeable surfaces (e.g. roads, driveways and roofs) and as population numbers and building densities increase. The conventional response to increased pressure on drainage capacity and other problems such as receiving water pollution from CSOs has traditionally focused on 'end-of-pipe' or 'at the point of the problem' solutions using storage and increasing the size of sewers.

In most developed countries the use of 'separate sewer systems', in which foul sewage and surface water are handled in two separate sets of pipes or channels, is becoming more common. There are moves in some parts to convert the extensive existing combined sewer systems into foul only systems, with separate provision for surface water. However, there has been a growing appreciation that separated surface water runoff, which is often discharged directly to a watercourse, is an important carrier of pollutants. In the USA for example, where separate sewer systems are prevalent, stormwater has become regarded as one of the primary sources of watercourse pollution.

Many of the adaptation approaches applied to 'conventional' and combined sewer systems, such as increasing storage provision at CSOs, tend to create yet more challenges, including increased pollutant discharges, downstream treatment plant inefficiencies and consequential air pollution. The use of high resource, energy and carbon intensive 'solutions' as required for new storage in combined sewer systems, lead to other problems, including increased greenhouse gas emissions and 'lock-in' to the use of non-abandonable or unadaptable assets for decades into the future. These approaches are still used, however, because they have the advantages of being centralised, have well-defined ownership (providing clearly defined tangible assets), draw on existing skill and knowledge availability and can be implemented without having to engage with too wide a group of stakeholders.

In contrast, in a number of places, such as in Philadelphia, the alternative of retrofitting stormwater source controls to manage surface water nearer to the source is now the main policy option for managing excess stormwater flows that result in CSO spills.

Hence, there is a shift away from combined piped urban drainage systems to systems where stormwater is managed (ideally) as close to the point at which it is generated and then successively at a local level and only at a regional or 'end-of-pipe' position when other alternatives are not feasible. This approach is known as 'sustainable (urban) drainage systems/sustainable drainage systems' or 'SUDS' in the UK. In the USA, the alternative terms 'Best Management Practice' (BMP) or 'Low Impact Development' (LID) have been used; and 'water sensitive urban design' (WSUD) is used in Australia.

Urban drainage should be seen as just one component of the management of the water cycle as a whole. Hence, appropriate drainage systems need to be set within the context of the urban water cycle as a whole.

8.2 MAJOR AND MINOR FLOWS

The management of stormwater in urban areas is considered to be effected by 'natural' rivers, streams and watercourses plus 'artificial' drains, pipes, channels and other measures such as SUDS (see above).

Urban stormwater drainage systems typically comprise two separate components:

1. a surface, or 'major' system composed of streets, gutters, ditches, and various open natural and artificial channels;
2. a subsurface storm sewer network or 'minor' system.

Planning of stormwater runoff systems must consider both a normal or typical rainfall situation and less frequent but larger rainfall events. If the minor system is overloaded or blocked, there must be a drainage system on the surface in which stormwater can drain away without unacceptable adverse effects. Typically, the minor system is designed to fully contain and convey runoff generated during a storm of 2–10 year return frequency, while the surface system is designed to handle events of 25–100 year return frequency. According to the European Standard EN 752, urban drainage systems should be designed to safely convey discharges from events with return periods of 10–50 years. Dual drainage principles, which involve both the major and the minor systems, should be considered in an integrated way in the planning of urban stormwater runoff systems.

These systems are linked via street curb inlets which are designed to convey certain amounts of stormwater runoff into the underground storm sewer system. Manholes also link the two systems during surcharge. Street flooding is a special case of sewer surcharge in which when water cannot be accommodated within or enter the sewer system, it is allowed to either discharge from the manhole and pond on the surface or flow down the street in response to topographic gradients.

The major system must provide proper gradients and pathways so that it has the capacity for a safe conveyance of surface flows.

The traditional approach to modelling the complex interactions between the major and minor systems consist of two stages. In the first stage of designing an urban drainage system for say, a 100-year flood event, the minor system is designed to convey the discharge from a ten-year event. Sewer pipes are sized to an acceptable level of performance. Next, the ten-year hyetographs are subtracted from 100-year design event hyetographs and the subsequent rainfall is input to the hydrologic model. The resultant surface flows are used as the design capacity for the major drainage system.

Storm sewers can overflow via manholes, and the excess volume can either return via the original manhole or travel overland to another manhole in the storm sewer system. They may also overflow due to separate storm sewer overflows (SSOs) or combined sewer overflows (CSOs) when the storm and sanitary flows are conveyed in the same pipe. A dual drainage system may consist of low-lying areas in the city e.g. parks, being used to detain and delay flows from the major system to further reduce the peak discharge from an urban area. Specific inlet control devices are recommended to restrict major system flow into the minor system so as to avoid flooding of basements and overloading the storm sewers.

Floodways, see Figure 8-1, or exceedance flow pathways should be included in urban planning at both the overall level and at the detailed level. Floodways should be dimensioned to be able to take all drainage water from the sub-catchment area, and should at least be considered being sized to accommodate a 1:100-year flood event. When selecting the floodway's capacity, in terms of a return period, it is necessary to weigh the estimated damage and disadvantages against the cost of construction of the floodway, in other words carry out an appropriate cost–benefit analysis. The optimal capacity of a floodway is when the total sum of investment costs, operation and maintenance costs, costs of damage from flooding and flood disadvantages are either at a minimum (total life costs) or when the benefits outweigh the costs. The calculated capacity of floodways and the corresponding return period of the flood should be shown on maps. Downstream areas should be checked to ensure they can handle the added water flows from upstream floodways. Floodways should preferably follow public space and areas, like roads and street surfaces, parks and recreational areas. This requires special justification as well as approval from the proper authority. If floodways must follow private ground, such plans should be implemented by agreements with the property owners or an acquisition of the ground. Floodways should generally be built in such a way that they delay and detain the flood, so that large flow peaks will not propagate in the

FIGURE 8-1
Floodway integrated in high–density urban area, Phoenix, USA

Source: Chris Zevenbergen, 2004.

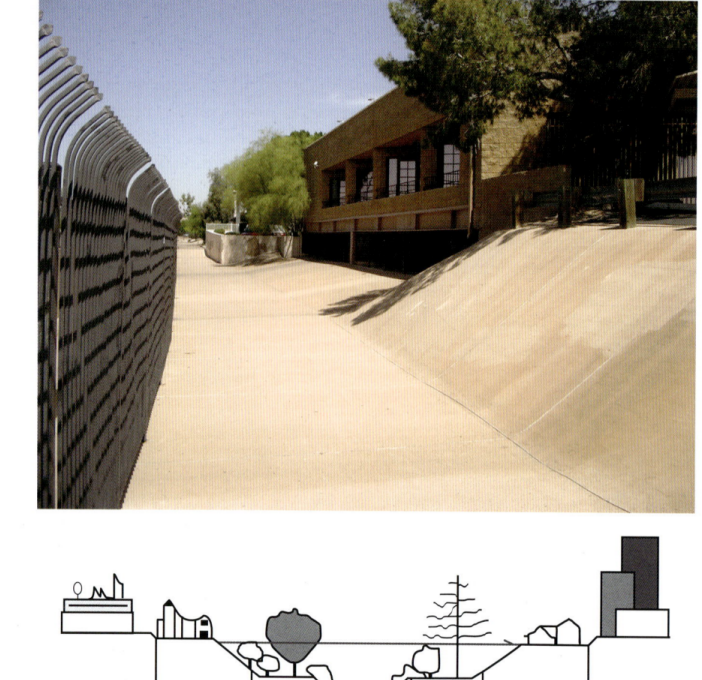

landscape below. GIS tools and terrain models are of great benefit in calculating the direction and capacity of floodways on the surface. This identifies the lowest points and profiles in the terrain and indicates how high the water surface will rise and which areas will be flooded.

8.3 SUDS/LIDS

In non-urban catchments, typically up to 50 per cent of precipitation is contained in natural depression storage and 40 per cent is intercepted and taken up by trees and vegetation in the form of evapo-transpiration. Only 10 per cent is turned into runoff. With urbanisation these relative percentages change and Figure 8-2 illustrates this for various percentages of paved area, reflecting the increasing levels of urbanisation. When 75 to 100 per cent of the ground is impervious only 15 per cent of runoff is detained and can infiltrate and some 55 per cent is converted to runoff. Various standards for

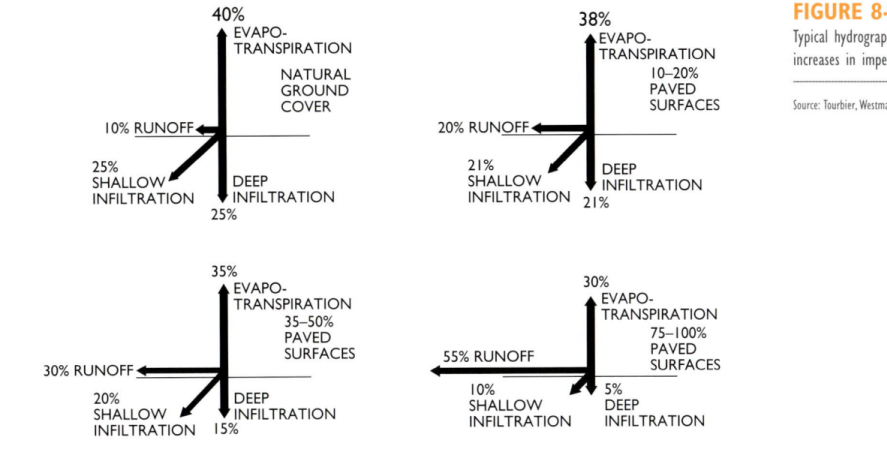

FIGURE 8-2
Typical hydrograph changes due to increases in impervious surfaces

Source: Tourbier, Westmacott 1981 in EPA, 1983.

developments have been defined to try to ensure that measures are incorporated into developments and urban design to maintain pre-development evapo-transpiration, infiltration rates and temporary storage and slow release.

Controlled discharge from a development site can be achieved through a detention basin, for example, that reduces runoff peaks after development through a slow release over a longer time period in order to mitigate downstream effects and possible flooding.

The 'common enemy rule' in managing stormwater in urban areas has traditionally meant that rainwater should be shed away from buildings as fast as possible, and it is common practice for it be collected through curbs and gutters to be directed to sewer inlets and sewer systems leading to an 'end of pipe treatment' with massive discharges. Alternatives are decentralised systems, which attempt to handle the volumes locally. Porous pavements, as shown in Figure 8-3, are examples of components of decentralised systems.

Low Impact Development (LID) was developed in Prince George's County, Maryland, USA, in the early 1990s as a site design strategy to maintain and replicate the pre-development hydrologic regime. This was achieved through storage, groundwater recharge and runoff volume and peak control including the lengthening of flow times and times of concentration. This approach could be practised on individual development sites to reduce or eliminate the need for centralised stormwater management practices. LID practices include the increase of flow path length, increased roughness, minimised disturbance, infiltration swales, vegetative filter strips, disconnected impervious areas, reduced curb and gutter, rain barrels, rooftop storage, bioretention and revegetation. In the US, the State of Maryland is considering site design criteria that consider 'woods in good condition' as the 'pre-development' hydrology condition. Such standards have implications for design as no single, end of pipe treatment measure can achieve them, but a 'treatment train' of practices of combined measures does offer this possibility.

In the UK in 2001, the first planning advisory note on Sustainable Drainage Systems (SUDS) was published in Scotland, followed by the consultation document *Framework for Sustainable Urban Drainage Systems (SUDS) in England and Wales* in 2003 (NSDW 2003). It was followed in turn by an *Interim Code of Practice for Sustainable Drainage Systems* (NSDW 2004), published by the National SUDS working group in 2004. In the UK, the Construction Industry Research and Information Association (CIRIA) has

8.3 SUDS/LIDS

taken up SUDS as a programme that fits into its development objectives of improving the built environment through networking, information services, conferences, training and publications, and a design manual and construction guide was produced in 2007.

In Germany, the concept of 'Natural Drainage' is being advocated, not only for new developments but also to retrofit existing cities in the Ruhr Valley, where a policy of 15 per cent runoff reduction has been set. Throughout Germany, cities are charging fees to those who are connected to separate stormwater systems, thus providing an incentive for the implementation of SUDS. The city of Dresden, for example, charges property owners around €1.60 per year per square metre of impervious surface, unless there is an on-site disposal for runoff, such as infiltration. Natural Drainage efforts are also being

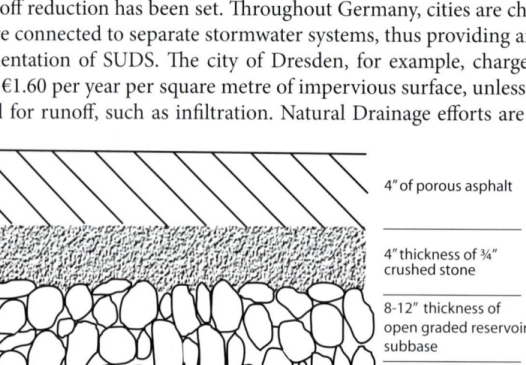

FIGURE 8-3
Porous asphalt to lessen runoff and provide for infiltration

Source: Joachim Tourbier, 1983.

TEXTBOX 8-1

Bioretention as a 'Low Impact Development (LID)' at the Beltway Plaza Shopping Mall in Greenbelt, Maryland, USA

The Prince George's County Department of Environmental Resource encouraged the development of a bioretention facility as a LID initiative on the parking lot of the Beltway Plaza Shopping Mall in Greenbelt, Maryland, USA. The parking lot is graded to direct runoff to enter the heavily planted median strips through curb cuts. It is planted with shrubs, trees and decorative grasses. Here drop inlets are set at a height that permits ponding of the first 1.5 inches of runoff which has been found to have the highest levels of pollution. Ponded runoff inundates not only the median strip but also the lower portion of parking stalls before it infiltrates to be cleansed through a 2–4 inch layer of mulch, planting soil and a sandbed containing an underdrain at a depth of 45 inches. The runoff is then fed into a stormwater drain.

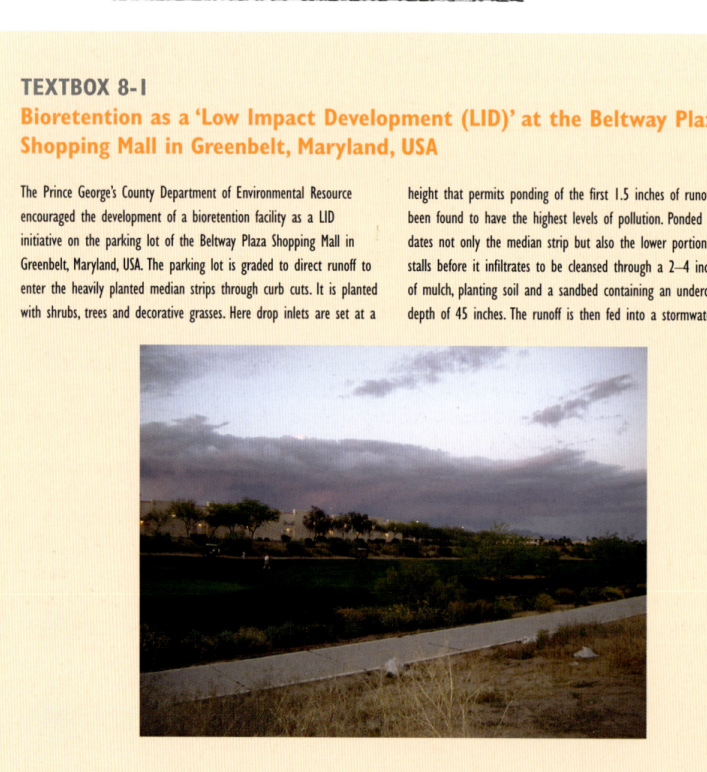

FIGURE 8-4
Low Impact Development (LID) at the Beltway Plaza Shopping Mall in Greenbelt, Maryland, USA

Source: J.T. Tourbier, 1983.

made at the Federal UBA (Umweltbundesamt) to control stormwater runoff at new construction sites by replicating runoff volumes and peaks found in woodland conditions, including evapo-transpiration.

SUDS, BMPs and LID practices are often sized for design storms of 1:30-year return frequency. During a major storm, they overflow and exceedance flows will find a course into urbanised areas that may result in flood damage. In response, the concept of SUDS should be expanded to become Sustainable Drainage and Conveyance Systems (SUDACS) to also consider the safe design of overflow floodways. During extreme rainfalls, the capacity of sewers and covered watercourses will also be exceeded and stormwater will travel above ground. How to convey such flash floodwaters through a city without causing damage is becoming a particularly pressing question especially given the likely impacts of climate change.

8.3.1 Hydraulic design of SUDS

The hydraulic design of SUDS involves the determination of the main parameters characterising the functioning of SUDS, which are hydraulic load (inflow, overflow, infiltration, outflow) and dimensions (volume, area, height). The processes in SUDS are interdependent where, for example, the processes in the unsaturated soil layer influence the groundwater level. For a full simulation of processes and their interdependences, hydrodynamic models are necessary, but they are complex, time and resources intensive and are usually not readily available. In hydrologic approaches, the processes can be simplified such that SUDS are regarded as reservoirs where storage effects (retention) dominate over water movement (translation). For these reservoirs, the continuity equation is solved, whereas the retention and infiltration of water in the layers is balanced. For the practical application and dimensioning of SUDS structures, the hydrologic approach is further simplified where SUDS are described as an uncoupled system composed of layers for which the balance equation is solved. Although the processes are not fully simulated here, a set of design criteria describing the main geomorphologic conditions (slope, distance to groundwater table and permeability) is defined, to ensure that they do not have any negative impact on the SUDS performance.

Design procedures for SUDS are defined within a number of standards (e.g. in Germany in DWA- A138 and in the UK in CIRIA, 2008). The definition of the hydraulic system and design procedures differ for different SUDS techniques. In this chapter, they are given for green roofs, pervious pavement, swales and swales with filter drains, assuming that all design criteria are fulfilled.

Independently of the type of SUDS used the following steps should be performed.

- definition of the design storm event;
- definition of runoff from drained area;
- hydraulic design of SUDS for design storm;
- performance proof and design for exceedance.

8.3.1.1 Definition of design storm events

SUDS are typically dimensioned for a defined 'design storm event'. It is given as precipitation height (h_N in mm) for a defined frequency of occurrence (n) and a precipitation duration (D in min). Based on those values the rainfall intensity ($r_{D,n}$ in [l/(s*ha)]) of a storm event can be calculated as:

$$r_{D,n} = 166.7 * \frac{h_N}{D} \quad [l/(s*ha)]$$ (8-1)

The design storm event is not a unique value and differs across Europe. In Germany, the standard DWA A 138 (2005) postulates a storm event, which has at least the frequency of occurrence of 0.2 which means a storm event with a return period of once in five years. In the UK, the standard BS EN 752 (2008) requires protection against pluvial flooding for a 30-year event. Kellagher and Laughlan recommend protection against a 100-year event and the need to use continuous simulation with time series rainfall. For German catchments, precipitation data can be obtained from KOSTRA (KOordinierte STarkniederschlags Regionalisierungs—Auswertungen). It contains a raster-based representation of precipitation heights that depends on duration and frequency. Alternatively, precipitation data can be processed and prepared according to DWA-A 121.

For the calculation of a design storm event, the precipitation of the constant intensity is usually assumed (block rain). This is a simplification of the real situation, but for storm events of short duration (1 minute to 60 minutes) this approximation is acceptable. For longer duration, a non-uniform temporal distribution has to be assumed. Guidance on this is to be found in DWA-A 118.

8.3.1.2 Definition of the runoff from sealed areas

Within urban catchments, the whole area does not contribute equally to the formation of runoff. Specific losses have to be taken into account, which reduce the runoff. For a defined area A_{total}, discharge Q_{runoff} can be calculated as:

$$Q_{runoff} = r_{D,n} \times \Psi_m \times A_{total} \times 10^{-4}$$
(8-2)

where:

Q_{runoff} Runoff (l/s)
A_{total}, Catchment area (m²)
$r_{D,n}$ Rainfall intensity [l/(s*ha)]
Ψ_m Runoff coefficient (–)

The runoff coefficient ψ_m for an area A_{total} defines the percentage of precipitation that results in surface runoff and as such depends on the material and the structure of the surface as well as on the gradient. It can be calculated by summation of the specific surface types and corresponding specific runoff coefficients as.

$$\Psi_m = \sum_{i=1}^{n} \frac{\psi_i \times A_i}{\sum A_i}$$
(8-3)

The runoff coefficients are still a matter of research and usually the experiential values are taken, Table 8-1 gives some of the recommended values. Further coefficients are defined in the standards ATV DVWK- A 117 and ATV DVWK- M 153.

Those values might in some cases be too high and, in reality, the runoff coefficient never reaches the maximal value of 1 as there is always a specific loss either by spraying the rainwater on the surface or its detention in depressions. It mostly depends on the terrain slope in the area. Such a dependency between the loss of rainfall contributing to runoff and the slope is given in Table 8-2.

8.3.1.3 Hydraulic design of SUDS

Definition of the hydraulic system and concrete design procedures are described here in detail for green roofs, pervious pavement, swales and swales with filter drains.

TABLE 8-1
Recommended values for runoff coefficient (Geiger et al., 1995)

Surface type	Runoff coeff. Ψ
Wooden elements and roofs	0.50 to 0.70
Asphalt and ways	0.85 to 0.90
Pavement	0.75 to 0.85
Open pavement	0.10 to 0.20
Park green spaces	0.05 to 0.10

TABLE 8-2
Runoff loss to slope

Cat	Slope	Wetting loss	Depression loss
1	<1%	0.5 mm	2.0 mm
2	$1 < x < 4\%$	0.5 mm	1.5 mm
3	$4 < x < 10\%$	0.5 mm	1.0 mm
4	>10%	0.5 mm	0.5 mm

8.3.1.3.1 Green roofs:

- Hydraulic System (HS)

The main processes for designing a green roof are the inflow through precipitation (Q_r [l/s]), drainage discharge from the drainage layer (Q_{drain} [l/s]) and discharge in case of exceedance flow through an emergency spill (Q_{ex} [l/s]). The hydraulic system of a green roof is depicted in Figure 8-5.

- Hydraulic design for design storm event

In the hydraulic design for design storm event, the height of the soil layer should be determined. It is calculated considering the following processes:

$$Q_r = r_{n,D} \times 10^{-4} \times A_r$$
(see Equation 8-3)

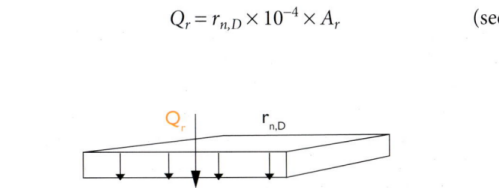

FIGURE 8-5
Hydraulic system

Source: Joachim Tourbier, 1983.

Q_{drain} can be calculated applying the POLENI Formula:

$$Q_{drain} = \frac{2}{3} \times \mu \times \sqrt{2g \times h_{drain}^{3}}$$ (8-4)

In practice, this discharge is usually set to a certain value (e.g. 0.5 l/s) beforehand. To prove whether this discharge in the drainage layer can be reached, Darcy's law should be applied.

The volume of the soil layer is determined for the design storm event as:

$$V_s = A_r \times h_s \times n_{pors}$$ (8-5)

$$V_s = (Q_p - Q_{drain}) \times 10^{-3} \times D \times 60$$ (8-6)

The height of the soil and substrate layer (hs) results in:

$$h_s = D \times 60 \times 10^{-3} \times \frac{r_{n,D} \times A_r \times 10^{-4} - Q_{drain}}{A_r \times n_{pors}}$$ (8-7)

where:

V_s Volume of the soil layer (m³)

h_s Height of the soil and substrate layer (m)

$r_{D,n}$ Rainfall intensity [l/(s*ha)]

D Duration of the design storm (min)

Q_{drain} Drainage discharge (l/s)

A_r Storage area of the roof (m²)

n_{pors} Porosity of the soil layer (–)

• Hydraulic design and proof for exceedance

Green roofs, although designed for precipitation events of T = 5-30a, have to be equipped for the case of an extreme event. For that purpose, emergency spills are used and designed for at least 100-year precipitation events.

For calculation of the height of the storage above the soil layer of the roof and under the height of the emergency spill, the following approach is recommended (see Figure 8-5).

$$\Delta h_{sl} = \Delta t \times 60 \times 10^{-3} \times \frac{r_{100,D} \times A_r \times 10^{-4} - Q_{drain} - Q_{ex}}{A_r}$$ (8-8)

where:

Δh_{sl} Height above the soil substrate layer (m)

$r_{100,D}$ Rainfall intensity [l/(s*ha)] n = 0.01

Q_{drain} Drainage discharge (l/s)

Q_{ex} Discharge through emergency spill (l/s)

A_r Storage area of the roof (m²)

The emergency spill discharge is generated when the water level on the roof reaches a specific critical level and ensures that the roof is not 'overfilled', or:

If $h_{sl,i} < h_{crit}$ (with $n_{pors} < 100\%$) $Q_{ex} = 0$

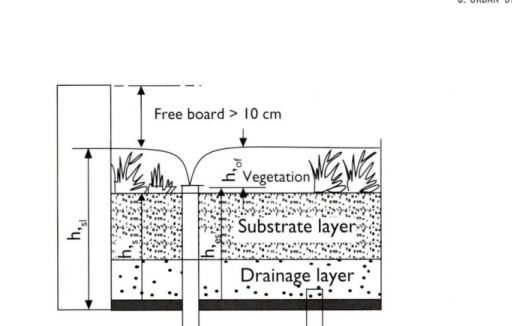

FIGURE 8-6
Cross section of a green roof

Source: Joachim Tourbier.

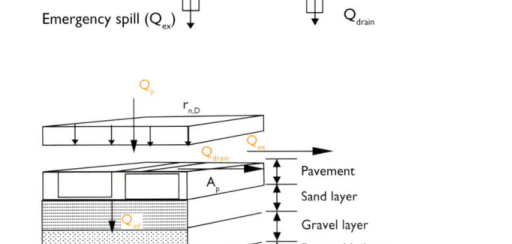

FIGURE 8-7
HS of a pervious pavement

Source: Joachim Tourbier.

If $h_{sl,i} > h_{crit}$ (with $n_{pors} = 100\%$) then $Q_{ex} = l_u \times \frac{2}{3} \times \sqrt{2g} \times \mu \times \sqrt{h_{of}^3}$ (8-9)

where:

- l_u Perimeter of emergency spill (m)
- g 9.81 m/s²
- μ Overflow coefficient = ca. 0.7
- h_{of} Overflow height (m)
- $h_{sl,i}$ Actual water level on the green roof (m)
- n_{pors} Porosity of the soil layer on the green roof (–)

The calculation of the critical overflow height is an iterative process as shown in example in Textbox 8-2).

8.3.1.3.2 Pervious pavements

The main processes for design of pervious pavements are direct inflow through precipitation (Q_p [l/s]), runoff from the pervious pavement (Q_{drain} [l/s]) and discharge into ground (infiltration) (Q_{inf} [l/s])

- Hydraulic System (HS)
- Hydraulic design for design event

For the case that the permeable pavement should only improve the infiltration characteristics of the soil, the improved runoff from the area and the infiltration discharge can be calculated as follows:

$$Q_{drain} = \psi'_m \times Q_p$$ (8-10)

where

ψ'_m is the modified runoff coefficient calculated for the unsealing plan (see Equation 8-4).

$$Q_p = r_{n,D} \times 10^{-4} \times A_p \qquad \text{(see Equation 8-3).}$$

and

$$Q_{inf} = Q_p - Q_{drain} \tag{8-11}$$

In case that an additional sealed area drains onto the pervious surface, the area A_p required to receive the runoff from the connected sealed surface can be calculated as:

$$A_p = \frac{A_{red}}{\frac{10^7 \times k_{f,sat} \times (1 - \psi_{m'})}{2 \times r_{D;n}} - 1} \tag{8-12}$$

where:

A_p Area of pervious pavement (m^2)

A_{red} Drained sealed area (m^2)

$k_{f,sat}$ Permeability coefficient (m/s): if the infiltration area is used as parking place, the kf-value is reduced by the compaction of the soil.

$r_{D;n}$ Rainfall intensity [l/(s*ha)]

• Hydraulic design and proof for exceedance

Pervious pavements are designed for a design storm event, but in the case of an exceedance flow, the Q_{ex}, has to be considered for design of the conveyance systems.

8.3.1.3.3 Swales

Three main processes are decisive for the design of swales. These are: inflow from the drained area (Q_{in} [l/s]) during a storm event with the duration D from the area A_e; inflow into the swale during a storm event with the duration D from the area A_e (Q_s [l/s]); and infiltration rate from the swale into the soil (Q_{inf} [l/s]).

- Hydraulic system of a swale
- Hydraulic design of a swale for design event

The objective of the hydraulic design for design event is to calculate the area A_s and height (z)

$$Q_{in} = (\Psi_m \times A_e) \times r_{n,D} \times 10^{-4} + Q_{rill} \tag{8-13}$$

where can be calculated using Equation 8-4.

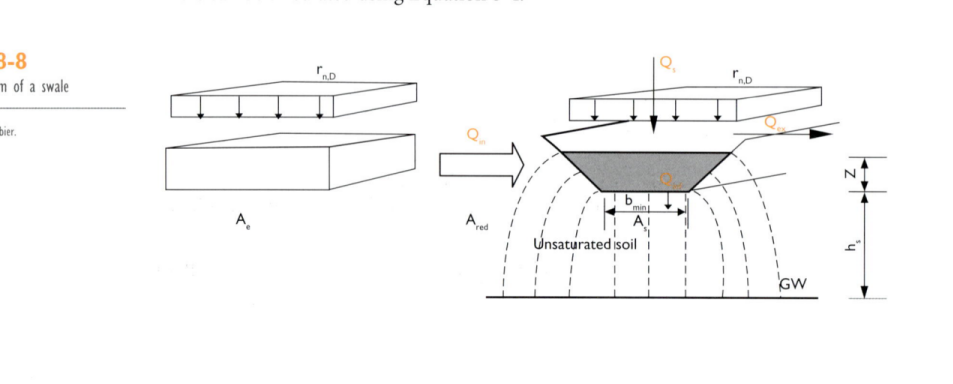

FIGURE 8-8
Hydraulic system of a swale

Source: Joachim Tourbier.

Q_{rill} is the discharge coming from rills (e.g. conveying water from the green roofs).

$$Q_s = r_{n,D} \times A_s \times 10^{-4} \tag{8-14}$$

where:

$r_{D,n}$ Rainfall intensity $[l/(s*ha)]$

A_s Area of the swale (m^2)

The infiltration into the ground is described by Darcy's equation:

$$Q_{inf} = k_f \times I_{hyd} \times A_s \tag{8-15}$$

whereby:

Q_{inf} Infiltration discharge from the swale into soil (m^3/s)

K_f Permeability (Darcy's) coefficient (m/s)

I_{hyd} Hydraulic gradient (m/m)

A_s Infiltration area of the swale (m^2)

The infiltration into ground takes place on both horizontal and inclined sidewalls. The hydraulic gradient is defined as (DWA-A 138):

$$I_{hyd} = \frac{\Delta h}{\Delta l} = \frac{(h_s + z)}{\left(h_s + \frac{z}{2}\right)} \tag{8-16}$$

where:

h_s Distance between the bottom of the swale and the GW level (m)

z Water level within the swale (m)

When the infiltration through the sidewalls can be neglected in comparison to the infiltration through the bottom of the swale, the hydraulic gradient becomes: (DWA-A 138):

$$I_{hyd} = \frac{\Delta h}{\Delta l} = \frac{\left(h_s + \frac{z}{2}\right)}{h_s} \tag{8-17}$$

In swales with a small depth, the hydraulic gradient is almost 1. The infiltration rate within the swales is calculated as:

$$Q_{inf} = v_{f,u} \times A_s = \frac{k_f}{2} \times A_s \tag{8-18}$$

where:

kf Permeability coefficient (m/s) in the saturated soil

Q_{inf} Infiltration rate of the swale (m^3/s)

A_s Infiltration area of the swale (m^2)

$v_{f,u}$ Infiltration (m/s)

with the assumption: that $k_{f,unsat}$ equals $k_f/2$ that of the saturated soil.

The infiltration area A_s, is related to the water level within the swale (DWA-A 138) as:

$$A_{s,average} = \frac{A_{s,min} + A_{s,max}}{2} \tag{8-19}$$

or

$$A_{s,average} = \frac{b_{max} + b_{min}}{2} \times L_s \tag{8-20}$$

The storage volume of the swale is calculated as (DWA-A 138):

$$V_s = (Q_{in} + Q_s - Q_{inf}) \times D \times 60 \times f_z \times f_A \tag{8-21}$$

where:

- V_s Required storage volume in swale (m³)
- Q_{in} Inflow into the swale during a storm event with the duration D from the area A_c (m³/s)
- Q_s Inflow into swale due to precipitation (m³/s)
- D Duration of storm event (min)
- f_Z Safety parameter (1.1 to 1.2 according to DWA-A 117)
- f_A Attenuation parameter runoff (≤1.0 according to DWA-A 117)

Determination of the swale volume is an iterative process and depends on the local conditions and available space for swales. In general the water depth in swales (h_s) should not exceed 0.3 m (DWA-A 138).

To achieve this requirement, the swale area (A_s) has to be maximised. Replacing V_s in Equation 3-9 gives the following equation:

$$z = \Delta t \times 60 \times 10^{-3} \times \frac{Q_{in} + Q_s - Q_{inf}}{A_s} \tag{8-22}$$

where:

- z Water level in the swale (m)
- Δt Time step (min)
- Q_{in} Flow into the swale (l/s)
- Q_s Inflow into swale due to precipitation (l/s)
- Q_{inf} Infiltration rate from the swale (l/s)
- A_s Area of the swale (m²)

• Hydraulic design and proof for exceedance

Swales are designed for a design storm event, but in the case of an exceedance flow, the $Q_e x$, has to be considered for design of the conveyance systems.

8.3.1.3.4 Swales with filter drains

In case that the soil permeability is not sufficient for the required infiltration rates, it can be improved by applying a filter layer below the swale and a perforated drainage pipe.

• Hydraulic system

The system is composed of two main elements: swales and filter drains. The design principle of those systems is as depicted in Figure 8-9.

• Hydraulic design

Swales can be designed as shown in the section above.

Filter drain:

The volume of the filter drain can be calculated as:

$$V_f = A_f \times h_f \times n_{pors} \tag{8-23}$$

TEXTBOX 8-2
Example of hydraulic design of a green roof

For a commercial building in an urban area, design a green roof, for design storm event $T = 10a$. The roof should provide an emergency spill to drain the storm event with a return period $T = 100a$.

Design values:

$A_r = 1712 \text{ m}^2$ $\Psi = 1$ $n_{design} = 0.1$ $n_{overflow} = 0.01$

$Q_{drain} = \text{const} = 0.5 \text{ l/s}$ $n_{pors} = 0.25$ $\mu = 0.5$ $h_{N,100,240} = 61.4 \text{ mm}$

1. *Definition of design storm event*
   The $r_{10,D}$ is taken as the design rain which controls the maximal height of the soil layer $h_{s,max}$.

2. *Definition of runoff from sealed area*
   With this design rain, the following runoff results:
   $Q_{runoff} = r_{10,D} \times A_r \times \Psi_r \times 10^{-4} = r_{10,D} \times 1712 \times 1 \times 10^{-4}$ [l/s]

3. *Hydraulic design*
   Determination of thickness (height) of the soil layer:

$$h_s = \frac{D \times 60 \times 10^{-3} \times (10^{-4} \times r_{D,n} \times A_r \times \psi_r - Q_{drain})}{A_r \times n_{pors}}$$

D (min)	**$H_{10,D}$ (mm)**	**$R_{10,D}$ [l/(s*ha)]**	**Q_{DRAIN} (l/s)**	**$H_{S,MAX}$ (m)**
5	10.6	353.404	0.5	0.042
10	14.3	238.381	0.5	0.056
60	31.7	88.073	0.5	0.123
480	47.9	16.635	0.5	0.158
720	49.1	12.039	0.5	0.157
1080	56.6	8.736	0.5	0.150

$h_{s,max} = 0.158$ m adopted value: 16 cm crest height and capacity of emergency spill.

Flow through an emergency spill is an unsteady process and has to be iteratively solved for different durations. An example is given for the $T = 100$, $D = 240$ min.

T (min)	**ΔT (min)**	**$R_{100,90}$ [l/(s*ha)]**	**Q_{DRAIN} (l/s)**	**Q_{EX} (l/s)**	**$DELTA$ $H_{OF,i}$ (cm)**	**H_{SL} (cm)**	**H_{ES} (cm)**	**H_{OF} (cm)**
(0)	(1)	(2)	(3)	(4)	(5)	(6)	(7)	(6-7)
5	5	42.65	0.5	0.00	0	0	17	0
10	5	42.65	0.5	0.00	0.5	0	17	0
...	...	...	...	...			17	
200	5	42.65	0.5	0.117	0.1	17.5	17	0.517
205	5	42.65	0.5	0.173	0.1	17.6	17	0.636
...	...	...		...		...		
235	5	42.65	0.5	0.608	0.1	18.3	17	1.305
240	5	42.65	0.5	0.692	0.1	18.4	17	1.412

The water level over the soil layer on the green roof for $T = 100$, $D = 240$ reaches 1.4 cm. This means that the roof edge should be constructed in a way not to allow the overflow (including 10 cm free board). Further calculations with different durations of storm event are required to find the critical storm duration, which gives the maximum height of the emergency spill. Also, the exceedance flow Q_{ex} should be considered for the design of conveyance systems.

FIGURE 8-9
Hydraulic system of a swale with filter drain

Source: Joachim Tourbier.

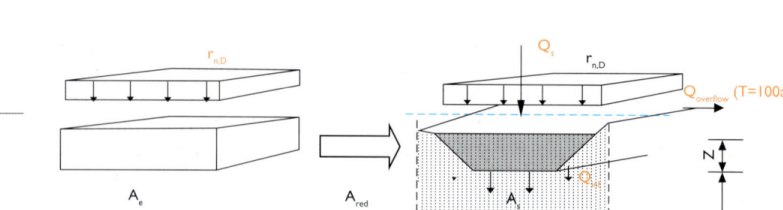

assuming that $A_f = A_s$.
Further, the volume of the filter drain can be defined as:

$$V_f = (Q_{inf} - Q_{GW,recharge} - Q_{drain}) \times D \times 60 \tag{8-24}$$

where:

Q_{inf}	Infiltration rate from the swale (m^3/s)
$Q_{GW,recharge}$	Recharge from the filter drain into the ground water aquifer (m^3/s)
Q_{drain}	Discharge through the drainage pipe (m^3/s)
A_s	Area of the swale (m^2)
D	Duration of storm event (min)

Recharge from the filter drain into the groundwater aquifer can be calculated as:

$$Q_{GW,recharge} = k_{f,saturated} \times I_{hy} \times A_f \tag{8-25}$$

with the following assumptions:

$$I_{hy} = 1 \quad \text{and} \quad A_f = A_s$$

Discharge through the drainage pipe can be calculated according to POLENI as:

$$Q_{drain} = \frac{2}{3} \times \mu \times \sqrt{2 \times g} \times h^{2/3}_{drain} \tag{8-26}$$

- Hydraulic design and proof for exceedance

Swales with filter drains are designed for a design storm event, but in the case of an exceedance flow, the Q_{ex}, has to be considered for design of the conveyance systems.

8.4 PRACTICES IN WATER SENSITIVE URBAN DESIGN

Flood management is often viewed as sizing and building engineering flood abatement structures. However, it also involves an additional factor: people and the related socio-economic, legal and stakeholder aspects. A project that has shortfalls with these aspects will fail to be implemented, no matter how well it is engineered.

Urban design is an element of urban planning with emphasis on physical improvements and their perception by users. It relates to spatial arrangements and their appearance, but also to their functionality. Urban design is an element of architecture, city planning and landscape architecture, and has recently been termed landscape urbanism. Urban design theory has been advocated by writers like Kevin Lynch, Christopher Alexander, Edmund Bacon, Andrew Duany, Robert Venturi and others. It relates to elements like public spaces, edges, gateways, accessibility, activity areas, and character and meaning of urban places. Bodies of water, along which cities have often been founded, offer special opportunities to add character and identity to a site.

In a post-carbon, climate-responsible world, Water Sensitive Urban Design (WSUD) takes on a new meaning. In a resilient city, many of the practices that we have become accustomed to, like planning around the automobile, cheap water supplies and wastewater disposal, no longer apply in the form we are used to. WSUD has become an approach, originating in drought-stricken Australia, that is now being used more widely, especially in Australasia. The approach is globally applicable, as climate change and its associated extremes potentially affect every country. Protecting natural water systems and integrating stormwater management into the landscape should be an underlying and primary consideration in every urban design.

WSUD places emphasis on the integration of the three urban water streams—drinking water supply, wastewater and stormwater. Its objectives are reduced water demand, reduced discharge of sewage and decentralised stormwater management, by applying the principles of LID (Section 8.3). Urban design, however, particularly applies to urban form and the perception people have of a city. In a textbook on design, Vitruvius, a Roman architect of antiquity stated that a good design needed to provide for structure, commodity and delight. The latter implies aesthetic appeal and quality of life. It has been suggested that there are six types of aesthetic emotional reactions: attraction, distraction, stress, relief, disillusionment and aversion. Attraction is experienced when the human mind makes a match with the present aesthetic stimuli and conjures up related positive memories. Distraction occurs when stimuli appear with great intensity, leading to 'over-optional processing' as negative experience. Stress occurs when the mind cannot process the information fast enough and retrieve related experiences. Relief occurs when sub-optimal processing that leads to stress ceases. Disillusionment has to do with unfulfilled anticipation. Aversion is, unlike the preceding experiences quite subjective, and as a learned experience is very personal. In addition, the ensemble of human senses: sight, smell, hearing and touch affect aesthetic experience, which can be positive or negative. Of those senses, sight is the most determining. As Kevin Lynch said, a city needs to look good and be memorable. A city planner's role is to enhance the positive and minimise the negative.

In the EU project Urban River Basin Enhancement Methods (URBEM), five elements were recommended to be considered as elements of urban design. These are spaces, boundaries or transition forms, focal points (and landmarks), activity nodes and access. These are based on the work of Kevin Lynch^1. Water in urban places, be it in the form of rivers, drainage swales or stormwater management structures, relates to all of these. Spaces may take the form of rivers, streams, parks, squares, recreation areas and LID practices. Boundaries or transition forms create these spaces. They may be streets, topography, vegetation and the shoreline as a transition between land and water. Focal points and landmarks have a prominence of location and make cities memorable. Activity nodes are centres of social activity, meeting places and sites of recurring events. Access or the pathways of a city permit us to experience it while in motion, be it on foot, bicycle, bus, tram or automobile. When we plan an activity, we make a mental map of how to get there and back, and places to visit en route. Pathways are the elements by which we mentally organise the city. All the above are design resources to be used and emphasised through WSUD. Special design opportunities are offered by LID, BMP or SUDS measures, conveyance measures for exceedance flows and stream rehabilitation measures.

^{1} Kevin Andrew Lynch (1918–1984) was an American urban planner and author. Lynch's most famous work, The Image of the City published in 1960, is on how users perceive and organize spatial information as they navigate through cities. Using three disparate cities as examples (Boston, Jersey City, and Los Angeles), Lynch reported that users understood their surroundings in consistent and predictable ways, forming mental maps with five elements:

* paths, the streets, sidewalks, trails, and other channels in which people travel;
* edges, perceived boundaries such as walls, buildings, and shorelines;
* districts, relatively large sections of the city distinguished by some identity or character;
* nodes, focal points, intersections or loci;
* landmarks, readily identifiable objects which serve as external reference points.

LID, BMP or SUDS (see also Section 8.3) measures are normally decentralised and primarily practised on private lands. The mere inclusion of these practices does not constitute urban design. Single purpose detention basins, surrounded by a chain link fence, that required the removal of the last remaining piece of woodland on a development site is hardly an improvement. Private developers prefer facilities that enhance the appearance of a site (curb appeal), offer multiple uses and upgrade property values. WSUD has not yet achieved mainstream acceptance and currently there are few excellent design examples.

Conveyance measures for exceedance flows are open floodways that are incorporated into urban settings to transport runoff without causing damage (SUDACS, Section 8.3). They transport the exceedance flows from conventional underground drainage facilities and the overflow from surface drainage systems during major storm events that exceed their design performance. A network of public open spaces in the form of squares, streets, alleyways, promenades and other open spaces with non-erodible surfaces can be designed to function as conveyance routes for flash flows. Conveyance measures should be combined with dry and wet flood proofing of individual structures. Open conveyance is an alternative to conventional levees and floodwalls that back up floodwaters and reduce the flood storage capacity of natural floodplains. Figure 8-10 shows an example of flood proofing the emergency access to buildings, waterproofing of manholes and sewers, anchorage for underground oil tanks and flood protected elevator equipment.

8.4.1 Urban stream restoration

Streams in the city have, in the past, often been structurally confined to maximise navigation and dispose of floodwaters at the city limits. The EU Water Framework directives require that the ecologic potential of such 'Heavily Modified Water Bodies' be realised. This encourages restoration practices and suggests that streams that have been reinforced with flood defence walls, covered streams and streams that have been turned into culverts should be rehabilitated and 'daylighted'. Daylighting entails exposing pre-

FIGURE 8-10
Elevated structures and floodways to floodproofing the buildings in the Flood Control District of Maricopa County, USA

Source: Chris Zevenbergen.

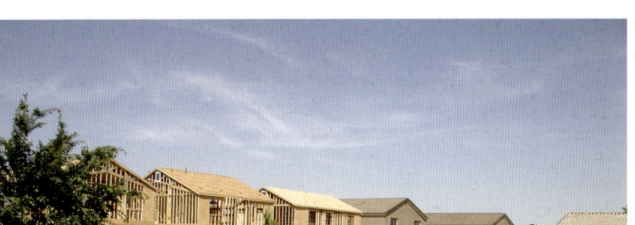

FIGURE 8-11
Swales with filter drains and green infrastructure in new urban developments in the Flood Control District of Maricopa County, USA

Source: Erik Pasche, 2009.

viously covered culverts and this necessitates making sure that the 'recovered' open watercourse is aesthetically as well as functionally effective. Urban waterways can be the core of multi-purpose green spaces and park systems that convey exceedance flows, and extend into development sites.

WSUD has become an international movement. In Hamburg, Germany, where WSUD is part of the SWITCH project^2, it has been defined as 'interdisciplinary cooperation of water management, urban design and landscape architecture which considers all parts of the urban water cycle, combines water management functions and urban design approaches and facilitates synergies for ecological, economical, social and cultural sustainability'. In Australia, the concept is being advocated by the Commonwealth Scientific and Industrial Research Organisation and its practice is supported by national guidelines for evaluating WSUD. Its objectives in regard to stormwater have been listed as:

1. protect natural systems: protect and enhance natural water systems in urban developments;
2. integrate stormwater treatment into the landscape: use stormwater in the landscape by incorporating multiple use corridors that maximise the visual and recreational amenity of developments;
3. protect water quality: protect the water quality draining from developments;
4. reduce runoff and peak flows: reduce peak flows from urban developments by local detention-on measures and minimising impervious areas;
5. add value while minimising development costs: minimise the drainage infrastructure costs of development.

Urban runoff should be seen as a resource, rather than as a problem. It can enhance the social and environmental amenity of the urban landscape through multi-purpose green space, landscaping and enhancement of visual, social, cultural and ecological values. Recommended steps in implementing WSUD are the application of Best Planning Practices (BPP) for achieving defined management objectives, and to conduct a site assessment of physical and natural attributes. In a parallel effort, Best Management

^{2}SWITCH is the name of an action research programme, implemented and co-funded by the European Union and a cross-disciplinary team of 33 partners from 15 countries around the world.

Practices (BMPs) are to be defined, alternative practices to be selected and investigated in a feasibility assessment and arranged in a site layout. It is anticipated that WSUD concepts and technologies, when planned and implemented correctly, not only offer an opportunity for the water cycle to compliment the development, but also that the development in turn would complement the water cycle.

8.4.1.1 Stakeholder participation

The Flood Risk Management Directives and Water Framework Directives of the EU widen areas of concern to include protection of nature, recreation and land use management. WSUD has implications for a wide variety of user groups. EU Directives explicitly call for stakeholder identification and stakeholder participation. Stakeholders are not to be involved in an 'after the fact' public review of completed plans but are meant to be a component of the entire planning process, starting with the definition of goals, up to the formulation of plans and post project evaluation of outcomes. It is recommended that a learning alliance be formed with stakeholder groups of similar interests. Table 8-3

TABLE 8-3
Stakeholder groups

Level A Stakeholders Community Groups	Level B Stakeholders Property Managers	Level C Stakeholders Government Agencies
Residents	Developers	City executives
Landowners	Insurers/mortgage agencies	Highway agencies
Allotment garden tenants		Department of natural resources
Voluntary nature associations	Facility mangers	City engineering department
City and neighbourhood associations		Water and Sewer Agency
Educational institutions and students		City planning commission
Sport clubs		Parks and recreation department
Business and industrial corporations		Rivers and harbours agency
Political parties		

Source: Tourbier, Ashley, 2007.

TEXTBOX 8-3
Daimler–Chrysler Project at Potsdamer Platz, Germany

A prominent example of WSUD is the Potsdamer Platz in Berlin, designed by Atelier Dreiseitl for the headquarters of the Daimler–Chrysler Company. The architectural design was the result of a competition that included 'reflection pools'. Herbert Dreiseitl converted that into a WSUD concept that starts with 44,000 m^2 of roof surface and a water surface of 12,000 m^2, receiving an annual rainfall of 21 inches. Only 0.2 per cent of the precipitation received is discharged from the site. Water is being harvested through

12,000 m^2 of green roofs and temporarily stored for flushing toilets, irrigation and fire fighting. Excess water flows into a system of narrow pools and canals that run along streets and walkways throughout the entire site and a larger pond with a total water volume of 12,000 m^3. Water is circulated through purification biotopes of 1,900 m^3 surface area to remove algae and impurities. Water bodies have a permanent surface elevation to insure maximum access and amenity. During large rainfalls, underground cistern provide for

2,000 m^3 of storage. Water disposal occurs through the evaporation of around 11,570 m^3/year, use of around 10,800 m^3/year for toilet flushing and use of around 114 m^3/year for irrigation.

The Potzdamer Platz is one of the most visited places in Berlin and the water features at the Daimler–Chrysler project contribute to that. It had been completely destroyed during World War II and later divided by the Berlin wall. When the wall came down in 1989, the redevelopment of the area provided one of the largest building sites in Europe at that time.

FIGURE 8-12
Excellence in urban design for highly functional urban water features at the Potsdamer Platz in Berlin, Germany

Source: Atelier Dreiseitl, Berlin, 1998.

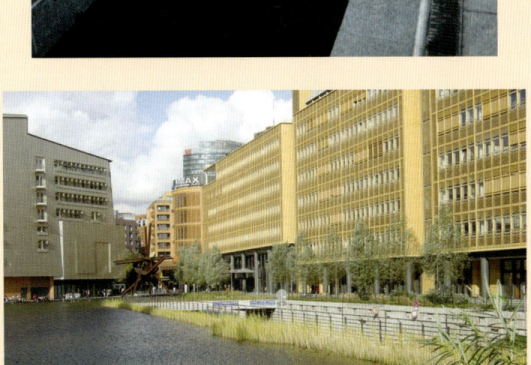

shows a listing of three groups of stakeholders, divided into level A: community groups, Level B: property managers and level C: government agencies. Stakeholder engagement is a process of consensus building through conviction and learning, often via learning alliances. Special methods, such as architectural programming, brainstorming, and Metaplan can be used in the process.

8.4.1.2 Post project evaluation

All projects should be subject to monitoring and to a post-development evaluation. This requires a comparison of what has been achieved to what was intended. In order to make these goals possible, objectives and performance criteria need to be defined early in a project. Evaluation of the results should be made accessible, as post project evaluations are essential for an advancement of the state of the art.

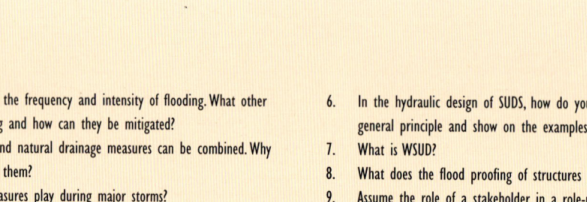

Key Questions

1. Climate change is increasing the frequency and intensity of flooding. What other factors contribute to flooding and how can they be mitigated?
2. Individual SUDS, BMPs, LID and natural drainage measures can be combined. Why would you want to combine them?
3. What role do the above measures play during major storms?
4. What are the main steps of hydraulic design? What are the main assumptions and simplifications taken for practical design of SUDS?
5. What are the main processes considered for hydraulic design of SUDS for design storm event? How are they calculated?
6. In the hydraulic design of SUDS, how do you consider exceedance flow? Explain the general principle and show on the examples of different SUDS structures.
7. What is WSUD?
8. What does the flood proofing of structures mean?
9. Assume the role of a stakeholder in a role-play related to flooding. What are the key questions that affect you?

FURTHER READING

Balmforth, D., Digman, C., Kellagher, R. & Butler, D. (2006). *Designing for Exceedance in Urban Drainage—Good Practice.* CIRIA, Publication C635 c CIRIA 2006 RP699 ISBN 978-0-86017-635-0 British Library Cataloguing in Publication Data.

Coffman, L.S. (2001). *Low Impact Development creating a storm of controversy.* Water Resources Impact, 3 (6), 7–9. http://www.awra.org

EPA (2006). "*Fact Sheet: Low Impact Development and Other Green Design Strategies.*" June 1, 2006.

Lynch, K. (1960). *The Image of the City*, MIT Press, Cambridge, MA.

Martin, P., Turner, B., Dell, J., Payne, J., Elliott, C. & Reed, B. (2001). *Sustainable Urban Drainage Systems: Best Practice Manual for England, Scotland, Wales and Northern Ireland C523.* London: CIRIA.

NSDW (2003). *Framework for Sustainable Drainage Systems (SUDS) in England and Wales.* National SUDS Working Group.

NSDW (2004). *Interim Code of Practice for Sustainable Drainage Systems,* National SUDS Working Group, ISBN 0-86017-904-4, ISBN 0-86017-904-4.

Tourbier, J.T. & Westmacott, R. (1981). *Water resources protection technology,* US Environmental Protection Agency (EPA) 1983.

Tourbier, J. & White, I. (2007). *Sustainable Drainage and Conveyance Systems (SUDACS),* in Ashley, R.; Garvin. S., Pasche, E., Vassilopoulos, A. and Zevenbergen, C. (Eds) Advances in Urban Flood Management, Taylor and Francis, London, pp. 13–28.

Wilson, S., Bray, R. & Cooper, P. (2004). *Sustainable Drainage Systems. Hydraulic, Structural and Water Quality Advice C609.* London: CIRIA.

Flood proofing the urban fabric

Learning outcomes

Urban flood-risk management encompasses a wide variety of possible responses and actions. In this chapter, we will identify and focus on managing the flood risk through site design and detailed design of individual buildings, infrastructure, services and public spaces.

9.1 MANAGING FLOODING THROUGH SITE DESIGN: BASIC PRINCIPLES

Both site selection and site planning have a major impact on the occurrence and impact of floods on any facility being planned. Site selection includes consideration of issues such as the historical flood record, terrain topography and ground level, and impacts on the hydrology (storm water flows, wetlands, etc.). Decisions made during site planning will impact the local hydrological situation as well as those downstream. The latter falls beyond the scope of this chapter and will therefore be discussed elsewhere (see Chapters 3 and 5).

Opportunities to maximise the flood resilience should be acted on as early as possible in the site selection and site planning process to ensure that site issues and temporal features are integrated into the design process (see Textbox 9-1, LifE project). Consequently, extreme climate change scenarios and future storm events should be considered when planning in areas at risk of flooding, in particular areas where the impact from climate change would be the greatest, such as low-lying and coastal land.

Carefully site design should:

- safeguard, where possible, the natural floodplain;
- minimise storm water runoff;
- compensate for any loss of floodplain space;
- reduce the risk of erosion;
- be adaptable; enable shifts in space planning and buildings to accommodate change in future flood levels and volumes.

TEXTBOX 9-1

Long-term initiatives for flood-risk environments (LifE)¹

¹ The LifE Project (Long-term Initiatives for Flood-risk Environments) is one of six projects funded by Defra's Flood and Coastal Erosion Risk Management—Innovation Fund.

The LifE approach is an integrated design approach that uses long-term and adaptable non-defence flood-risk management measures, which provide wider benefits to the development and the community; and therefore reduces the economic obstacles to delivering responsible and sustainable development. The means of managing flood risk that were explored were designed to have alternative uses for the majority of the time that they were not required to cope with flooding.

The LifE approach

The LifE approach embraces the following principles:

- Integrated Design. An integrated design approach seeks to use land and water assets for multiple functions, such as selective development, recreation, renewable energy production, local food production, water storage and flood alleviation.
- Non-Defensive Flood-Risk Management. A non-defensive approach to flood-risk management that works with natural processes can help to reduce the ongoing maintenance costs associated with defences, reduce the residual risk and increase awareness. Instead of keeping water out, developments are designed to allow floodwater and rainwater into, over or around the sites in a controlled and predetermined manner. The intention is to create more adaptable and intuitive landscapes, improving awareness of flood risk and potentially reducing risk to other areas.
- Long-term Initiatives. A development that intrinsically provides flood resilience, through good design and planning, should give insurers and financiers the confidence to provide affordable, long-term policies and investment. A versatile and adaptable flood-risk management strategy is required, which, through an explicit response to specific risk and vulnerability, can cope with the increasing unpredictability and severity of flooding resulting from climate change. Secondary and tertiary mechanisms to cope with flooding need to be integrated into the design and planning so that protection is not just reliant on a single line of defence. Furthermore, the concept of continuity of daily life, where possible, before, during and after a flood, should help to avoid the detrimental economic and social impacts that can result from flooding.
- Wider Benefits. Flood-risk management measures can be used to provide wider benefits for the development and the community, such as space for outdoor recreation, space for renewable energy and housing and job provision. Identifying the potential to create wider benefits at all scales of design can help to improve the quality, success and sustainability of developments.
- Reducing Economic Obstacles. Using available resources efficiently and combining functions may reduce implementation costs.
- Responsible Development. This should encourage greater environmental improvements to be achieved, higher quality design and more sustainable developments to be built.

LifE project case studies

The three study sites are located within three distinct river catchment locations within the UK. For each study site, a conceptual master plan was developed following the methodology set out below:

Site analysis

Information on each site was gathered from a range of sources to determine:

- location and character;
- planning policy;
- flood risk and flood zones;
- river characteristics;
- climate change implications;
- available resources.

Quantified brief

The site analysis was used to establish a detailed schedule for each site, quantifying a range of development requirements as follows:

- accommodation mix for domestic and non-domestic buildings;
- area of roads and pavements required;
- energy and water demand (based on the accommodation and occupancy schedule);
- public and private outdoor amenity requirement;
- surface water drainage required for a 1 in 100 year rain storm;
- floodplain compensation required.

Development objectives

Site-specific development objectives were established in combination with the project-wide objectives to (amongst others):

- reduce reliance on flood defences;
- use non-defensive flood-risk management measures;
- create adaptable plans for future climate change;
- provide surface water runoff rates equivalent to greenfield;
- use land for multiple purposes.

Iterative design process

An iterative design process, involving exploring and refining options, was carried out for each site to determine the optimal solution. These were reviewed by the local authorities and by the Environment Agency. The capital cost of the designs was calculated and the designs further refined.

Case Upper catchment: Let rain slow

Flood types: fluvial, typically flash floods; they also include surface water, sewers and groundwater.

Climate change: increased amount of flash flooding from rivers, increased flooding from land and sewers.

Responses: let rain slow. Slow rainwater runoff to reduce pressure on the drains and delay rainwater entering the river. Green roofs and underground storage may be suitable in the floodplain. Ground

Upper Catchment > Rain Courtyards

FIGURE 9-1
Rain courtyards

Source: Baca Architects, 2009.

Roofs used for solar PVs and solar hot water.

Planted communal 'rain gardens' provide buffer to local areas of play.

Greenswales provide threshold to buildings.

Rainwater is collected within communal harvesting system.

Middle Catchment > Stream Corridors

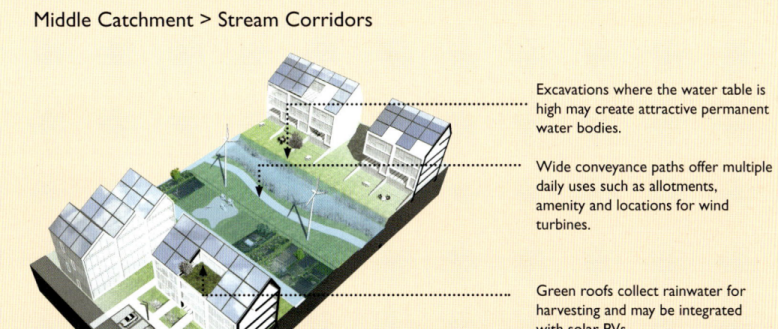

FIGURE 9-2
Stream corridors

Source: Baca Architects, 2009.

Excavations where the water table is high may create attractive permanent water bodies.

Wide conveyance paths offer multiple daily uses such as allotments, amenity and locations for wind turbines.

Green roofs collect rainwater for harvesting and may be integrated with solar PVs.

Gravel or planted swales attenuate rainwater and make building thresholds.

measures, like swales and rain gardens may also be suitable in the floodplain.

Case Middle catchment: Let rivers flow

Flood types: fluvial flooding dominant; they also include surface water, sewers and groundwater flooding.

Climate change: increased flood frequency, extent, depth and duration, increased risk of flooding from land sewers and groundwater.

Responses: let the river flow. They allow the river to flow during a flood by reducing obstacles in the floodplain conveyance. Provide paths for floodwater to return to the river afterwards.

Let rain slow: allow greenroof and underground storage may be suitable in the floodplain. Ground measures, like swales and rain gardens are suitable in the floodplain.

Case Lower catchment: Let tides go

Flood types: tidal flooding dominant; fluvial flooding, surface water, surcharged sewers and groundwater flooding can be also present.

Climate change: increased flood frequency, extent and depth. Increased duration of flooding from land, sewers and groundwater. Increased residual risk behind defences.

Responses: let tides go. Avoid high tides by letting water pass around the site into dedicated flood storage areas. Build in resilience behind defences and emergency escape routes.

Let rivers flow. Allow the river to flow during a flood by reducing obstacles to floodplain conveyance. Provide paths for floodwater to return to the river afterwards.

Let rain slow. Attenuate rainwater runoff to reduce risk of surcharged sewers or surface water flooding.

FIGURE 9-3
Design behind defences

Source: Baca Architects, 2009.

Lower Catchment > Behind Defences

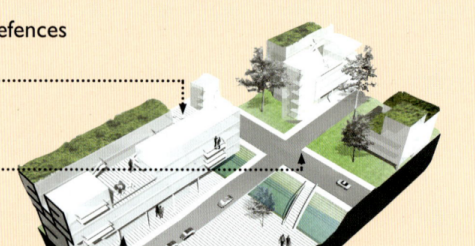

Generous elevated balconies or walkways can provide safe access and egress.

Level variation and street layout can slow flood flows, even behind defences.

Concrete frame buildings designed to allow floodwater to pass through help to resist structural collapse.

Rainwater needs to be attenuated and potentially stored, particularly during high tides, when drains can be blocked.

Lessons learned: LiFe project principles

Provide space for the river
Create areas that flood, to compensate for any loss of floodplain and to direct floodwater away from homes. Protect the natural floodplain, where possible.

Provide space for rain
Create areas that slow and store rainwater out of the floodplain to reduce surface water runoff rates to at least equivalent to greenfield rates.

Create space amenity
Locate play and recreation areas in areas designed to flood. Use permeable surfaces to reduce runoff and help prevent waterlogging.

Integrate with community needs
Plan proposals that could complement the area through provision of amenities and services that are lacking.

Design to be adaptable
Identify space to accommodate more frequent and higher volume floods in the future. Allow for retrofitting of resilience measures and new renewable technology.

Provide backups
Design fail-safe mechanisms, such as 'safe haven buildings', that are built to a higher resilience standard to provide sanctuary.

All images are by Baca Architects and all references to the LiFE project as by Baca Architects, with BRE and its consultants.

9.2 MANAGING FLOODING THROUGH DETAILED DESIGN (INDIVIDUAL PROPERTIES/BUILDINGS)

A building that resist flood loads over a period of decades, perfectly exhibits the following characteristics:

- any flood damage will be minor and easily repairable;
- the foundation will remain intact and fully functional following a design flood;
- the building envelope will remain sound;
- utility connections will be intact or easily restorable after a design flood;
- the building will be accessible and usable after a design flood.

9.2.1 Flood proofing

Flood proofing is the process of making a building resilient to flood damage, either by taking the building out of contact with floodwaters or by making the building resistant to any potential damage resulting from contact with floodwaters. Flood proofing can be subdivided into several categories:

- Raising or moving. Raising or moving a structure so that floodwaters cannot reach damageable portions. One technique is to raise the structure in place so that the lowest floor is above the expected level of floodwaters (DFE) (see Figure 9-4). This is commonly referred to as 'elevation'. In areas where flooding is likely to have high velocities, elevation on piles or columns is a recommended flood proofing technique. A second technique is to move the building to another location where floodwaters cannot reach it. This technique is commonly referred to in flood proofing literature as 'relocation'.
- Floating and amphibious structures. Floating and amphibious homes both rise with floodwater and are built on hollow concrete pontoons, on polystyrene blocks, or a combination of the two to give them buoyancy. In normal conditions, amphibious homes rest on piles or concrete slab foundations (see Figure 9-5 and 9-6).
- Dry flood proofing, or resistance measures, require the use of special sealants, coatings, components and/or equipment to render the lower portion of a building watertight and substantially impermeable to the passage of water.
- Wet flood proofing, or resilience measures, allow the uninhabited lower portion of a building to flood, but uses materials that will not be damaged by flooding.
- Active flood proofing, sometimes known as contingent (partial) or emergency (temporary) flood proofing, requires human intervention to implement actions that will protect a building and its contents from flooding. Successful use of this technique requires ample warning time to mobilise people and equipment and flood proofing materials.
- Passive flood proofing, sometimes referred to as permanent flood proofing, requires no human intervention—the building (and/or its immediate surroundings) is designed and constructed to be flood proof without human intervention.

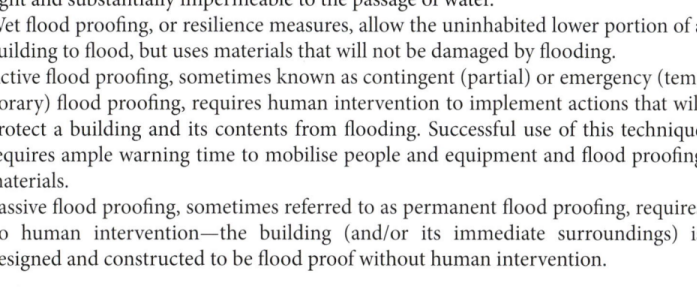

FIGURE 9-4
New elevated houses of modern architecture in Lower Ninth Ward, New Orleans, Spring 2010

Source: Chris Zevenbergen, 2009.

FIGURE 9-5
Amphibious house of modern architecture in Lower Ninth Ward, New Orleans, Spring 2010

Source: Chris Zevenbergen, 2009.

FIGURE 9-6
Amphibious house in classical style in Lower Ninth Ward, New Orleans, Spring 2010

Source: Chris Zevenbergen, 2009.

In the UK, the government have also provided guidance in the form of flowcharts which describe a design decision support methodology for avoidance, resistance and resilience measures. In addition to this, guidance is given on a resistance which can be implemented when the water depth is less than 0.3 m. This is similar to a dry proofing strategy and focuses on minimising water entry and using materials and construction techniques which allow drying and cleaning. Also included in this UK guidance is a water entry strategy or resilience which is similar to wet proofing which is implemented at higher flood water depths (greater than 0.6 m).

TEXTBOX 9-2
Thai traditional architecture

The traditional Thai houses are made from a variety of woods or bamboo. One universal aspect of Thailand's traditional architecture is the elevation of its buildings on stilts (Figure 9-7a, b and c). The area beneath the house is often used for storage, crafts, lounging in the daytime and sometimes for livestock. The houses were raised as a result of heavy flooding during certain parts of the year, and, in more ancient times, predators. Another type of traditional architecture in Thailand is the floating house (Figure 9-7d). A floating house is built on bamboo rafts or pontoons and moored in the rivers and canals. It was very common for a family to move from one floating community to the other.

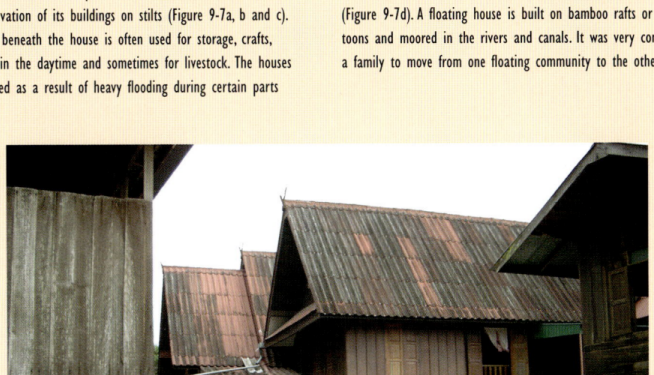

FIGURE 9-7A
Traditional Thai houses built on stilts

Source: Chris Zevenbergen, 2005.

FIGURE 9-7B
Traditional Thai houses built on stilts

Source: Chris Zevenbergen, 2005.

9.2 MANAGING FLOODING THROUGH DETAILED DESIGN (INDIVIDUAL PROPERTIES/BUILDINGS)

FIGURE 9-7C
Traditional Thai houses built on stilts

Source: Chris Zevenbergen, 2005.

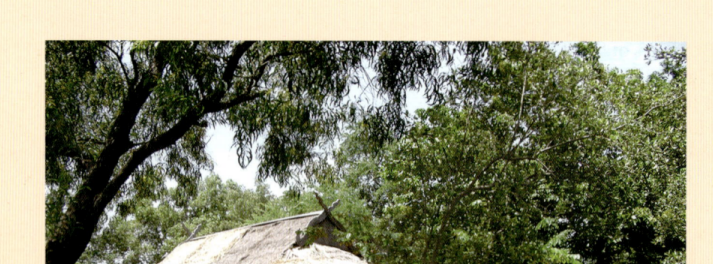

FIGURE 9-7D
Traditional Thai houses built on potoons

Source: Chris Zevenbergen, 2005.

Much of the UK guidance is based on knowledge of materials and how they react during a wetting and drying cycle. This is useful information and can be used to determine the most suitable method of construction to minimise the damage and disruption resulting from a flood event. It is more difficult to predict, however, how Modern Methods of Construction (MMC) and the use of materials with high thermal performance will react in an environment where the threat of urban flooding is increasing.

Examples of flood proofing methods are shown in Table 9-1 and Figure 9-8 below. Figure 9-8 illustrates that flood proofing involves more than the building envelope, it also involves penetrations through the envelope and associated utility systems.

Flood shields, panels, doors and gates are typically used to close medium to large openings in building walls. They can be temporary closures that are installed only when a flood threatens, or they can be permanent features that are closed manually or automatically. Key design parameters of these barriers are their height, their stiffness (and resistance to hydrostatic forces), their method of attachment or installation, and their seals and gaskets.

As a general rule, flood shields, panels, doors and gates should not be attached to building windows, glazing or doors. Given the potential for large flood loads, they should be attached to exterior walls or the structural frame. Designers planning to incorporate flood shields, panels, doors or gates into a building design are advised to consult with engineers and vendors experienced with the design and installation of these components.

Backflow valves are often installed to prevent drains and sewers from backing up into a building, and are an important component of flood proofing. They are relatively inexpensive to install and simple to operate.

TABLE 9-1
Flood proofing methods

	Dry	**Wet**
Active	Temporary flood shields or doors (on building openings) Temporary gates or panels (on levees and floodwalls) Emergency sand bagging	Temporary relocation of vulnerable contents and equipment prior to a flood, in conjunction with use of flood-resistant materials for the building
Passive	Waterproof sealants and coatings on walls and floors Permanently installed, automatic flood shields and doors Installation of backflow prevention valves and sump pumps	Use of flood-resistant materials below DFE Installation of flood vents to permit automatic equalisation of water levels Elevation of vulnerable equipment above DFE

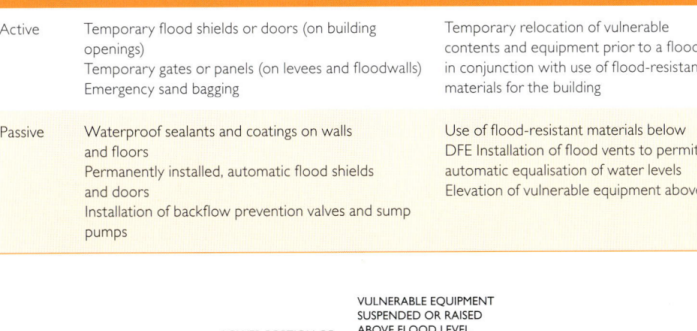

FIGURE 9-8
Typical dry flood proofing techniques

Source: FEMA, 1993.

9.2.1.1 Wet flood proofing

Wet flood proofing is employed commonly in parking garages, building access areas, crawlspaces and similar spaces. It is not employed in spaces where offices, commercial activity, residential uses and/or similar uses take place.

There are two principal methods used in wet proofing areas below the DFE, as follows:

- in enclosed areas, installation of flood openings or vents in walls to allow the automatic equalisation of flood levels on both sides of the walls (and to prevent wall failures due to unbalanced hydrostatic loads);
- use of flood-resilient materials.

9.2.1.1.1 Flood openings

Unless dry proofed, enclosed areas below the Design Flood Elevation (DFE) level must be equipped with flood openings capable of equalising water levels and hydrostatic loads. Since owners usually want to control temperature and moisture in these enclosed areas (and prevent rodents, birds and insects from entering), opening covers are often employed. These covers must not interfere with the equalisation of water levels in the event of a flood, and should be selected to minimise potential blockage by debris. There are a variety of commercially available covers, such as grates, louvers and grills (see Figure 9-9) that allow for control of the enclosed space and the passage of floodwaters.

9.2.1.2 Flood-resistant materials

Floodplain management regulations and model building codes both require materials used below the design flood elevation to be flood-resistant. But, what exactly are flood-resistant materials? Flood-resistant materials are those materials that are capable of withstanding direct and prolonged contact with floodwaters, without sustaining significant damage.

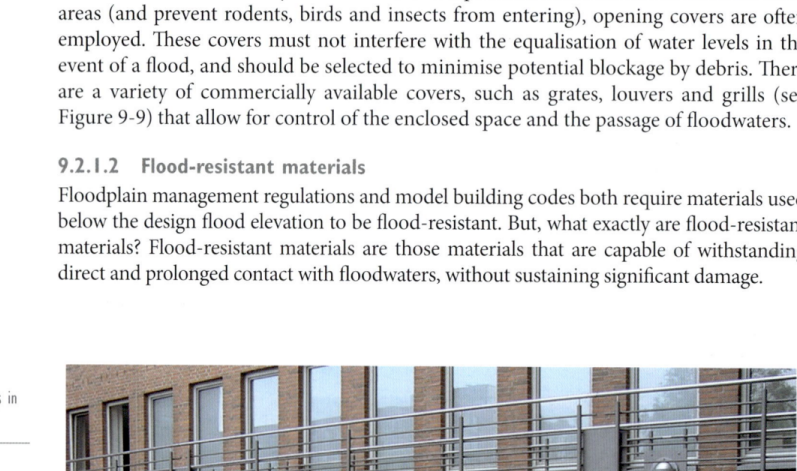

FIGURE 9-9
Active flood proofing of buildings in Hamburg, Germany

Source: Erik Pasche, 2004.

9.2.1.3 Structural factors

When floodwater surrounds a building, it creates pressure on the structure and the foundations. The loads which are exercised by the floodwaters must be determined in order to design an adequate flood proof, resilient building.

9.2.1.3.1 Walls

All structural walls must de designed to withstand the hydrostatic, hydrodynamic and debris loads. The walls must be able to withstand the predicted depth and the velocity of floodwater. To withstand the vertical loads of floodwater and the increasing groundwater level, the walls must be safely anchored to the foundation.

9.2.1.3.2 Floors

The floors must have enough stiffness to withstand the vertical pressure linked to floodwater. The floor must be thick enough and have suitable reinforcing to resist the floodwaters, and have enough weight to prevent floatation. The connections between walls and floors must be safe and sealed to prevent displacement or leakage.

9.2.1.3.3 Foundations

The foundations require special consideration in areas prone to floods. They should be sufficiently deep and resting on a suitable soil layer, in order to be resistant against floodwater pressure, namely vertical pressure. In some cases, this can require extra anchorage using piling or extra concrete as ballast. Furthermore, it is necessary to protect the foundation against erosion.

9.2.1.3.4 Openings

The most vulnerable components of buildings are openings such as doors and windows. Such openings must be designed above the expected water level or must be temporarily or permanently watertight.

9.3 FLOOD RESILIENT REPAIR AND RETROFITTING

The flood resilient repair and retrofitting of buildings have increased in significance as urban floods have become more common and the amount of loss and costs of repair have increased substantially in recent years. For the purposes of this book, resilience includes all measures to improve the flood performance of buildings.

Building owners and indeed building professionals should be aware that flood resilient repair and retrofitting can be limited by the type of property that has been affected by a flood. For example, historic buildings in cities are often 'listed' by planning authorities, therefore, limiting the potential to use methods that could alter the appearance, performance or function of the building. Other significant factors are the costs involved in the resilient repair of buildings, which are typically higher than the cost to return the building to its non-resilient pre-flood condition. The issue of who pays for the additional cost is often the most difficult to resolve in delivering flood resilient repair.

The various steps involved in recovery and repair of the building include the following:

- cleaning;
- drying;
- assessing damage to walls, timber, metals and decorations;
- repair methods.

The flood resistance of buildings can be accomplished by three types of strategies, using waterproof materials, sealing and shielding the building. In the first strategy, the floodwater is not held back, but the impact of water on fabric and fixtures is minimised. This strategy requires that all household products and furniture are removable. The second and third strategies try to keep the floodwater out of the building.

9.3.1 Flood water and damage

Buildings will allow water to enter during a flood through the following routes:

- mortar joints;
- brickwork and blockwork;
- cracks in external walls;
- vents, airbricks and flaws;
- windows and doors, including joints to the walls;
- door thresholds;
- gaps around wall outlets and voids for services such as pipes for water and gas, ventilation for heating systems, cables for electricity and telephone lines;
- party walls of terraced or semi-detached buildings;
- damp-proof course, where the lap between the wall damp-proof course and floor membrane is inadequate;
- underground seepage directly rises via floors and basements;
- sanitary appliances (particularly WCs, baths and showers) caused by backflow from flooded drainage systems.

The amount of damage caused to buildings will depend, amongst other factors, upon the water depth of exposure, as follows:

- Below ground floor: basement damage, plus damage to any below ground electrical sockets or other services, carpets, fittings and possessions. Minimal damage to the main building. Deterioration of floors may result if the flood is of long duration, and/or where drying out is not effective.
- Above ground floor: in addition to the above, damage to internal finishes, saturated floors and walls, damp problems, chipboard flooring destroyed, plaster and plasterboard. Services, carpets, kitchen appliances, furniture, electrical goods and belongings are all likely to be damaged to the point of destruction. Services such as water tanks and above ground electrical and gas services may be damaged.

Generally, short duration flash flooding will be quickly remedied, and will be less costly to repair than a flood of longer duration. A flood of longer than 24 hours duration can potentially cause serious damage to the building elements. The nature of the floodwater is also significant, for example, saltwater from coastal floods results in corrosion to metal components, and water containing sewage requires extensive cleaning and decontamination.

9.3.2 Post flood

The process of flood resilient repair commences immediately after a flood has subsided and the building is once again accessible to professional recovery and building workers. All information gathered and work undertaken from this point should concentrate on ensuring future floods are mitigated by resilience measures.

Making safe, decontamination and drying should be undertaken as soon as possible after the floodwater recedes and prior to the post flood survey. It is likely that the building fabric will absorb moisture; the degree of saturation depends on the duration of the flood and the properties of the materials. Although it is desirable to reduce the moisture content in all locations as quickly as possible, care is required so that the process is controlled and the situation is not made worse, e.g. by causing damage to materials through rapid drying.

A post flood survey should be undertaken to identify damage as soon as the floodwaters have receded and decontamination and initial drying has been completed. The following list describes the necessary assessment:

- structural damage;
- settlement damage;
- surface material condition of the wall (internal and external);
- material damage;
- internal condition of the wall;
- staining;
- measure moisture content;
- corrosion;
- blocked ventilators, or other openings in the building.

9.3.3 Repair

Flood resilient repair methods include interventions in the design, methods of construction and materials used. However, flood resilient repair may also require the use of flood protection products, household products and temporary barriers; these methods of protecting buildings may be used where there is a high risk of flood of a building in subsequent years.

There are a number of factors that contribute to resilient repair. The performance of materials in a flood situation is, however, one of essential requirements. Opportunities should be taken to reduce the vulnerability of buildings to future floods where possible, including changing the type of materials used in walls, floors or fenestration. Table 9-2 summarises the flood performance of common building materials.

TABLE 9-2
Materials resilience

	Uses	**Flood resilience**	**Immediate impact**	**Long-term effects**
Masonry and concrete	Walls, foundations, floors	Good to excellent	Saturation of bricks and some concrete blocks, leakage through mortar joints, contamination of surfaces	Efflorescence from drying out, frost damage in winter, cracking on drying out
Timber	Floors, wall frames, windows, doors	Poor to good	Swelling and warping of unprotected timber, poor quality materials or poor condition, loss of coatings	Timber decay if moisture content remains high for long periods

Wall finishes— plaster, plasterboard, render	Internal and external finishes	Poor to good	Gypsum-based materials deterioration, saturation of porous materials	Efflorescence, frost damage in winter, cracking on drying out
Metals	Cladding, fixings, window frames	Adequate to good	Corrosion, loss of coatings	Corrosion
Insulation	Cavity, external, internal and floors	Wool—poor Closed cell—good	Wetting of wool-based material, slumping of wool-based material on wetting	Difficulty to dry out materials, wall cavities, etc. Absorption of water by some closed cell types (phenolic)
Glazing	Windows, doors, curtain walling	Single panes—poor to good Double-glazing— adequate to good	Breakage of glass panes, water leakage through glazing	Breakdown of sealed units

TEXTBOX 9-3

Shearing layers concept and pro-active retrofitting

Shearing layers is a concept invented by architect Frank Duffy which was later elaborated by Stewart Brand in his book *How Buildings Learn: What Happens After They're Built* (Brand, 1994). It is increasingly recognised that dynamic societies require flexible architecture and that the built environment is the product of an ongoing and never-ending design process. Much theory and practice on the transformation of buildings such as the 'open building concept' have been developed to understand and exploit these dynamics to create buildings that can provide the capacity for changing functional requirements,different urban conditions, standards of use and life-styles. One of the fundaments of the 'open building concept' is the principle of distinct Levels of Intervention: a building can be conceived as a collection of several layers of longevity of built components. Parts of a building can be removed and replaced such as the entire façade of a building, revealing a layer which is independent of its structure.

From an economic point of view, a building can be considered as an aggregation of stocks services with different life spans. The latter implies that a building should be deconstructed into separate layers with different life cycles and depreciation rates. For example in the following layers:

- the structural layer of the building (Group C components) which generally has a life span of 50 yrs or more (European context);
- the space plan, services and skin layer (Group B components) with life spans between 10 and 25 years;
- the mobilia (Group A components). At a constant price level and without renewal the total economic value of the property (land and building group $B + C$ components) depreciates continuously down to the level of the land value.

Urban renewal of buildings and infrastructure is one of the means which allow individual buildings, neighbourhoods and even cities to adapt to long-term changes, to correct old errors and to increase flood resilience. The inclusion of pro-active retrofitting in regular urban renewal schemes and decreasing lifetime cycles of new buildings are probably sound and effective strategies to increase the flood resilience of the urban fabric. If a pro-active strategy is aimed for, often a decision will need to be made on whether to undertake flood proofing measures now or at some future point in time during the 'normal' renewal scheme of the property (or to do nothing). It is arguable that considerable cost savings are possible if retrofitting is combined with 'normal' renewal. Two main momenta can be distinguished to improve the resistance or resilience of buildings, namely the replacement of structure (Group C) $\rightarrow$ major adjustments, and the substitution of built components (Group B) $\rightarrow$ minor adjustments. In this sense, the type of appropriate flood proofing measure strongly depends on the type of renewal action: Group B or C. The structural components (Group C) are important for resistance measures where the floodwaters are prevented from penetrating the property. Resilience measures that can be carried out inside the property to minimise the damage caused by floodwaters entering the building, may be combined with the replacement of shell components (Group B). The decision whether or not to invest in pro-active flood resilient retrofitting should be based on life cycle costs and include potential benefits arisen from reduced potential flood damage.

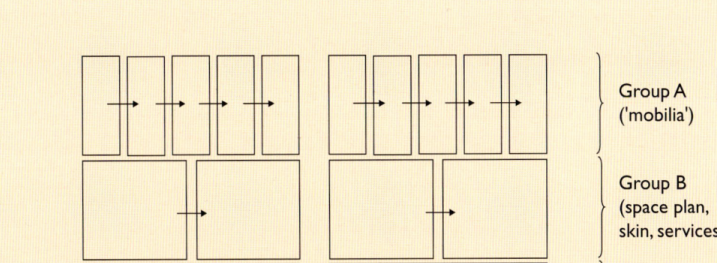

FIGURE 9-10
Duffy's shearing layers of change (Group A; mobilia, Group B; space plan, services and skin, Group C; structure (new building)

Source: Brand, S., How Buildings Learn. New York: Viking, 1994.

9.4 URBAN FLOOD DEFENCES AND BARRIERS

Conventional flood defences are barriers constructed away from a building as a means of preventing floodwaters from reaching the building. Permanent barriers are typically constructed of earth, concrete or metal. Temporary barriers are typically installed using sand bags or water-filled tubes or may comprise door boards. Permanent barriers typically require gates or doors at any locations where normal building access (vehicular and pedestrian) is maintained. Drainage systems are required inside the flood defences to remove rainfall that becomes trapped behind the barrier, and to remove floodwater that enters through leaks in the barrier or seeps through the ground.

The function of a flood defence structure is to protect people and their assets against floods. Each government has its own policy as to what extent these structures must provide safety. This results in different types of structures and dimensions of these structures. However, all structures need to be inspected to determine if they work correctly, need to be maintained so that they can work properly, and need to be improved if the hydraulic boundary conditions or policies alter. Moreover, all structures need space. These actions are often more difficult to accomplish in urban areas than in rural areas.

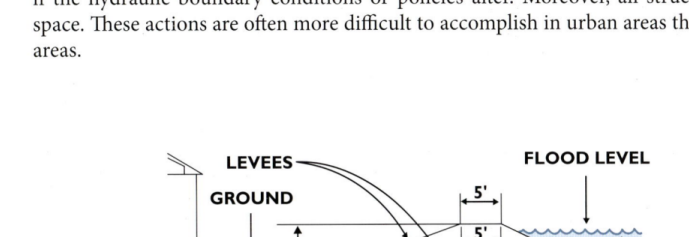

FIGURE 9-11
Cross-section of typical flood defences

Source: Levee Safety Program, U.S. Army Corps of Engineers.

9.4.1 Flood defences and failure

Traditionally structural flood 'defences' have been designed in a narrow (probabilistic) way, using the limit state equation which in its simplest form is:

$$PF = P(F \leq S) \tag{9-1}$$

where PF is probability of failure, P is cumulative probability, R is resistance (strength) and S is loading. Both R and S are combinations of several statistical variables. Equation 9-1 says that failure occurs when S exceeds R, but this is really only failure of the PES. Real failure occurs when the SES cannot bear the consequences (see Textbox 1-5). To underline this point, we denote structural failure as 'PES failure' in this section.

Traditional responses to risk, using Equation 9-1, consist of ensuring a suitably low value of PF by increasing R. The typical engineering reaction to PES failure is therefore also to strengthen the resistance, the strength of the structures. This response to PES failure is called *resistance*. But provision of a 'suitably low PF' is not always possible, because structure strength is often proscribed by social, economic, aesthetic or other considerations. Some examples: structures may need to be built so high that they block the view of the sea and cause public objections (aesthetic consideration); provision of a totally safe environment against the probability of direct hit by a hurricane, earthquake or tsunami will be too costly (economic consideration); people in other parts of the country do not want their tax money to provide a suitably low PF for a project in a far away location (social consideration).

If the PF cannot be made low enough, we must learn to accept PES failure and manage its consequences. This means we must consider risk, which is defined as

$$Risk = \Sigma \ PF * C \tag{9-2}$$

where C is consequence of failure and (*) represents some combination of PF and C. The usual, but not necessarily correct, combination of the two is simple multiplication. Equation 9-1 represents the design criterion to provide a suitably low PF. Equation 9-2 presents a design criterion—design for minimum risk.

Equation 9-2 has been successfully used in the field of design of structures, where PF and C refer to the same structure. But there are serious problems when Equation 9-2 is applied to a system consisting of (PES + SES). First, PF and C are innately different; PF is by design; C is not by design, but develops by historical evolution (such as population expansion in vulnerable areas). A second problem with applying Equation 9-2 to a system with an SES is that individuals in the system simply want to be safe; they want the lowest possible PF, but the collectives (the communities or governments within the system) want to design for minimum total cost (to them). It will be seen that these contrasting expectations result in conflicting requirements that will produce totally different responses. Ultimately, all the stakeholders must somehow agree and the contemporary decision-making process is the tool to reach this level of agreement through the SES input into the design process.

9.4.1.1 Minimum cost

Because the collective (government or community) normally pays for both the protection (PF) and the consequences (C) of any PES failure, the collective will normally insist on minimising its cost, which is the sum of the risk it takes plus the costs of designing and building the PES. This extension of Equation 9-2 is shown schematically in Figures 9-12 and 9-13.

The minimum cost in Figure 9-12 occurs near $PF = 2 \times 10^{-3}$. This design PF, based on the collective minimum cost criterion, may be higher than the individuals within the system are prepared to accept and this will result in difficult negotiations during decision-making.

Figure 9-13 shows the same schematic as Figure 9-12, but the costs of the consequences are high, relative to the structural costs of the PES. Such is the case, for example, in the Netherlands, where much of the country consists of high-density

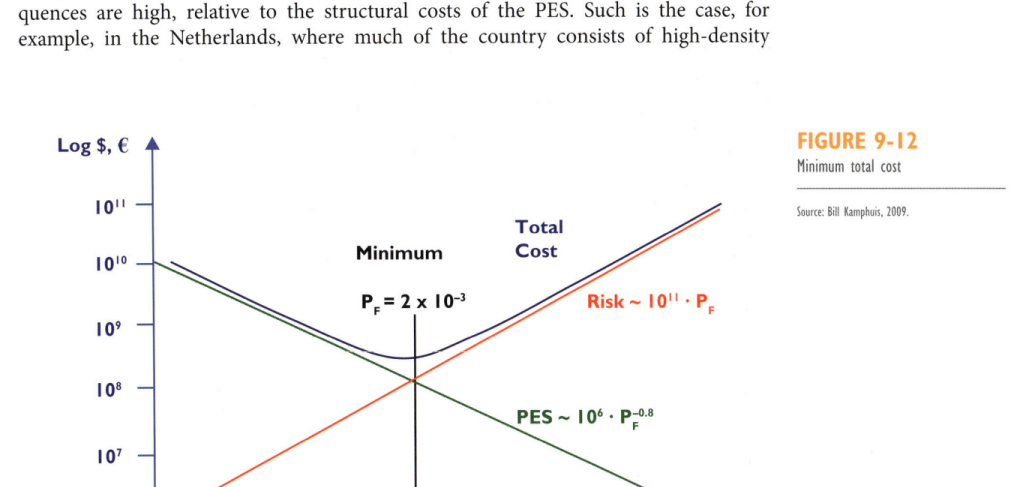

FIGURE 9-12
Minimum total cost

Source: Bill Kamphuis, 2009.

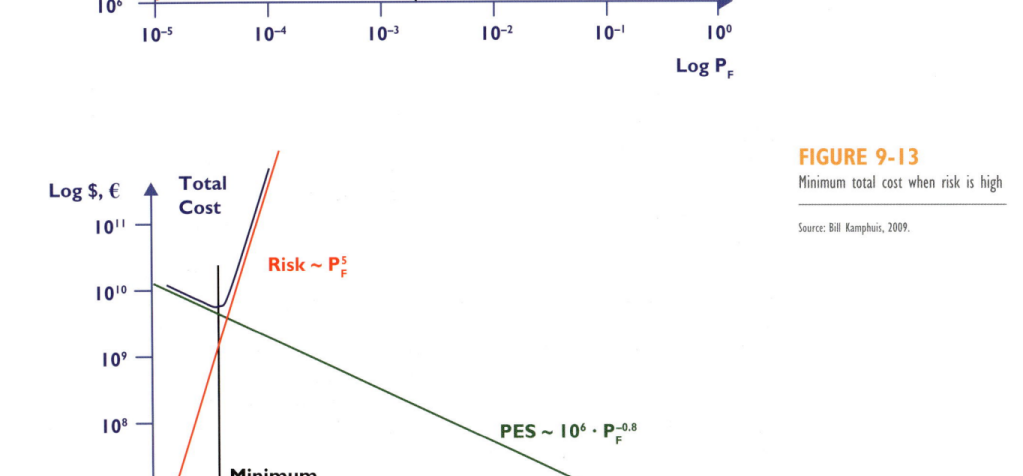

FIGURE 9-13
Minimum total cost when risk is high

Source: Bill Kamphuis, 2009.

populations living several metres below sea level. The minimum cost in Figure 9-13 is at $PF = 4 \times 10^{-5}$. Here governments achieve minimum cost and individuals will be satisfied with the low *PF* value. Subsequent decisions concerning safety should be relatively easy and approvals should be relatively simple to obtain. Figure 9-13 also comes close to the traditional engineering solution in which it is assumed that safety must be provided at all cost.

Figures 9-12 and 9-13 are also representative of two common situations; Figure 9-12 represents rural areas, while Figure 9-13 represents urban areas. In most designs, several distinct areas must be combined. If a region consists of a combination of urban and rural sections, a minimum total cost calculation for the combined area would result in a design *PF* for the rural sections that is higher than the design *PF* for the urban parts. Thus, the rural areas would experience PES failure earlier than the urban areas. In other words, rural areas will be flooded before urban areas. This inevitable (and potentially anti-social) outcome of minimum cost calculations will obviously also lead to difficult negotiations in the decision-making process.

If the rural and urban areas interact physically, if for example they are both located in a common river basin, then permitting upstream rural areas to flood earlier than downstream urban areas will by itself decrease *PF* for the urban areas. Flooding of upstream rural areas is, therefore, even more beneficial than strict minimum cost calculations would indicate.

Conversely, a political decision to protect all areas to the same design *PF* value (equal protection for all people) would result in an increase in the value of *PF* for the urban areas (not acceptable to individual citizens) or a decrease in *PF* for the rural areas (expensive for the collective).

9.4.2 Some design concerns

Recent disasters show that events with a low design *PF* value do occur. Let us first examine *PF*, its meaning and some caveats.

PF is a statistical quantity, which must be based on an adequate and appropriate database. In practice, however:

- *PF* is normally based on extrapolation of a database of a century or so of relatively quiet weather. These values are inappropriate for the present and future climates, which will experience elevated temperatures, higher wind speeds and more frequent storms. For this reason we can expect a greater frequency of PES failures in the future.
- There are no adequate databases for extreme events, such as direct hits by powerful cyclones or by tsunamis. In the past, the design *PF* values for such extreme events have mostly been based on surrogate databases, for example, flooding by hurricane storm surge and waves is based on 100-year river basin flood levels. This is not sensible.
- Synthesised databases for such extreme events may be the only approach, but they contain high uncertainties. Using inappropriate *PF* values makes subsequent sophisticated risk analysis and minimum total cost calculations meaningless.

Apart from the above concerns about *PF*, there are three other serious concerns in connection with failure and the need for resilience. These are 'secondary' processes, infrastructure concerns and rampant real estate development.

'Secondary' processes include:

- land subsidence;
- sea level rise;
- effects of climate change (such as accelerated sea level rise and increase in frequency and intensity of storms).

Such 'secondary' processes have not been routinely taken into account in the structural design of flood defence systems, since they were generally thought to be small. But they cause PF to increase with time, e.g. a design PF of 10^{-4} may become $PF = 10^{-2}$ with time. Since a return to the original PF is costly, maintenance and upgrading are often delayed, which causes the structures to be vulnerable.

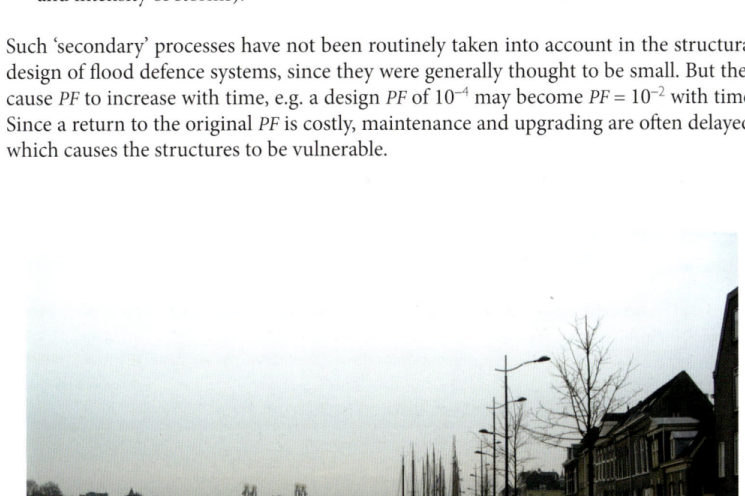

FIGURE 9-14
Flexible quay wall in Kampen, The Netherlands

Source: Bianca Stalenberg, 2006.

TEXTBOX 9-4
Cities of the Hanseatic League

Cities that were of importance for the Hanseatic League in the past, have a large trading history and this resulted in the construction of a quay. Along the quay, a fortification was built that not only protected the cities from the enemy but also, together with the quay wall, protected the cities against natural floods. By the end of the twentieth century, the fortification could be demolished due to the cancellation of the fortification legislation. These cities developed stately avenues with beautiful houses along the quay. An open city front with allure was designed.

Examples are the cities of Kampen and Zutphen which are both located at the river IJssel. Today these quay walls are too low and too weak to protect the cities against floods. In Kampen, this has been solved partly by constructing a quay wall with stop logs and partly by altering house fronts into flood defences. The quay in Zutphen was improved with a short sheet pile in combination with cuts. The sheet pile was embellished with stones of granite at the land side.

Source: Bianca Stalenberg & Cornelia Redeker: Urban flood protection strategies. p. 70–71 in: Proceedings NCR-days 2006; 2007. (ISSN 1568-234X).

FIGURE 9-15

Sheet pile with stones of granite in Zutphen, The Netherlands

Source: Bianca Stalenberg, 2006.

Key Questions

1. Under what circumstances is flood proof building design and construction viable?
2. Flood proofing is more viable for new structures than for existing buildings. Why?
3. What is the difference between the design of residential and industrial buildings in flood-prone areas?
4. Which consideration is correct for flood proofing a new building? With or without a basement or crawlspace and why?
5. Which parties and contractors are involved in the design process of flood proofing buildings?
6. How large are the uplift and lateral forces (in kg) against a foundation slab and walls, that should be caused by infiltration of floodwater through the foundation backfill in a typical house in your country?
7. When will flood proofing be economically justified?
8. Which flood proofing methods, as described in Textbox 9-1 of this chapter, are required for an adequate flood warning system and why?
9. How are the buildings illustrated in Textbox 9-1 different?
10. Why have they been designed that way?
11. What will happen if the global water level rises?

FURTHER READING

Baca Architects and BRE (2009). *The LifE Report: Long-term Initiatives for Flood-risk Environments.* IHS BRE press. ISBN: 978-1-84806-101-9.

Bowker, P. (2007). *Flood Resistance and Resilience Solutions: An R&D Scoping Study.* London: Department for Environment, Food and Rural Affairs.

BRE (2006). *Repairing flooded buildings: an insurance industry guide to investigation and repair of flood damage to housing and small businesses.* EP69, BRE press.

CIRIA (2005). *C623, Standards for the Repair of Buildings Following Flooding.* London: CIRIA publication C623 (authors Garvin, S.L., Reid, J. & Scott, M.).

Communities and Local Government (2007). *Improving the flood performance of new buildings: flood resilient construction*, CLG, UK.

ICPR (2002). *Non Structural Flood Plain Management—Measures and their Effectiveness*. Koblenz: International Commission for the Protection of the Rhine.

Kamphuis, B. *Coastal Resilience*. Kingston ON, Canada: Queen's University. K713N6.

Kreibich, H., Thieken, A.H., Petrow, Th., Müller, M. & Merz, B. (2005). Flood loss reduction of private households due to building precautionary measures—lessons learned from the Elbe flood in August 2002. *Natural Hazards and Earth System Sciences*, 5, 117–126.

Penning-Rowsell, E., Parker, D., Harries, T. & Werrity, A. (2008). *Systematisation, Evaluation and Context Conditions of Structural and Non-structural Measures for Flood Risk Reduction: FLOOD-ERA Report for England and Scotland*. R&D Technical Report FD2602/TR.

Thurston, N., Finlinson, B., Breakspear, R., Williams, N., Shaw, J. & Chatterton, J. (2008). *Developing the Evidence Base for Flood Resistance and Resilience*. FD2607/TR1.

Enhancing coping and recovery capacity

Learning outcomes

In this chapter you will learn about the following

- The objective of flood forecasting, warning and response, and how this non-structural approach can contribute to the reduction of flood risk.
- Why an effective forecasting, warning and response system must be embedded within the community to which the warnings are provided.
- How flood insurance operates around the world and the underlying conditions for successful functioning of flood insurance.

10.1 FLOOD FORECASTING, WARNING AND RESPONSE

10.1.1 Some general notions

This chapter will provide an overview of the concept of flood forecasting, warning and response. It is clear that this can be an effective approach in reducing damage and loss of life due to flooding, not, as with structural measures, through attempting to remove the flood risk, but rather through ensuring people have time to move themselves and where possible their property out of harm's way. The process of flood forecasting, warning and response is, however, complex. Without real time observation data for flood risk on the one hand, and an effective response of a well-prepared and supported public on the other, the objectives will not be attained.

The provision of early warning to a forthcoming event is increasingly being recognised as an effective approach to reducing losses due to floods. In response to such a warning, simple yet effective measures can be taken to reduce damage to property and loss of life. Depending on the scale, this may include the evacuation of people from areas where flooding is predicted or local measures such as moving mobile items, such as cars, electronic equipment and furniture that may be damaged, out of harm's way. In some cases, emergency flood protection measures may also be employed, such as temporary flood defences using for example sandbags or erectable defences. To be effective, such a warning should not only be reliable but also timely enough to provide sufficient lead time for communities to respond, allowing the measures described to be implemented in time, before the onset of the flood event.

10.1 FLOOD FORECASTING, WARNING AND RESPONSE

Flood warning systems must be reliable and able to operate during the most severe floods. It is during severe events that the benefits from flood warning systems are the greatest. The implementation of an end-to-end flood forecast, warning and response system consists of many components, which will be described below. For the system to function, each component must be linked for successful operation and interact properly with each other, see Figure 10-1. An integrated flood warning system focuses on three factors: relevant institutions; technology used in flood forecasting and issuing warnings; and the response of inhabitants receiving the warnings.

A good example of how losses can be reduced through provision of warning was shown in the Meuse basin. Financial losses due to flooding in the Netherlands during the event of December 1993 greatly exceeded those of the event of January 1995, despite the two events being very comparable in magnitude. The reduction was partly attributed to the provision of an early warning and subsequent response by the authorities and the population affected.

Sometimes a distinction is made between passive and active flood warning systems. Passive systems make inhabitants aware of flood hazards through actions such as marking flood lines and transferring responsibility to communities for taking action. Active warning is more formal and usually tries to meet three conditions: immediate action when a warning is issued; it must be possible to warn all the people about the flood hazard in a timely manner; and an appropriate centre of control available that can receive information, process it and undertake action.

In the design of a Flood Forecasting and warning system, there are a number of inter-related factors that have to be taken into consideration. These factors are amongst others:

- Basin characteristics—surface area, topography, geology, surface cover and use all help to determine the nature of potential flooding events and susceptibility to related hazards.
- Flood history—identifies areas that have been subject to flooding, seasonal characteristics and feasible warning times.
- Environmental factors—changes in morphology, nutrient and pollutant mobilisation, and effluent discharges can inform what the additional forecasting needs might be.
- Economic factors—past damages and potential future damages help identify priority areas for flood forecasting, warning and response.

FIGURE 10-1
Flood forecasting and warning management chain

Source: UNISDR—Guidelines for Reducing Flood Losses (2001) (website: www.unisdr.org).

TEXTBOX 10-1
São Paulo: Successful implementation of non-structural approaches

São Paulo Metropolitan Region is the largest urban conurbation in South America and the largest industrial complex of Latin America. Its 16 million inhabitants are spread over an area of approximately 8,000 km, and it has 950 km^2 of urbanised area, which is entirely within the Upper Tiete river basin. The upper Tiete River basin at Edgard de Souza dam has a drainage area of approximately 4,000 km^2. Gentle slopes (of the order of 0.17 m/km) characterise a meandering Tiete river. From its headwaters until Edgard de Souza dam, the Tiete river flows through 161 km, with the Tamanduateí and the Pinheiros rivers as its main tributaries. The Tiete River basin is being urbanised at a very high rate upstream of the Penha dam and is almost completely urbanised downstream of this dam. Flood hydrographs for the upstream basin show a slow rising limb with moderate peaks, typical of rural areas, while downstream of the dam, flood hydrographs are typically

urban. Due to the lack of adequate wastewater treatment, these urban watercourses are highly polluted, conveying all sorts of municipal and industrial wastes, thus imposing a serious threat to human health and so, complicating even more the usual flooding problems. Due to inappropriate land use in the basin, floods are becoming more severe in small creeks and the problem is being transferred to the Pinheiros and Tiete river basins. The situation has become extremely critical, since the main freeways of São Paulo are located at the banks of the Tiete and Pinheiros rivers. During the wet period (December to March), the population is frightened and local radio, television and newspapers give much importance to the problem. It is clear that any hydraulic work done in the Tiete or Pinheiros will be of little efficacy in the future, if flood control is not carried out in the contributing basins. A flood warning system that started in the late 1970s has proven to be a very

effective non-structural alternative for coping with urban floods in the basin. The system has evolved in recent years to a powerful Decision Support System which provides users with reliable information in real time via an internet channel. This is a very important social component in the sense that it provides the State and municipal civil defence means of mitigating the floods effects on the poor people of the region, many of whom live in the river corridor. An added problem, common to many South American cities, is that sewage flows in all the creeks and rivers in the region, with attendant risks to public health, making it extremely important to have a reliable flood forecasting system. The forecasting system has been in operation since 1976 and its performance was greatly improved in 1988 with the installation of a weather radar in the Ponte Nova dam.

- Communities at risk—combining the above factors gives some indication of the type of flood forecasting and warning system that would be suitable and effective.
- Existing capabilities—identifies where there are opportunities and where attention to strengthening will be required in terms of people, systems and capabilities.
- Key users and collaborators—knowing who the users and collaborators are enables user needs to be met so that forecasts are more likely to be acted upon during an emergency.

10.1.2 The flood forecasting, warning and response process

Provision of a flood warning that can lead to an effective response is often the result of a larger process, delivered through a Flood Forecasting, Warning and Response System (FFWRS). FFWRS is one component of various flood control options that can be installed to reduce flood losses and perhaps the most important of all non-structural measures. This FFWRS can be conceptually sub-divided into four components: a forecasting sub-system (preparation); a decision sub-system; a warning sub-system (warning distribution); and a response sub-system Figure 10.2. If one or more of these components does not operate properly then the whole process will underperform and the expected reduction in flood losses as a result of the provision of flood warning will not be achieved.

Monitoring is a key element of any operational activity, where observations and up to date information on phenomena leading to a flood event in the area of interest should be readily available. This will allow the imminent threat to be recognised and acted on. Relevant real time data primarily include monitoring of hydrological and meteorological conditions in the catchment through telemetry systems, climate stations, weather radar etc.

FIGURE 10-2
Principal components of a flood forecasting and warning system

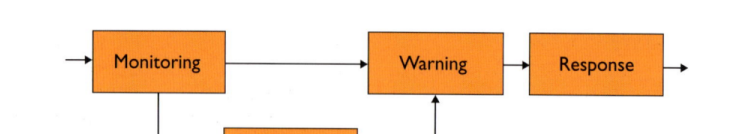

In the forecasting sub-system, predictions are made of levels and flows, as well as the time of occurrence of forthcoming flood events, using primarily information obtained from the monitoring stage. In several FFWRS, hydrological and hydraulic models are applied in the forecast stage to predict future levels and flows. These use the data gathered on the current conditions, possibly augmented with forecasts of short- to medium-term meteorological conditions such as rainfall and temperature. Predictions of future conditions are assessed and used to make the decision to provide warnings. Not all FFWRS, however, include a well-developed forecasting stage, and in some cases, the decision to warn is taken solely on the real time observations from the monitoring phase.

The warning and dissemination phase is key to the success of an effective FFWRS. It is in this phase that the decision is taken to act on the threat that has been recognised and information is disseminated to the relevant authorities and public to be acted on. This is a significant challenge, as the information provided must be timely, it must be targeted, and it must provide a message about the imminent threat that is as unambiguous as possible.

Once warnings have been received, authorities and the public must respond. Although the type of effective response will vary greatly between the different recipients of warnings, it is clear that without such a response all previous steps are in vain.

10.1.3 Lead time

For a response to be taken effectively it is clear that the warning must be disseminated in a timely way, such that an effective response can be taken before the flooding

---

TEXTBOX 10-2
Norwegian flood forecasting service

Norway has had a 24 hour flood forecasting service since 1989 with 10 hydrologists operating the flood warning office. The office has two main objectives: long term to provide information and undertake initiatives to reduce the probability of flooding; short term to issue flood warnings.

The Meteorological Institute supplies data to the forecasting service and this is supplemented by satellite imagery and information from hydropower companies. Stream flow data is uploaded twice a day from 175 gauging stations plus information from local observers. These data are used to update the hydrological models. Warnings are issued on the internet and distributed via other means to the Emergency division at the county level and this is cascaded down to the municipal level for further action by emergency services.

While the focus of much research is on the forecasting phase, it is the detection step together with the warning and response steps that are the most important for an operational flood warning service to attain its objectives. This is reflected in the criteria used by Parker and Fordham (1996) in assessing and comparing the maturity of various operational flood-warning systems in operation across Europe, with most criteria addressing the dissemination of the flood warning and the organisational embedment of the system. Failure of flood warnings to reach the public, or a large number of warnings issued to the public that prove to be false will lead to deterioration in performance of the flood warning service as a whole.

actually occurs. The longer the lead-time with which the warning can be provided, the more effective that response can be, and the more damage avoided. With thirty minutes warning, residents may, for example, move small electric appliances such as televisions and personal effects, but with four hours lead time larger appliances such as washing machines and even furniture and carpets can be saved. Parker *et al.* (2008) found in a survey following recent flooding in the UK that those not warned saved only some two-thirds of the items when compared to what those who were warned managed to move to safety. Those that received a warning with a lead-time of over eight hours were also found to have saved quite a bit more than those who received a warning with less lead-time.

Figure 10-3 shows a schematic time line of the flood forecasting, warning and response process. While the different steps are not drawn to scale, it is clear that if not done efficiently the time left for response (mitigation time) is too small to be effective. Each step should be as efficient as possible; an efficient process for gathering real time information, an efficient and dedicated forecasting and modelling system, and an efficient approach to the taking of a decision if action is to be taken. In particular, the decision-making process must be planned carefully. Clear procedures must establish how the decisions are made and who should make them and when, and these procedures must be developed well before any flood event happens.

Chapter 2 discusses different types of flood hazard. It is clear that the time available for each step may be very different depending on the type of flood hazard. For slow onset fluvial flooding, impacts may be widespread and the required mitigation time long, but sufficient time may still be available to extensively analyse data with several experts and decision-makers, and run complex hydro-meteorological models in the forecasting and evaluation phase. For flash flood events, such as pluvial flooding which is very common in urban areas, the time available may be very short, and the approach to an effective warning service may be very different, and may even need to be automated to a large degree.

10.1.4 Flood forecasting systems

Operational forecasting agencies often employ computerised flood forecasting systems to reduce the time required in the evaluation and forecasting phase. Such systems manage data and models in the operational environment, and provide decision-makers, and/or flood forecasters with ready access to the information they require to recognise a threat at the location(s) of interest, or forecasting point(s). In most cases, these systems are tailor-made, and may vary quite considerably depending on the characteristics of

FIGURE 10-3
Time-line of the flood forecasting, warning and response process (after Carsell, 2004)

Source: Kim Carsell, 2004.

FIGURE 10-4
Elements of a flood warning system

Source: Micha Werner, 2009.

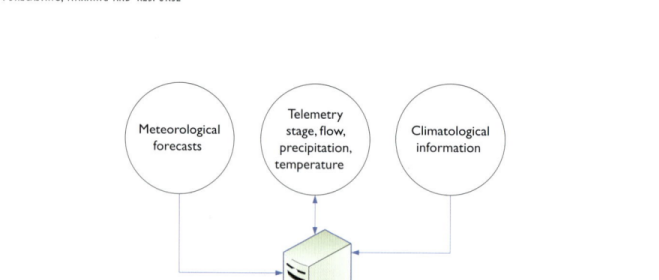

the flood hazard at the forecasting point(s), as well as on the available data, the available hydrological and modelling expertise, and even the organisational structure of the forecasting agency itself. Madsen *et al.* (2000) list the primary elements of a flood forecasting system: (i) a real time data acquisition system for observed meteorological and hydrological conditions; (ii) hydrological and hydraulic models for simulation; (iii) a system for forecasting of meteorological conditions; and (iv) a system for updating and data assimilation. This list can be extended to include a dedicated and efficient interface through which forecasters can interact, as well as the ability to create suitable output products such that information relevant to the decision-making and warning phases can be provided to those who require it.

The first computerised flood forecasting systems were developed in the early days of computer technology in the early 1970s and range from fast responding local warning systems in the headwaters of a river, to flood warning systems for lower reaches of large river basins or for all rivers in an administrative region. In several countries, a national approach is taken to operational flood forecasting, with forecasting centres in different administrative regions using a common approach to data, modelling and operational forecasting procedures.

Traditionally the development of such operational systems has been specific to each operational forecasting centre, but while this usually leads to adequate forecasting capabilities for the river systems for which these were developed, it has been found that often such systems are inflexible to changing requirements through new models and data becoming available. Recently more modular approaches to the development of operational forecasting systems have emerged. These can be configured to provide a tailor-made forecasting solution to the forecasting centre where these are used, while being flexible to incorporating state of the art modelling techniques and data as these become available.

A variety of techniques are used to predict future levels and flows within operational forecasting systems. In many cases a suite of models is used, which are run in real time using real time monitoring data where available. A detailed analysis of the types of model used would extend beyond the scope of this chapter, but includes regression and transfer function type, hydrological models relating rainfall to runoff, and hydrodynamic models to determine the propagation of flow through the river

TEXTBOX 10-3
Example: Flood forecasting for the city of Glasgow

The city of Glasgow in Scotland lies on the Clyde estuary on the west coast of Scotland. As with many cities across the world, flooding in the city may be due to coastal flooding, fluvial flooding or a combination of both as well as pluvial flooding from rainfall. The most severe coastal and fluvial events in recent history were the Strathclyde floods of January 1994 when during an exceptionally wet 48-hour spell, many rivers reached record levels, resulting in widespread disruption to infrastructure and the loss of three lives.

While various flood defence schemes have been commissioned, it was recognised that these alone could not mitigate the impact of flooding across the region, and a comprehensive flood early warning system has been established to provide warnings to those at risk.

Forecast of tidal surge levels in the estuary are provided through a real time operational two-dimensional hydrodynamic model of the Clyde estuary using predictions of tidal levels, surge levels in the Irish Sea and wind velocity forecasts provided by the UK Meteorological Office. This model in turn provides forecast boundary conditions for the River Clyde and other tributaries that flow into the estuary. Hydrological forecasting in this catchment is a particular challenge, as while the River Clyde is relatively slow in responding to runoff there are several tributaries that are prone to flash floods, with response times in the order of 2–3 hours that flow through the dense conurbation. A conceptual hydrological rainfall–runoff model is used to forecast river levels and flows, where the response of the river to rainfall data obtained both from a network of river gauges and radar data is calculated. Flood hydrographs are routed through the network using a hydrodynamic routing model, which provides levels and

flows at the forecasting points of interest. Forecasters interact with the system through a dedicated user interface. A more complete description of the system can be found in Cranston *et al.*, (2006).

Since the 1994 floods, various agencies responsible for flood management have pursued flood alleviation schemes to mitigate future flooding impacts, such as the White Cart, Kilmarnock and Kirkintilloch schemes. In addition, the Clyde Flood Management Strategy has investigated the flood management options for the Clyde catchment and Glasgow conurbation. However, it is recognised that these schemes alone cannot fully mitigate the future impact of flooding across the region and the alleviation schemes themselves require some form of early warning system to allow time for various gate closures.

network. Together with the observations available, the model results are disseminated to the decision-making and warning process.

10.1.5 Warning and response

The way in which warnings are provided is perhaps more diverse than the methods used for flood forecasting. Methods of warning will be different depending on the hydro-meteorological characteristics of the events that are warned for, but will also differ depending on administrative and organisational structures. In the Rhine basin, for example, there are some twenty-five operational flood-warning centres within the trans-boundary basin. These have evolved primarily as a consequence of national and regional boundaries as civil protection is vested in national and regional administrations, and not as an approach that is best set up to cope with control using hydro-meteorological criteria.

Dissemination of flood warning to the public and professional partners can be through various methods. Mostly information to professional partners involved in civil protection such as the police and fire departments is disseminated along established lines of communication, using for example specially created reports, telephone and two-way radio. Dissemination to the wider public is often more passive, including radio, television and internet. More direct methods include automated telephone messaging, where registered users receive an automated call in case of a warning coming into effect. With the advent of modern methods of communication, multimedia warning dissemination systems are coming into operation that can deliver a larger number of messages using a wide range of methods, including telephone, internet, e-mail, mobile phone and text messages. An example is the Floodline Warnings Direct system in use by the UK Environment Agency in England and Wales. Figure 10-5 gives an example of flood warnings posted on the Agency's website during an event in 2008. Interesting to note is that the information provided on the website available to the

10.1 FLOOD FORECASTING, WARNING AND RESPONSE

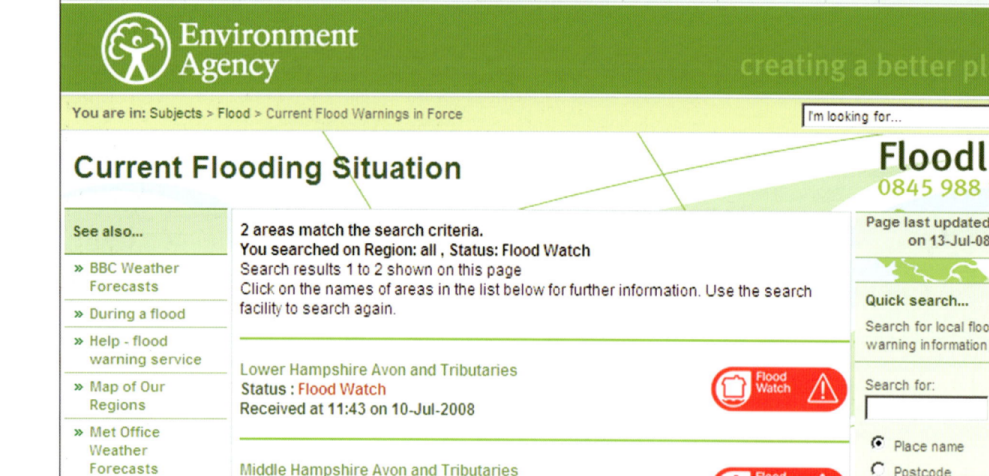

FIGURE 10-5

Example of flood warnings published on the internet (www.environment-agency.gov.uk)

Source: EA (UK), 2009.

general public is at a very different level of detail to the data and forecast model results considered by the forecasters. This is because generally forecasts are made at river locations the public are generally not familiar with, while the information provided as a warning must be understandable.

In order to make FFWRS more effective there have been improvements in the process of dissemination and communication of warnings and in the process of responding to warnings in appropriate ways. In the past, there were problems associated with these two aspects and efforts to overcome them have led to innovations in flood warning services. One of the changes in approach has been to look at what the social needs rather than just the technical needs are; by considering whom the system is designed for and what the requirements of these users are.

FFWRS and flood risk information provides the key information for effective risk management by individuals and authorities. Therefore, the information should facilitate the actions taken to minimise risk to life, property and infrastructure. Hence improving the appropriateness of responses to warnings is an important issue in improving FFWRS as a whole. Possible responses to a flood warning can include small-scale responses such as individual homeowners moving furniture to the first floor, or wide-scale evacuation as seen in, for example, New Orleans in 2004. In Japan, evacuation is in many cases one of the options regularly considered in the approach to reducing losses due to floods. Each urban area has flood proof evacuation centres that are available for the local population to find shelter in, that function even if individual properties have been inundated.

Although these measures can reduce losses due to flooding, their success depends very much on the participation of the public who are at risk. As with the decision-making that leads to the issuing of a flood warning, those required to respond should be aware of how to respond when receiving a warning. Preparing communities for flooding and how to respond is key in a successful FFWRS, and may include clear directions on how to reach evacuation centres and/or higher ground. Good examples of this are seen in Japan and in tsunami hazard areas on the west coast of North America, and through the distribution of leaflets to homeowners on how to prepare for flooding, Figure 10-6. Apart from a failure to receive a warning, the most important reason why no action is taken by people is that they do not know what to do.

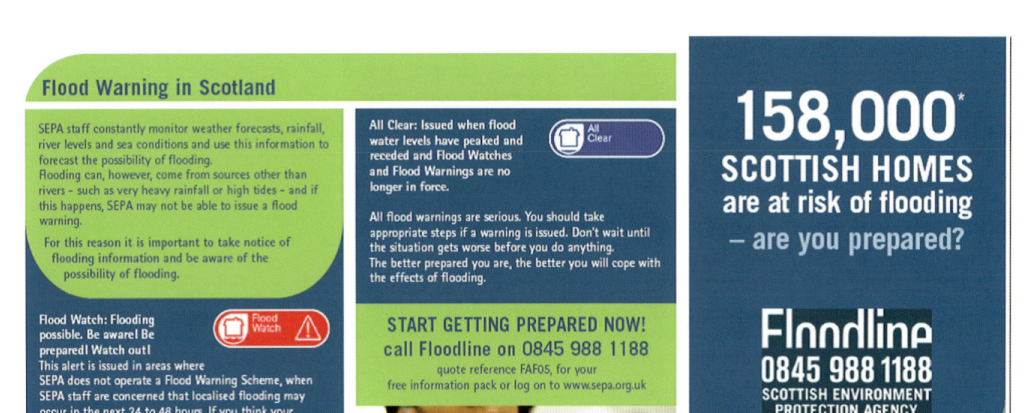

FIGURE 10-6
Example of a leaflet informing how to prepare for flooding distributed to homeowners by the Scottish Environment Protection Agency (www.sepa.gov.org)

Source: SEPA (UK), 2009.

The development of forecasting models used within forecasting and warning is very much a challenge, but while this often gains much attention, a truly effective forecasting, warning and response system must be embedded within the community to which the warnings are provided.

10.2 EMERGENCY PLANNING, MANAGEMENT AND EVACUATION

Planning and responding to floods and the consequences of flooding is usually dealt with within the broader framework of emergency (or 'contingency') planning, where it is just one of a number of potential sources of emergency. Emergency planning is about being prepared to cope with situations that have the potential to cause a significant amount of harm and disruption. The harm is often suffered by those who are 'at risk' and have a limited ability to secure themselves against that harm. Preparedness encompasses all aspects of emergency management but first and foremost, it must focus on mitigation—taking pre-emptive measures to help avoid emergencies in the first place but also to be better equipped so that when they do occur the consequences are lessened.

10.2.1 Emergency planning

Emergency planning can be best considered as a process: a continuing cycle of analysis; plan development; and the acquisition of skills necessary to implement the outputs of planning. This latter can be achieved through training, exercises, drills, etc. Although arrangements may differ, there is usually some formal national requirement for emergency planning and this may be accompanied by allocation or assignment of responsibility. Often emergencies involve organisations that do not routinely cooperate with each other, who work in different ways and who now have to coordinate their activities to provide an integrated response. The level of financial support available for this may also vary. Similarly, emergency response plans and procedures may be, but are not always, written down and formalised in a set of guidance documents, which must be flexible enough to ensure that special needs and circumstances can be accommodated. In developing countries for example, the lack of resources and access to effective IT systems often means that procedures, with key contact details are kept in printed lists that are often several years old. These need to be continuously updated, which without adequate IT systems, is often extremely difficult.

Approaches to emergency planning may differ among organisations (public or private) but there are two features that can serve to turn an emergency into a disaster:

- no one thought that it could happen or would happen to them;
- those who were prepared saved lives and livelihoods.

Without a plan, there is very little chance that there will be an effective response. The reasons why an organisation should produce and maintain emergency plans are:

- There is a legal requirement: there are statutory requirements for the preparation of emergency plans.
- It is a public protection activity.
- It is an organisational management function.
- It provides for business or operational continuity.

These reasons apply equally to public as well as private organisations. In most cases where the effects of an emergency maybe widespread there will need to be coordination and cooperation between public and private organisations.

The planning process can be broken down into different sequential steps. The first requirement is information in order to make informed decisions. Here historic data can provide inputs for identifying the broad parameters which the emergency planning has to consider. The next step is to identify, analyse and evaluate potential risks, their likelihood and consequences. Here it is important to consider the different types of hazard that will be faced: natural hazard, technological and environmental. Flooding, for example, can give rise to all three, in other words it is almost certain that multiple hazards will be faced by organisations and communities and this has to be planned for. The extent to which mitigation measures, preparedness and responses can impact on

TEXTBOX 10-4
Emergency planning principles

Over the years, eight principles of emergency planning have been identified. These are:

- Anticipate active and passive resistance to the planning process, and develop strategies to manage them. This can arise through apathy or an unwillingness to acknowledge that an emergency could happen. Active resistance might come about over questions of allocating resources that might be put to other uses. Overcoming these requires strong institutional support.
- Address all hazards to which there might be exposure. This requires that hazard-vulnerability analyses be carried out to identify the levels of exposure that might arise. Different hazard agents will make different demands on the emergency response organisations. Thus, a flood that affects a hazardous chemical works in a densely populated urban area will require a different response to that of flooding in a less densely populated suburb. Hazards can include technological, social and health risks as well as natural ones. Assessing vulnerability to hazards involves asking, risk:

From what?
Of what?
To what?
So what?

Although hazards can be external to an organisation, the assumption that its organisational infrastructure is unaffected has to be questioned. It may well not be realistic to assume that all the resources of the organisation will be accessible, as they too will be affected. Severe flooding will affect the ability of emergency personnel to get to their operational base or to vehicles and equipment. Most significantly, disruption of electricity supplies will impact on the use of almost every other function including equipment and facilities unless there are back-up generators available.

- Include all response organisations, seeking their participation, commitment and clearly defined agreement. In order to be effective, emergency planning and response must promote inter-organisational coordination, as has been noted above. This requires building trust and having in place mechanisms that enable participation and commitment. The range of organisations should reflect those who are active or have responsibility for the affected area and the communities and individuals in that area. This can include those who are the source of

potential hazards and organisations that must protect sensitive populations as well as the emergency services. Coordination is required because although the various parties have differing capabilities they will all have to work together to perform the four major functions of responders—emergency assessment, hazard operations, population protection and incident management. As has already been pointed out, it is not just government bodies that have a responsibility for emergency planning, private companies and operations also have a responsibility, sometimes a statutory one. In some cases, emergency planning and management arrangements are elements of a number of environmental management system standards (e.g. ISO 14001; EMAS; OHSAS 18001). They may deal with internally generated emergencies, with effects outside of their own boundaries or externally generated emergencies that they have to respond to and cope with. In all cases, coordination with the authorities has to be a requirement.

- Base pre-impact planning on accurate assumptions about the threat, about typical human behaviour in disasters, and about likely support from external sources. This includes knowledge of the affected area, its population and demographics and the activities and facilities in that area. Part of knowing what the threats are means understanding the basic characteristics of the hazards, such as speed of onset, scope and duration of the impacts and the potential for producing damage and casualties. It has been noted time and time again that disaster victims will act rationally within the limited amount of information that they have about a situation. Furthermore they are more likely to contact informal (friends and local groups) and trusted (fire services) sources rather than official government agencies.
- Identify the types of emergency response actions that are most likely to be appropriate. Responses should be based on the likely disaster demands but, at the same time, planning should encourage improvisation on the basis of continuing emergency assessment and the appropriate responses. Thus, planning should emphasise the principles of response rather than over defining what the particular responses should be.
- Address the linkage of emergency response to disaster recovery. There should be no clear demarcation

between emergency response and disaster recovery; one should blend into the other. So accompanying an emergency response plan there should be a disaster recovery plan.

- Provide training and evaluation of the emergency response organisation at all levels. A plan is no good if it hasn't been tested to see how well—or not—it works. Also, as an emergency will require those involved to take on tasks that differ from their day-to-day tasks, they need practice to become proficient. This includes the at-risk population, as it increases awareness and knowledge, which in turn promotes effectiveness. Multi-functional exercises, those that test emergency assessment, hazard operations, population protection and incident management facilitate inter-organisational contact and foster better understanding between parties and individuals. Hence, there is a role in evaluation for assessing effectiveness and identifying what lessons can be learnt.
- Recognise that emergency planning is a continuous process. Conditions change, conditions in a community change, people in organisations move, requirements evolve and the nature of hazards alters. These changes need to be taken into account and hence the need for emergency planning to be a continuous process.

With respect to floods, these are different from other disaster types for the following reasons:

- Floods can usually be forecasted even though unexpected flash flooding does sometimes occur.
- Floods can be relatively slow, evolving in time and it can take several days before the full impact of the event manifests itself. For example, in the Katrina flood in 2005, a hospital in New Orleans flooded two days after the start of the levee breaches to the surprise of all the people in the hospital. The reason for this was that a ridge—natural sand elevation—overtopped only after two days into the event;
- Floodwaters often do not flow out of the flooded area automatically, and have to be pumped out, which can take months. (In New Orleans in 2005 it took six weeks to dry the area, with much added pump capacity.)

the level of risk should also form part of any risk scenario evaluation. For example, what might be the impact of implementing inadequate plans or delaying evacuation from flood-risk areas? The identification and evaluation of emergency scenarios is perhaps the most difficult and onerous part of the whole emergency planning procedure as it calls for both rigour and imagination.

Having identified the potential scale and consequences of the problems to be faced, the operational and management considerations have to be matched to them. This will also highlight the potential gaps and organisationally weak areas that would militate against successful emergency planning and implementation. Having gone through all of this, the collective knowledge and wisdom developed needs to be recorded and communicated to provide a set of guidance for all parties involved, whatever the emergency situation they find themselves facing. But this has to go further than just a set of documents, it has to be put into practice and this is where the lessons to be learnt from simulations and exercises can be incorporated.

Emergency services such as the fire brigade or hospital emergency services often have difficulties in being able to respond adequately. For example, even if a flood is forecast, there will be uncertainties concerning the timing of the flood, which areas might be flooded and the numbers of people who are likely to require assistance. Emergency plans for dealing with flooding events have to pay particular attention to these uncertainties.

10.2.2 Evacuation

Evacuation and escape routes are a major issue in the emergency planning and management of floods. Evacuation entails transferring people from an area that is likely to be or is already flooded and which if they remained there they would be at risk of incurring harm, to an area in which that risk is removed or at least minimised. Evacuating people from urban areas has been said to be highly complex but also routine. It is routine in as much as virtually every urban area experiences evacuation twice a day—this is usually known as 'commuting to and from work'. However, the process is also complex as it involves thousands of people trying to do the same thing at the same time. Timing provides a key insight into successful evacuation. It must begin early enough to ensure that everyone who wants to leave is able to leave.

One of the problems that occurs when evacuation becomes a possibility or necessity connected with flooding is that often there is a breakdown in the flow of information to those moving around as to what is going on and the ensuing impacts on the various evacuation means and routes. Experience from the United States has shown that great care has to be taken at the planning stage to ensure that evacuation routes are not overwhelmed by sheer numbers. Failure to do so increases the risk of harm. Apart from the logistics involved in moving large numbers of people, the other factors that have to be taken into consideration in planning evacuations are special needs populations, failure to heed evacuation warnings and those individuals or communities that due to their location may not be able to evacuate. However, it is known that the effectiveness of evacuation is greater if those involved have had some prior experience of an evacuation, either real or simulated.

Emergency planners now make a distinction between horizontal evacuation and vertical evacuation. Horizontal evacuation moves people out of an at-risk area. This relies on having sufficient time and the available means to move and, to a lesser extent, an alternative location to move to, where basic facilities of food, shelter and security are available. Increasingly, the role of vertical evacuation is being considered as part of an integrated flood evacuation strategy. Vertical evacuation involves transferring people to a more elevated location where they can be located above the level of flooding. To be effective, vertical evacuation solutions must be capable of receiving large numbers of people in a short time period.

Vertical evacuation refuge structures can be stand-alone or part of a larger facility, and they can be single or multi-purpose. In some areas, they may also be multi-hazard facilities that can cope with more than one type of hazard. Refuges range between the use of designated high ground areas that are set aside for emergencies and artificial high ground areas to new and existing structures that meet the requirements to provide a vertical refuge. In the case of structures these can be: parking garages that have been designed to the required building standards; community facilities—community centres, libraries, museums, police stations, sports complexes taking into account the limitations that their uses might have on their suitability; commercial facilities; and school buildings. Hospital facilities also need to be designed to serve as vertical refuges, depending on their location. This is especially important given the nature of the activities they are called on to perform and the special needs of those in them.

TEXTBOX 10-5
Evacuation planning

Floods can be large or small-scale events. Large-scale flooding events can have a dramatic impact on human life, health and welfare, so the evacuation of areas at risk can reduce the number of victims dramatically. It is therefore important to plan evacuation routes and procedures as a precaution alongside other preventative measure. For example, in 2005 the mayor of New Orleans ordered that all inhabitants of the City had to be evacuated. Some 80 per cent of the population followed that instruction, but 20 per cent (some 100,000 persons) did not. During the flood around 1,000 people drowned, many as a result of not following evacuation instructions. However, if the 80 per cent of the population had not been evacuated, the loss of life in New Orleans during the Katrina event would have been much higher. One way of testing plans is to evaluate actual evacuation events. However, particularly in the case of large-scale flooding, mass evacuation may be a once-in-a-lifetime occurrence because of a country's high level of safety and emergency preparedness standards. As a result, any prior experience is likely to be outdated by the next incident, because social structures, public perceptions, decision-makers and infrastructure all change over time. There are a number of tools available to test evacuation plans. For example, the evacuation exercise tool 'SPOEL' has been developed to train, prepare and evaluate personnel for a potential mass evacuation. The tool can be used as a gaming-style simulation for emergency responders and decision-makers. From practical experiences with the use of SPOEL, it has been concluded that it is possible to develop experience of carrying out evacuations through exercises using SPOEL and evaluate emergency planning procedures. It is therefore possible to compensate for the lack of real life mass-evacuation experience through such means.

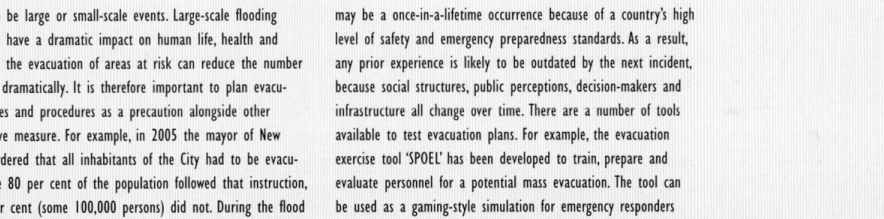

FIGURE 10-7
Scheme for disaster simulation using models that combine the decision-making processes by public and by government with the use of available infrastructure, as well as information regarding the possible threat

Source: B. Kolen, B. Thonus, K.M. Zuilekom, E. de Romph. SPOEL: An instrument for training, simulation and testing of emergency planning for mass evacuation—user experiences. Paper presented at the Evacuation Symposium, Delft. HKV Consultants, 2009.

10.3 COMPENSATION AND FLOOD INSURANCE

In many middle to high income countries, some form of insurance market operates and where it does it can make a difference to recovery from flooding. Unfortunately, for an overwhelming number of the world's population, access to most forms of insurance, let alone flood insurance, is not within their grasp. For natural hazards such as floods, most have to rely on government relief and international aid. This is because the viability of an insurance scheme is based on cost. In general, insurance schemes work by pooling the funds contributed (premiums) by members of the pool. It works best where there is a significant number of randomly occurring independent losses to members which, although large for the individual, are in total small relative to the size of the total insured value covered by the pool. So the premium is relatively small and thus affordable. The problem with flood damage is that although the average annual risk may be small, a large number may occur simultaneously. Thus, there can be a high variability of annual losses to be paid out from a funding pool and this can make a very large demand when a major event occurs. It is for this reason that in some countries the risk is considered too much for the commercial sector and government-based schemes are in place (e.g. Spain, Norway and France). But where governments readily come in to cover the costs of a catastrophe (so-called 'insurer of last resort') this reduces the attractiveness of insurance. In France for instance, where there is a national tax on all insurance premiums ('social solidarity'), even for motor vehicles, that is then redistributed to cover flood-risk recovery, many at-risk enterprises are blasé about being located in a floodplain as they hold the view that they will be adequately compensated should they be affected. In this way, preparation for flooding in e.g. the Loire River Basin is not as well advanced as it is in other countries where such insurance cover is not so readily accessible.

In order to function, a flood insurance scheme must successfully meet a number of challenges related to the supply of and demand for insurance as well as market and government-related factors. On the supply side, insurers may find it difficult to assess flood risks, people's vulnerability and potential damage. All of which influence the design of the insurance package. On the demand side, there might be low demand and on the government and market side, there can be deficiencies in the institutional framework, especially in developing and low income countries. These challenges are summarised in Table 10.1.

10.3.1 Reinsurance

When there are no insurance payouts arising from events such as floods, premiums accrue but when they do payout the loss may exceed the annual premiums several times over. This introduces an unwanted degree of volatility into the insurance market. To be effective this calls for very large levels of reserves, which is not efficient from a business perspective. The way around this is through reinsurance. This takes the risk from one country and spreads it around the world on the assumption that most hazardous events have a limited geographical extent. Thus, by spreading the risk geographically, the annual volatility relative to annual losses is reduced. Thus, insurance against flood risk along with that of other natural hazards is dependent on the availability of reinsurance. In the USA, reinsurance is of less importance due to the internalised geographic spread of risk available within the country's own borders.

TABLE 10.1
Potential challenges in adopting flood insurance

Supply side factors	• Difficulty in assessing risk and vulnerability before the floods
	• Difficulty in estimating damage after the floods
	• High administrative costs
	• Limited access to reinsurance markets
Demand side factors	• Global climate change
	• High premium due to limited risk collective
	• Limited awareness and information
	• Moral hazard—relying on government disaster relief efforts
Market and government factors	• Low income
	• Lack of relevant legislation and policies
	• Lack of clear partnership scheme between government and the private sector

10.3.2 Mitigation

One of the important contributions insurance can make is to reduce the human stress associated with coping with losses. However, there are difficulties, and direct incentives through insurance schemes to mitigate risk disasters have been limited. Incentives imply options, a voluntary scheme and preparedness to incur up-front costs that may pay off in the long run. Making insurance subject to a certain level of mitigation measures being in place will only work where schemes are voluntary and where there is a large desire for insurance—such as in the United States. Another form is to use risk rated premiums. This is more complex but it has the advantage that it does not preclude insurance cover and can be used together with compulsory schemes. The Turkish Catastrophic Insurance Pool is an example of this approach which takes into account aspects such as building standards and zoning.

To be effective the provision of insurance for catastrophes such as flooding should:

- provide adequate funds for rehabilitation;
- be affordable;
- be sustainable over a long period of time;
- be administratively efficient;
- be linked to mitigation actions;
- have high penetration;
- be politically and culturally acceptable.

Lack of insurance or low market penetration usually reflects an inability to satisfy the above conditions. For large parts of Asia, Africa and Latin America and the Caribbean, flood insurance is practically non-existent.

In those countries where relatively sophisticated insurance schemes exist, engineers and architects alongside insurance companies have an important role to play in helping society to become more resilient. Some of the leading insurance companies have considered climate change adaptation very seriously with respect to their potential financial impacts, especially following Hurricane Katrina. Under climate change projections, the problems associated with different forms of flooding are projected to increase. For flood management to be sustainable, other, more natural flood management is needed. The insurance industry will have an increasingly important role in helping society to adapt and become more resilient. Some ways in which insurers can help are:

- assistance with identifying areas at risk;
- catastrophe modelling;
- economic incentives to discourage construction in the floodplain;
- collection of data on the costs of flood damage to feed into benefit cost appraisals for flood management schemes;
- promotion of resilient reinstatement techniques;
- promotion of temporary defence solutions.

The extent to which insurers can help society depends very much on how flood insurance cover is arranged, and this varies depending on the country. It also depends on how sophisticated the country's insurers are in mapping flood risks and how much the insurers are regulated by government. The more regulation, the less insurers can use market forces to manage the risk.

10.3.2.1 How insurance operates around the world for residential properties

There are many different approaches to insurance in different countries around the world. Experience shows that in almost all countries that have flood insurance programmes, governments are involved to a greater or lesser extent. There are three types of flood insurance depending on the level of government involvement.

- Market-led flood insurance schemes are where the private insurance industry is the main coverage provider, while government provides appropriate conditions for insurance cover (protection measures, infrastructure and flood-risk mapping). A market-based system has to operate in partnership with government to establish the responsibilities of each of the parties and their various roles.
- Government-led flood insurance schemes are operated and subsidised by government, whilst it still provides flood protection measures, compensation to flood victims and control of hazard through zoning. France's 'social solidarity' scheme is an example, which was mentioned earlier and is discussed further on. This can be a public programme but administered through the private sector and financed through mandatory contributions. Contributions are not based on risk and the government may cross-subsidise low-income households.
- Mixed flood insurance schemes are operated by the private sector with substantial involvement from the government (e.g. information provision, subsidised premiums, reinsurance and zoning). The public role is supplemented by mandatory insurance based on flat-rate premiums plus voluntary risk-based insurance for those who want a higher level of coverage. Here the government's role can be quite complicated.

Where private flood insurance cover is available (and it is not available in all countries, for example, The Netherlands) such different approaches can be categorised into just two basic types, what have been called the 'option' and the 'bundle' systems.

10.3.2.2 The option system

Under the option system, insurers agree to extend their policy to include floods, on payment of an additional premium. This system can be found in Belgium, Germany and

Italy for example, and in the North West Territories of Australia, but the take up rate is very low.

There are a number of problems with optional cover. Apart from the problems of defining what 'a flood' means so it can be excluded, the biggest problem is adverse selection. Adverse selection means that insurers tend to select against customers by only making the cover available in areas they consider to be safe, while customers select against insurers by only buying it in areas they deem to be risky. The result is that cover, when it is available at all, is expensive, and has very low market penetration. Such insurance is unlikely to be sustainable because a big enough 'book' of business cannot be achieved. In countries where the government will step in to compensate flood victims, this further reduces the effective demand for insurance.

10.3.2.3 The bundle system

In this system, cover for flood is only available if it is 'bundled' with other perils, such as fire, storm, theft and earthquake. This system is used in Britain, Japan, Israel, Portugal and Spain, for example.

With the bundle system, insurers have the freedom to charge differential rates, but excessive rate increases can be mitigated because the risk is not only spread over time, but also across perils and across rating areas. People living in areas safe from flood still have to buy flood cover if they want to get earthquake cover, for example (as in Portugal), and vice versa. This system is characterised by much higher market penetration. Because everyone is paying for flood insurance, whether they think they need it or not; this reduces the opportunities for adverse selection by customers.

10.3.2.4 State involvement in compensation for flood survivors

There are four categories of state involvement in compensation for flood survivors and these are outlined below.

- No state compensation for citizens, although there may be grants for infrastructure repair—Argentina, Germany, Israel, Japan, Portugal, UK and Ireland. The lack of financial compensation can slow recovery, but it does encourage private insurance.
- Procedures to provide compensation in hardship cases—Australia, Canada, Holland and China. The State may not be well geared up to assess how much compensation to pay or to administer it efficiently. Residential flood insurance is not generally available in these countries, although in Australia and Canada cover is available for floods caused by drainage or sewage overflows.
- State reinsurance—Belgium, France, Italy and Spain. Here the cover is often based on individual private insurance policies and the State involvement is only for reinsurance and/or catastrophe situations. Sometimes state assistance for the survivors of a disaster may be dependent on the whim of a politician who could be influenced by many other issues, not always entirely objectively. In France the CATNAT system has been running since 1982 under which the government acts as a reinsurer. There is also the Bernier Fund which provides loss reduction incentives. This is funded by a levy on insurance premiums and, according to Mission des Risques Naturels, a body formed in 1999 to monitor and report on natural hazards, this causes some resentment because those who live in safe areas are in effect forced to pay for risk reduction measures for others. There is a complex relationship between insurers, the government reinsurer (Caisse Centrale), the state

insurance commission, the state guarantee and local authorities, which often leads to confusion and inertia on the part of those at risk.

- National Flood Insurance Program (NFIP) in the USA.
Here, some 95 private insurance companies provide the cover and claims handling service in almost all cases but pass the premium on to the government and recover the claims costs from the government. The premium is expensive and average take up is only around 50 per cent. Until Hurricane Katrina, the scheme was self-funding with the US Treasury giving a guarantee that claims would be paid. Annual claims costs never exceeded $2bn. Following Katrina with an estimated NFIP claims cost of $23bn, the scheme is under review.

10.3.3 The effects of the variability and increasing uncertainty of climate change on flood risks

Within Europe, in the latter part of the first decade of the new millennium, more than 10 million people lived in areas at risk of extreme floods along the Rhine, and the potential damage from floods amounts to € 165 billion. Coastal areas are also at risk of flooding. The total value of economic assets located within 500 metres of the European coastline, including beaches, agricultural land and industrial facilities, is estimated at € 500 to € 1,000 billion. In the UK, reports from the 'Foresight' programme of the Office of Science and Technology show that the flood hazard in Britain will increase significantly in the next 100 years with the number of people at risk increasing from 1.6 million to between 2.3 and 3.6 million by 2080.

Drains and sewers in towns and cities could be particularly affected by climate change and increased flooding. Short duration rainfall events will become more severe, resulting in more flooding from drains and sewers as illustrated in the summer of 2007. New European standards for urban drainage (EN 752) are based on flood frequency rather than rainfall return periods as in the past, and therefore drainage designs will in future have to take into account other factors such as pre-existing groundwater levels, or changes in urban runoff. Even so, the new standards only require drainage to cope with a 1-in-30-year event in urban areas which is hardly adequate in the context of climate change, although designing to cope with more extreme events may be extremely costly. Recent court cases in Norway have challenged this limit and allowed insurers to recover flood claims costs for drainage failure caused by events up to 1 in 100 years.

All mainstream scientists now agree that climate change is happening and the impacts will get worse. The latest projections of extreme events in Europe are detailed in a report from an EU project called 'the Prediction of Regional scenarios and Uncertainties for Defining EuropeaN Climate change risks and Effects' (PRUDENCE). The main conclusions are that by 2071–2100:

- Heatwaves will have increased frequency, intensity and duration especially over the continental interior of Europe.
- Precipitation will show major changes. There will be heavier winter precipitation in central and northern Europe and decreases in the South. Heavy summer precipitation increases in NE Europe and decreases in the South are projected. The summer floods in England in 2007 are consistent with this (see below). Longer droughts will happen in Mediterranean countries, but there could still be severe rainstorms with faster runoff.
- Extreme wind speeds will increase between $45°N$ and $55°N$ except over and south of the Alps, and become more north westerly. These changes are associated with

reductions in mean sea level pressure and will generate more North Sea storms, leading to increases in storm surges along the North Sea coast, especially in Holland, Germany and Denmark.

It has been suggested that winter storm tracks may move south: this would affect areas such as France where buildings may be less resilient than in Scotland or Scandinavia, as happened in 1999. Indeed, the director of research at the French national meteorological research centre has warned of a 50 per cent increase in the number and severity of storms in France by 2080. Even in Scotland, where it is commonly assumed that the country is uniformly 'wet', the north east coast, which often has water shortages, has experienced several unprecedented high intensity rainfall events that will change the

TEXTBOX 10-6
The summer 2007 floods in England

In 2007, England and Wales had their wettest summer since records began in 1766. By 25 July, the Meteorological Office reported that 387.6 mm of rain had fallen since the beginning of May. There was unusually severe rainfall on 25 June and 20 to 22 July.

The record downpour in June flooded 27,000 homes and 5,000 businesses in the Midlands, Yorkshire and Northern Ireland. Sheffield, Doncaster and Hull were particularly affected, because drainage systems could not cope; costing insurers an estimated £1.5bn. Hull was the worst affected. Unfortunately, in Hull the flood was made worse by the 80-year-old storm drainage system which had not been cleaned and was blocked by tree roots, grass cuttings and leaves and had been enclosed by culverting by the City Council in the 1970s and 1980s.

The rain in July resulted in rivers overflowing and flooding 10,000 homes in Gloucestershire, Oxfordshire, Warwickshire, Worcestershire and Bedfordshire. It is estimated that this event could have cost insurers a further £2bn.

Ironically, a flood event can stimulate GDP owing to increased sales of replacement building materials and household goods, and increased purchase of insurance so the overall economic losses were not great. The human costs and suffering, however, were huge, even for people not actually flooded. The July 2007 flooding left at least 350,000 homes without running water and 50,000 without power. There were fears that a further 250,000 would lose power and water if Gloucester's Walham sub-station was flooded and emergency crews worked overnight to keep the floods out. There were serious concerns that not only might thousands of people suffer, but that the security of the Government's General Communications Head Quarters (GCHQ), a key anti-terrorism communication centre, might be compromised and a large contingent of army personnel were rushed in to help.

Questions were subsequently raised over the wisdom of locating housing, electrical sub-stations and water treatment plants in the floodplain. These floods triggered further insurance premium increases and flood excesses to premiums for properties located in English floodplains and will discourage people from living in such areas.

FIGURE 10-8
2007 United Kingdom flood

Source: Michael Wilson, 2007.

statistical records in the second half of 2009. This rainfall has led to widespread flooding due to drainage infrastructure being provided to cope with historical, statistical levels of intensity. If the same rainfall had occurred almost anywhere else in Scotland, the existing drainage systems would have readily coped.

Sea levels are rising as seawater expands due to rises in temperature and as polar ice melts. With a rising sea level, tsunami events and storm surges will become more damaging. With a warming ocean, tropical storms are likely to increase in frequency and intensity. Shallow slope coastal areas are particularly vulnerable as storm surges are higher there and run up further. Hurricane Anatole in December 1999 created an unprecedented five-metre storm surge on the West coast of Denmark, which resulted in coastal defences being overtopped or breached and thousands of properties damaged. Low-lying coastal areas in the Gulf of Mexico suffered from storm surges in excess of eight metres in 2005.

10.3.4 Flood-risk assessment and modelling

Insurers need to understand the components of risk in order to be able to assess and manage risks. There are many different definitions of 'risk' in use (see Chapter 6), but one of the simplest and most comprehensive is the 'Crichton Risk Triangle', Figure 10-9. With a changing society and climate, insurers can no longer rely solely on historical claims experience to predict risk. It is necessary to analyse each of the components of risk because they are each changing in different ways. In Figure 10-9 the area of the triangle represents the risk, so if any one of these components is missing, then there is no risk. Risk can be reduced by addressing all sides of the triangle, searching for the most cost-effective risk reduction measures. Each of these three elements is considered in more detail below.

Flood hazard can be managed to some extent by building flood defences such as sea walls and river embankments (see Chapter 9). However, the power of nature may overcome these, especially as the hazard grows with climate change. Exposure is already growing as more people live near the coast or on floodplains, but this could be controlled by land-use planning. For economic, historic or logistical reasons, building is still likely to continue in exposed locations. In the USA, for example, it is clear that a growing number of people wish to live and build in southern coastal areas. Even in hazardous areas, risk can still be managed by reducing vulnerability. This may mean building in a more resilient way. For example, in Holland there is an increasing use of floating houses, while in London's Docklands area, the ground floor of apartment blocks is often reserved for car parking (see Chapter 9).

Deterministic modelling requires consideration of the values of each of the three sides of the triangle (Figure 10-9) in given scenarios. Probabilistic modelling also requires an assessment of the probabilities of different frequency and severity situations. To the layman, frequency and severity is usually associated with the hazard—how often it floods and how severe the floods are. However, the best models must recognise that exposure and vulnerability can also vary in frequency and severity.

10.3.4.1 Hazard

Climate change will certainly increase the frequency and severity of flooding events but specific events are impossible to predict. Flood management policy in some cases has concentrated on dealing with the effects of flooding rather than the causes. Thus, the answer has often been seen as 'structural' solutions, such as walls, drains and reservoirs.

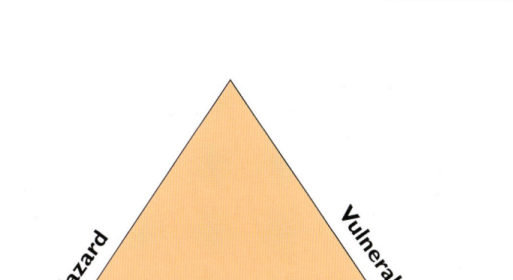

FIGURE 10-9
The Crichton Risk Triangle
© Copyright, Crichton 1999

Source: David Crichton, 1999.

10.3.4.2 Vulnerability

Different types of construction show different levels of vulnerability to flood or windstorm damage. All the major insurers in the UK have pooled flood insurance claims data in what is now the biggest database of flood damages in the world; the British Flood Insurance Claims Database, held at the University of Dundee in Scotland. This enables insurers to model different flood event scenarios, and assess total maximum losses, as well as calculate insurance premiums. The data shows the actual financial costs of a flood to insurers, rather than the theoretical economic losses. For example, if someone has a 20-year-old carpet, the economic loss is another 20-year-old carpet, whereas UK insurers will generally provide cover on a 'new for old' basis.

There have been attempts to develop similar databases in Germany and Italy, but these are relatively small, localised and under-funded. The UK database has demonstrated that if the property is likely to be flooded again, resilient reinstatement can often be cost justified. And it has been used to demonstrate how flood damage can be reduced by using more resilient building materials.

Urban vulnerability has increased considerably in the last 100 years. A good example of this is Paris, a region of 11 million people, of whom 800,000 are directly at risk from floods, and 2 million who could suffer indirectly. The last major Paris floods of 1910 occurred when the Seine rose eight metres as four billion cubic metres of muddy, sewage-laden, floodwater flowed through the city. Nonetheless, people were still able to get around the city thanks to horse-drawn carts and carriages. The 1910 Paris flood was a 100-year return period event, but less severe flooding was common in Paris in the latter half of the nineteenth century with floods in 1854, 1856, 1861, 1866, 1872, 1876, 1882 and 1883. Since 1910, four reservoirs have been built to store floodwater, and the people in Paris seem to have forgotten about flood hazards, but these reservoirs can only cope with a rise of a few centimetres in river level, thus giving only a false sense of security. As a result, Paris has become much more vulnerable compared with 1910. The Seine has been restricted in width, many new schools and hospitals have been built in the danger zone; 154 km of underground railway has been constructed; buses are now used instead of horse-drawn vehicles and people are dependent on electrical power. There are now 600,000 underground car park spaces with basement depths up to ten levels.

10.3.5 Exposure

Unlike perils such as windstorms or droughts, property exposed to flooding can be identified with reasonable confidence. It is usually specific to low-lying topography and dependent on the effectiveness of drainage. Flood mapping is therefore the obvious solution to assessing exposure. In the UK, concerns about possible adverse selection have caused a number of major insurers to invest substantial sums of money in better flood maps for their own use. Such research has been expensive, but commercial imperatives have meant that for several years, the leading insurers and reinsurers have been able to dynamically model various flood event scenarios around the world to an increasing level of sophistication. In the UK, insurance penetration is so high that one insurer for example, has spent £5m just on digital elevation models produced using the latest airborne 'synthetic aperture radar' (SAR) technology surveys in order to be able to produce high resolution flood maps. As flood maps improve, there is increased scope for governments to avoid exposing their citizens to flood risk. They can, for example, use the planning system to prevent developments in flood hazard areas. However, this does not help with existing properties already situated in exposed areas.

10.3.6 Compensation

Governments will budget financial resources for emergency relief and part of this risk management usually includes organisation of financial and humanitarian aid, which could be channelled through various government agencies and subsequent repair of damaged properties and provision of lifelines. Compensation mechanisms and hazard mitigation strategies vary considerably from one country to another because of differing national priorities and local cultures. Compensation may in some cases be made available to flood victims. The diverse national flood relief programmes include:

- systems with no state compensation (Germany, Japan, Portugal and UK);
- government policies providing extreme hardship compensation (Australia, Canada, Mexico, Slovak Republic and Turkey);
- government catastrophe programmes when a national disaster is declared (Belgium, France, Iceland, Italy, Netherlands, Norway, Poland, Spain, Switzerland and US).

Aside from post-disaster assistance, governments can subsidise pre-disaster insurance premiums, which is what the USA does. However, there are two issues that need to be assessed:

- the extent to which insurance premiums should be subsidised in order to provide sufficient and affordable insurance coverage to the intended group;
- whether the government can afford such subsidy.

The first relates to a household's ability to buy flood insurance, while the second involves the government's capability to provide flood insurance subsidies. In the US, NFIP was initiated to address the increasing costs of government-funded flood management and disaster relief for flood victims. However, flood claims were greater than the premium incomes collected from policyholders. This resulted in operating losses being experienced for some years because programme costs were greater than the income.

Thus, a planned flood insurance scheme should not be viewed as a direct replacement for government expenditure on flood management and disaster relief programmes. It requires an assessment of a country's present and medium term budget framework indicating the capacity to finance planned flood insurance schemes, subsequent recurrent spending and a good understanding of the impact of subsidies.

10.3.7 Micro-insurance

In low and middle income developing countries less than 3 per cent of households and businesses have some form of catastrophe insurance coverage and instead rely on other means (family and government support) to overcome or recover from the affects of a disaster such as flooding. And indeed, it is the poorest members of society that are often most at risk. A key concern is how the poor in developing countries can have access to insurance. At the macro level, there have been initiatives to facilitate better access to insurance, such as the World Bank's Global Index Insurance Facility that, among other services, will provide reinsurance to index-based insurance pools in developing countries.

Micro-insurance is attracting wide interest, as evidence suggests that there are potential benefits for low-income households and businesses that are traditionally excluded from conventional insurance services. The idea of micro-insurance is to provide easily accessible insurance cover for small-scale assets and livestock at affordable premiums by keeping transaction and other costs low. While micro-finance services over the last three decades have been offered on a wide scale, they have only recently included insurance for natural disaster losses. Micro-insurance provides indemnity for losses with respect to a pre-specified event in exchange for a premium payment. It is distinguished from conventional insurance by its provision of affordable cover to low-income clients who cannot be profitably insured by commercial firms, or who are not currently served by conventional insurance. Affordability, like conventional insurance, is usually secured by building and pooling groups of clients and, otherwise, minimising transaction costs, overheads and profits.

As already discussed above, disasters such as flooding events present a challenge because insurers must have a large capital reserve or reinsurance to cover infrequent but very high claims. Because of the high costs of capital and reinsurance, it is difficult therefore to offer low-cost catastrophe cover. However, a number of pilot schemes are emerging, for example, in India micro-insurance for sudden onset disaster risks is offered by NGOs in conjunction with insurance companies in two states. These schemes build on micro-insurance arrangements for independent risks, such as unemployment, fire and accidents, by extending cover to loss of life, property or livestock due to natural disaster events. Coverage for property losses due to floods, for example, is offered to groups such as women with a minimum group size of 250, or to community groups. In addition, clients can additionally engage in risk reduction training for a small fee.

Micro-insurance can also take the form of index-based weather derivatives. Such pilot schemes have been implemented in India and Ukraine with pilot projects underway in Nicaragua, Ethiopia, Malawi, Peru and Mongolia providing financial protection to farmers against weather risks, such as drought. Contracts are written against a physical trigger, say, severe rainfall measured at a regional weather station with the contracts designed by insurance companies. Since payouts are not coupled with individual loss experience, farmers have an incentive to engage in loss reduction measures. A physical trigger also means that claims are not always fully correlated with the actual losses experienced, but this 'basis risk' may be offset by the reduction of moral hazard and elimination of long and expensive claims settling. For such schemes, many of the considerations that apply to conventional insurance schemes also apply, e.g. sufficient funds to meet claims, means of risk estimation and reliable institutions.

Research from Bangladesh has shown that administrative implementation costs of micro-insurance play a significant role in determining the viability as well as long-term sustainability of micro-flood insurance schemes. In developing countries, costs can be minimised by offering policies to groups or communities through established micro-finance institutions. Furthermore, income is a major determinant on the demand side for setting up flood insurance schemes, although with respect to property, research has shown that a commercially viable market exists in Bangladesh. In Jakarta, Indonesia the insurer, Munich Re, teamed up with a local insurer to offer cheap index-based flood insurance to low-income families through a protection card system. A protection card guarantees a one-off payment should waters rise above a certain level at a particular measuring point. It has been claimed that this trigger-based micro-insurance product is the first to try to deal with flooding risks under these particular circumstances. Thus, for developing countries there are emerging insurance schemes that may enable them to better cope with recovery after flooding events. This is particularly important as disasters such as flooding make poverty worse.

Key Questions

Flood forecasting, warning & response

1. Explain the influence of the 4 components of a flood forecasting and warning process on the effectiveness of the whole process?
2. What is the difference between passive and active flood warning?
3. How is the process of flood forecasting, warning & response dependent on the type of flooding?

Emergency planning, management and evacuation

1. Which steps are involved in an emergency planning process?
2. Why are emergency planning processes for floods different from other natural hazard types?
3. How does a vertical evacuation strategy differ from a horizontal evacuation strategy?

Compensation and flood insurance

1. Explain the potential challenges in adopting flood insurance, from both a supply and demand side.
2. What are the characteristics of an effective flood insurance scheme?
3. What is the role of re-insurance in dealing with flood risks? And what is the role of micro-insurance?

FURTHER READING

Bouwer, L.M., Huitema, D. & Aerts, J.C.J.H. (2007). *Adaptive Flood Management: the Role of Insurance and Compensation in Europe.* Deliverable D 1.2.4 of the NeWater project. Institute for Environmental Studies, Vrije Universiteit Amsterdam.

Briscombe, N. (2002). *Flood Warning Dissemination Technologies—Technology Comparison Report.* FD2202/PR.

Cloke, H.L. & Pappenberger, F. (2009). Ensemble flood forecasting: A review, *Journal of Hydrology, 375(3–4),* 613–626.

Environment Agency (2007b). *Risk Assessment for Flood Incident Management: Management and Tools.* SC050028/SR1.

Mens, M.J.P., Erlich, M., Gaume, E., Lumbroso, D., Moreda, Y., van der Vat, M. & Versini, P.A. (2008). *Frameworks for Flood Event Management.* T19-07-03.

Parker, D. & Fordham, M. (1996). An Evaluation of Flood Forecasting, Warning and Response Systems in the European Union, *Water Resources Management, 10,* 279–302.

Sene, K., Huband, M., Chen, Y. & Darch, G. (2007). *Probabilistic Flood Forecasting Scoping Study.* FD2901/TR.

Wallingford, H.R. (2004). *Best Practice in Coastal Flood Forecasting.* FD2206/TR1.

V

Towards flood resilient cities

Urbanisation, both as a social phenomenon and physical transformation, is driven by processes that take place at varying temporal scales from relatively slow (e.g. migration, rising water demand, sea level rise and changes in laws) to rapid (e.g. natural disasters, changes in regulations and economic systems). While there is much that is uncertain about the urban future, some recent experiences described in Part V show that some urbanisation pathways are more desirable than others because they will likely lead to more (flood) resilient cities. These experiences highlight the need to take a completely new and different perspective on urban design, planning, and building. Creative thinking and innovations in socio-economic and technological systems are essential to change existing management structures and regimes. In Part V we will learn that responses that enhance resilience, can be implemented gradually and offer prospects for action in the short term in regional planning and development in cities. These interventions should operate in a mode of constant learning and experimentation. Those interventions do not only reduce flood impacts, but also create new opportunities.

In many Dutch polders the need for stormwater control, water supply and urban expansion resulted in competing land claims which necessitate the search for multi-functional land-use.
This impression of a flood resilient community at high water level shows how urban developments adapted to fluctuating water levels (up to 1 metre) and water storage can be combined in the same area.

Source: Dura Vermeer, The Netherlands, 2004.

Managing for resiliency

Learning outcomes

In this chapter you will learn about the following key concepts:

- How assets can be managed to minimise risk and improve performance.
- The application of the concept of resilience as a way of informing management approaches.
- The impact of previous courses of action on future options for flood-risk management.

11.1 ASSET MANAGEMENT, SOME BASIC PRINCIPLES

The Organisation for Economic Cooperation and Development (OECD) define asset management as:

A systematic process of effectively maintaining, upgrading and operating assets, combining engineering principles with sound business practice and economic rationale, and providing the tools to facilitate a more organised and flexible approach to making decisions necessary to achieve the public's expectations.

Good asset management has gained increasing recognition as an approach that gives the information and has associated analytical tools to manage assets more effectively and to plan for long-term needs. Planning and formulating asset management strategies that can meet long-term needs requires that factors such as the progressive deterioration of assets, changes in regulatory requirements, changes in stakeholder expectations and climate change issues are explicitly considered. Flood-risk managers need an asset management system that allows the information generated by the management systems to identify and assess the risks associated with an urban flood management system. The assets associated with urban flood management depends on the history of urban infrastructure provision, the geographic and climatic location of the urban area and the nature of the flooding to which the urban area has traditionally been exposed. Thus, the ensemble of flood management assets will be peculiar to the situation being considered. Once the potential risks have been identified a plan of management interventions to achieve performance and meet the level of service objectives can be formulated. Asset management is a process that involves four distinct phases: inventory and performance monitoring; risk assessment; decision-making; and implementation, Figure 11-1.

FIGURE 11-1
Asset management

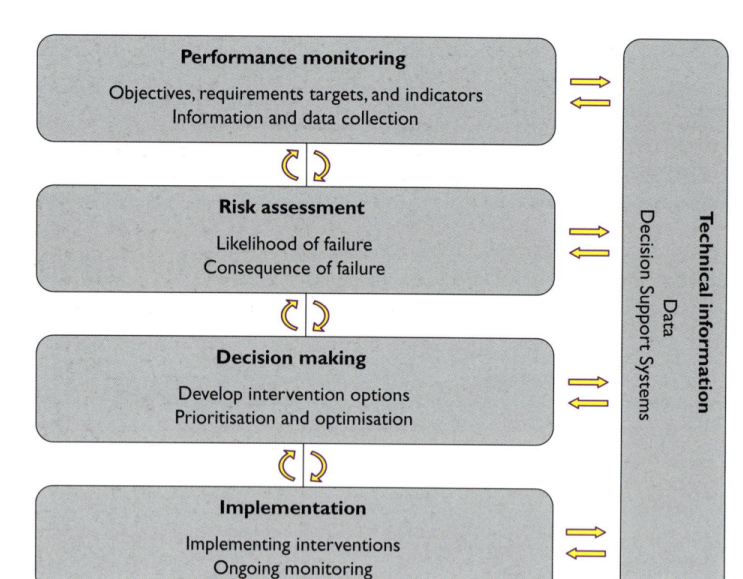

11.1.1 Performance monitoring

In order to improve the performance of the urban flood management system, it is important to quantify the performance of the system, and this should be done against some benchmark of the required level of performance. This should be driven by stated policy objectives, derived from organisational and stakeholder values as well as the design capabilities of the system. Performance requirements are also established that reflect the appropriate level of performance for a system. There are two alternative approaches: to adopt a fixed value or to use a variable performance criterion. The advantage of using variable criteria is that these allow the optimisation of the benefits and costs for each intervention. Finally, targets can be defined in terms of the proportion of assets that meet the performance requirements/goals and the threshold value.

The focus on meeting performance requirements and targets is supported by the use of performance indicators. In flood management, performance indicators can be expressed in terms of the risk associated with the failure of the urban flood management system or its components now and in the future. Risk assessment depends on the collection of information and data concerning the urban flood management system. The information and data to be gathered can be related to four aspects involved in the asset management and performance of these systems:

- Hydraulic information gathering, focusing on the hydraulic problems and effects (e.g. the consequences, the duration, the return period and the nature of the events associated with the lack of capacity and flooding effects). This aspect usually includes the development and use of a computational hydraulic model of the urban flood management system. Also, records of past flooding incidents should be collated.
- Environmental information gathering, focusing on possible environmental impacts (e.g. arising from sewer overflows). This aspect includes modelling the sources and

the passage of pollutants through the urban flood management system and into surface receiving watercourses or groundwater.

- Structural information gathering, focusing on the parameters associated with structural deterioration. This aspect requires detailed inspection to determine the physical condition of the assets using a range of investigative techniques, and systematic recording of defects observed as well as records of structural failures.
- Operational information gathering, focusing on risks associated with system maintenance, failure of support systems or emergency response. This aspect includes recording operational problems such as incident reports (e.g. sewer blockages, odour and noise complaints and pumping station failures). Further collection of operational information can take the form of inspecting flood defences, sewer systems or other assets.

11.1.2 Risk assessment

This phase involves assessing the risk associated with the failure of the urban flood-risk management system or its components. Risk can be defined as the product of the likelihood of failure occurring and the consequence of that failure should it occur. To assess the risk it is necessary to establish the link between the failure mode and the failure effects. The failure effects include:

- hydraulic effects (e.g. flooding);
- environmental and societal effects (e.g. pollution, health, traffic disruption, costs and damage);
- operational effects (e.g. operational costs).

To determine the consequences of a failure it is necessary to assess the effects.

For each failure effect, it is possible to identify the majority of the problems that would cause the effect. Problems can be grouped into three failure modes:

- hydraulic failure of the asset (e.g. inadequate hydraulic capacity);
- structural failure of the asset (e.g. sewer collapse);
- operational failure of the asset (e.g. blockage of a pipe).

To determine the likelihood of a failure it is necessary to estimate the likelihood of occurrence of the underlying problems of that failure. For example, consider flooding from sewer systems. This may be caused by both extreme rainfall exceeding the hydraulic capacity of the sewer system and/or as a result of the partial or complete failure of an asset (e.g. blockage or collapse of a pipe). This implies that risk assessment should consider both the probability of the occurrence of extreme rainfall as well as asset failure, among other factors.

The performance risk of flood defence systems can be described using 'fragility', which enables the strength assessment to be separated from the hydraulic loading regime. Fragility functions provide a representation of flood defence performance over a range of hydraulic loading conditions. The fragility can be estimated using a range of methods from expert judgement through to reliability analysis.

11.1.3 Decision-making

The aim of decision-making is to develop interventions that optimise the trade-offs between improvements in performance, risk and the cost of interventions to deliver those improvements. It involves the formulation of and technical and economic analysis of the interventions needed to reduce the risks of failure effects. It is important to consider all types of interventions, including those regarded as operations and maintenance as well as capital investment solutions. The types of intervention can be divided into four groups based on the underlying problems of failure:

- Interventions to improve the hydraulic capacity of the urban flood-management system or to otherwise reduce the risk of flooding. These options are also likely to be applicable to the mitigation of environmental problems (e.g. problems caused by sewer overflows).
- Interventions to improve the environmental performance of the system (other than by improving hydraulic capacity of the system).
- Interventions to improve the structural condition of the system.
- Interventions to improve the operation of the system.

It may be possible to mitigate several risks in a single intervention by considering the interventions in an integrated way (e.g. by replacing a dyke that is in poor structural condition rather than renovating it in order to provide both an enhanced amenity and also to improve the structural integrity at the same time), and to avoid transferring risks from one part of the urban flood-risk management system to another.

The analysis process to generate and evaluate options should be based on the need to recognise and manage uncertainty and also use a 'proportionality' approach. Proportionality is the principle whereby the level of detail, data and other studies such as modelling, need to be matched to the scale of the problem and potential responses. This is because carrying out a fully detailed analysis of each option from the outset is likely to be very expensive. Only once the main candidate responses have been decided upon, i.e. those with the highest potential level of benefits compared with costs, should the most detailed level of analysis be considered. More details on this and the overall appraisal process are provided in the 'Flood and Coastal Erosion Risk Management Appraisal Guidance' available from the UK Government Department of Environment Food and Rural Affairs (Defra) website. This guidance is periodically updated.

The selection of interventions should be based on an appraisal that includes an economic and performance analysis. Economic analysis methods can be divided into two categories. The first category is cost–benefit analysis, which can provide a quantitative assessment of the relative economic costs and benefits of interventions, expressed in a monetary measure. The second category of economic analysis is cost-effectiveness analysis, which deals with effects that are not so easily quantified or for which there are no easily defined monetary values. In cost-effectiveness analysis, the effectiveness of each intervention is expressed in some standard unit, with interventions then compared by a procedure analogous to that used for a benefit–cost analysis. Effects can be 'scored' with some scaling device and then aggregated. For asset management, these weighted effects would then be related to the costs of implementing each intervention. The intervention with the highest score per unit of expenditure is then selected.

When analysing infrastructure interventions, it is imperative that life cycle costs be considered. Life cycle cost analysis can provide the 'effectiveness' measure in cost–effectiveness analysis, just as it provides the 'benefit' measure in cost–benefit analysis if monetary measures are used.

In most evaluations, there are also formalised requirements to demonstrate that environmental impacts are managed appropriately and also in some cases, an assessment of the 'sustainability' of the intervention is required. In the EU, flood-risk management and other major schemes have to be considered in terms of the Environmental Impact

Assessment (EIA) Directive. Most authorities also have guidance and requirements in terms of a full 'sustainability assessment' that includes, for example, an evaluation of the carbon footprint of the intervention.

A major difficulty in application that applies to asset flood-risk management is caused by the boundaries of the problem. These include the alignment of all the beneficiaries (some of whom may not be local and may be downstream or interested in the watercourse for recreational or other reasons not directly related to flood risk) with those who have to fund any scheme and are typically not the same group, although they may all be taxpayers. This is an example of the need to consider secondary objectives such as environmental or other enhancement as well as the primary objectives related to flood-risk management.

11.1.4 Implementation

This step involves the implementation in the urban flood-risk management system of the approved interventions and any monitoring of the future system performance to check that the interventions remain effective and appropriate. The implementation should not be seen as a one-off activity but as part of a process that will semi-continuously produce feedbacks and updates to the plan of management interventions in order to facilitate continuous adaptation as necessary. New information and data, technological developments, and new or modified drivers of change may all lead to the need to revise and update the plan.

11.2 ASSESSING RESILIENCE IN FLOOD-RISK MANAGEMENT

The adoption of a systems approach is necessary for the effective management of uncertain flood risks (see Section 11.1). This approach has to be, to a greater or lesser extent, based on the principle of managing for resiliency. Resiliency refers to the self-organising properties of a system that provide the ability to re-organise itself so as to maintain functionality. The consequential effect of resiliency on the functioning of the system is dependent on the two complementary properties of the same system ability: restorative and adaptive resilience.

It is argued that resiliency can be assessed by looking at the system's reaction to perturbations and change. Applications of resilience to flood-risk management have so far used the concept in a way that corresponds directly only to the aspect of restorative resilience. Indicators to assess restorative resilience are determined by the fast-changing variables in the system. Dutch studies have assessed the resilience in flood-risk management (of lowland rivers) by studying three parameters that describe the system's reaction to flood waves: amplitude, graduality and recovery rate.

- Amplitude. The amplitude of the reaction indicates the severity of the expected damage resulting from a certain extreme rainfall event or peak discharge. The expected annual damage and the expected annual number of affected persons are used to describe the magnitude of the reaction to the whole regime of flood waves by only one number.
- Graduality. The graduality of the reaction is large when the increase in the disturbance is in relation to the increase in damage. A low graduality is found for systems in which a small increase in the disturbance results in a large increase in damage. This happens for example when a river flow depth just exceeds the top of

an embankment or when an embankment suddenly fails and extensive inundation occurs. In systems with a low graduality sudden disasters are more likely.

- Recovery rate. The recovery rate describes how rapidly the system will be able to react and recover to the impacts of the flood wave. It is assessed based on a mainly qualitative approach using indicators for physical factors and economic and social factors which influence the recovery capacity.

The resilience of a system is larger when the amplitudes (i.e. amounts of damage) are smaller and where the graduality is larger or the recovery rate is higher.

The aspect of adaptive resilience requires additional indicators that are dominated by the slowly changing variables in the system. Studies in Australia have suggested that adaptive resilience can be represented by the adaptive performance space open to the system. A comprehensive assessment of adaptive resilience therefore requires indicators for all aspects that describe the adaptive performance space of the system: scope, timeliness/cost; and the dependency on external and internal conditions.

- Scope. The qualitative scope is the range of response options that are available and also feasible. The qualitative scope of the flooding system includes flood barriers, embankments, pipe or river enlargement, storage, disconnection and source control. The quantitative scope that is provided by the set of response options available corresponds to the maximum realisable capacity (i.e. when all available response options have also been implemented). For example, the maximum realisable capacity of the flooding system may be described in m^3/h from a hydraulic perspective.
- Timeliness/cost. These aspects refer to the time and cost required to reach the various performance conditions in the adaptive performance space. The timeliness of implementing response options corresponds to the time necessary for the specific measure to be implemented. The time varies considerably between the different possible structural and non-structural measures, e.g. from less than 1 to 2 years for

TEXTBOX 11-1

Applicability of resilience indicators: Assessment of restorative resilience in the Mekong River

The lowland river part of the Mekong is situated in Cambodia between Kratie and the Vietnamese Cambodian border. In this case study, the current resilience of the system was determined and the influence of three strategies was assessed: continuation of the current strategy, a resilience and a resistance strategy. In order to calculate the resilience indicators the current socio-economic system and its relationships with the flooding regime were studied, extreme value analyses and flood wave volume analyses were carried out, flood patterns were calculated with the help of a hydrodynamic model, and a damage model was developed to assess flood impacts. To be able to develop sensible strategies the developments in the region which are expected to occur in the future were studied as well.

Since floods and society are strongly related in the lowland part of the Mekong, the systems approach which is required in order to be able to apply resilience proved to be very useful here. Agriculture and fisheries, transport and the way people build their houses and construct roads are all adapted to the frequent floods. Agriculture is that much adapted that it has become dependent on the floods. The current resilience of the Mekong River was expected to be very high, because the whole country seems to be adapted to the annual floods. Although the graduality was found to be very high indeed, the amplitude proved to be high also and the recovery rate low. The high amplitude is caused by frequent severe flood damages and the low recovery rate by relative poverty. The resilience of the system is expected to change in the future. Population growth, socio-economic development and climate change are expected to cause an increase of both the amplitude and recovery rate.

When developing flood-risk management strategies, the complex relationships between agriculture, fisheries and floods must be considered. The Mekong case study included a resilience strategy based on flood forecasting, flood regulation and agricultural development. This strategy significantly increases resilience: the resulting amplitude is lower than when the current strategy is continued and the recovery rate is similar. Next to the current strategy and the resilience strategy, a resistance strategy was also studied. This resistance strategy consists of the construction of embankments which prevent discharges with a probability larger than 1/100 a year to cause floods. This strategy results in a decrease of the amplitude, graduality and recovery rate. The evaluation of the strategies showed that the resilience strategy scores well on socio-economic development, on natural and land scenery values and on coping with unexpected events. The resistance strategy, however, scores higher on reducing flood impacts. Because the resilience strategy can be implemented stepwise and differs less from the current strategy, it seems more feasible than the resistance strategy.

Source: De Bruijn, 2005.

some source control options, such as regular drain inlet unblocking, to 10 to 20 years for measures such as pipe enlargement or building underground storage. The cost of implementing a response option in part depends on its inherent capital, operation and maintenance costs and also on the cost of empowering system managers and other actors involved to implement that option at the right time. The latter involves a significant investment in the capacity of the main actors and in continuously monitoring the system and its wider context and may also involve significant engagement with communities or property occupiers to build their capacity to respond.

- External (contingency) conditions. The achievement of the adaptive performance possibilities can, to a greater or lesser extent, depend on the existence of particular external conditions that are necessary for those possibilities to be realised. For example, the ability to implement certain source control options ultimately depends on the social and institutional acceptance and manageability of these measures. The associated external contingency conditions needed to enable the full adaptive performance space to be realisable can be referred to as the full external (contingency) conditions. The proportion of the external contingency conditions that is then satisfied determines the access available to the adaptive performance space.
- Internal (coordination) conditions. The internal coordination includes the capacities both to manage flood probabilities (and consequences) at each point in the adaptive performance space and to shift the evolution of the flooding system from one point to another. As with the external conditions, the range of internal coordination conditions that needs to be successfully managed to enable the full adaptive performance space to be realisable can be referred to as the full internal (coordination) conditions. The proportion of the internal coordination conditions that is then satisfied determines the access available to the adaptive performance space.

Scope and timeliness characterise the adaptive performance space inherent in the flooding system. It comprises those adaptive performance conditions that are technically and physically possible given the response options available. But the implementation of the response options may be facilitated or hindered by external and internal conditions, which also should be addressed. There a number of important constraints given by the socio-economic sub-system that relate to the accessibility of the adaptive performance space:

- awareness and concern about the potential impacts of changes in drivers for flood risk and other factors;
- belief that adaptation can be beneficial;
- susceptibility of the individual or group to being able to change;
- resources and capabilities of the individual or organisation to change;
- regulatory and market context within which the individual or group resides.

The adaptive resilience is increased when the adaptive performance scope is extended, when the rate at which it is possible to move from one point in the total space to another is increased, or when the achievement of performance possibilities is made less dependent upon external conditions.

11.3 TRANSITIONING FROM ENTRAPMENT TO RESILIENCE APPROACHES

The need for fundamental changes in the way in which flood risks are managed compared with historical practice is apparent if future climate and other changes are to be addressed. The 'non-stationarity' of future environmental drivers and lower confidence in the ability to predict and therefore plan for, flooding events, makes a change from the traditional approaches using large infrastructure solutions imperative. A major challenge in effecting this change in approach is the significant inertia caused by the legacy of existing flood defences and also a mindset by policymakers, professionals and others, that these are the only and sufficient solutions available. The historical traditions of large embankments and similar infrastructure are still being replicated as exemplified by the Dutch Delta Commission, which proposes building more coastal embankments to cope with rising sea levels and more intense storms over the coming decades. This reliance and belief in large technological solutions is known as 'technological entrapment', and whilst new embankments may be appropriate in certain circumstances, a reliance on these and other 'big' solutions exclusively risks a loss of flexibility, adaptability and ultimately sustainability in flood-risk management.

A major challenge is therefore to change current custom and practice to a more open, innovative and less tried-and-tested approach that entertains a wide range and diversity of options within a portfolio of 'no-regret' responses to changing flood risk. This is the approach advocated in the UK's Foresight Future Flooding studies carried out in 2004 and 2008. This change in custom and practice away from: 'we always do it this way because we know it works' to one that is more flexible and adaptable, requires a 'transition' in practice and also in professional and policymaker culture. It will also require a transition in the way in which society as a whole expects to be protected from risks in the future.

11.3.1 Transition theory

A transition is a structural change in the way a societal system (e.g. water management, energy supply and agriculture) operates, and can be described as a long-term process (25–50 years) that results from a co-evolution of cultural, institutional, economic, ecological and technological processes and developments on various scale levels. A transition can be described with the multi-stage concept and multi-level concept.

11.3.2 Multi-stage theory

Figure 11-2 shows the S-shaped curve of the multi-stage concept. The curve shows a change in society over time. At all times, the societal system is in a certain state and moves in a certain direction. It is only possible to give a rough indication of the state and direction of the system in the present time for a society in transition. Moreover, it is impossible to predict the state of the system after the transition has been completed. For example, the transition in urban flood management is directed towards more resilience, but it cannot be known 'a priori' what this more resilient system of the future will look like. However, transition theory is a useful tool to describe changes in society in the past. Knowledge about the state of the transition and the dynamics that are taking place in the multi-level concept (see below) can be used to accelerate the process of transitions by adapting to those developments and lessons from the past.

The transition is divided into four stages: pre-development, take-off, acceleration and stabilisation. During the pre-development stage, there is a stable situation that does not visibly change on the surface. Slow social change or events (crises) can cause the system to reach its threshold and suddenly start to shift in the take-off stage. This sudden shift is highly nonlinear and unpredictable. In the acceleration stage, socio-cultural,

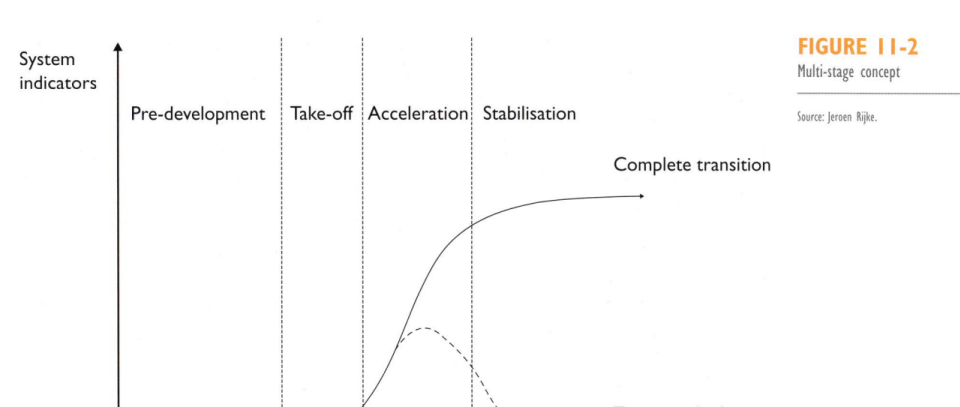

FIGURE 11-2
Multi-stage concept

Source: Jeroen Rijke.

economic, ecological and institutional changes reinforce each other. This is a period of much instability and uncertainty in the system. If the changes are a success, a new stable situation will occur in the stabilisation stage, where innovative processes and technologies have become mainstream. If not, the system could face a drawback that could result in a so-called lock-in.

11.3.3 Multi-level theory

Figure 11-3 shows the multi-level concept, which distinguishes three scale levels in society: the macro-level, the meso-level and the micro-level. At the macro-level, societal landscape is determined by relatively slow trends in economy, demographics, natural environment, climate, culture, politics and worldviews. At the meso-level, regimes are dominant patterns of artefacts, institutions, rules and norms that are being assembled to perform economic and social activities. The regime is often installed to preserve the status quo and to encourage optimisation, protecting investments rather than system innovations. At the micro-level, individual persons, organisations, projects and technologies are distinguished and innovation is apparent.

11.3.4 Combining the multi-stage and multi-level theories

The multi-stage concept can be integrated with the multi-level concept. In the pre-development stage, the regime, which is installed to protect the status quo, is often the inhibiting factor for change. Typically, it will seek to maintain social norms and belief systems and to improve existing technologies. This can be beneficial, for example in the maintenance of public health and hygiene by the promotion of sanitation systems. The 'take-off' stage of any innovation is often initiated by developments at the macro-level that are sometimes reinforced with developments at the micro-level. In the

FIGURE 11-3
Multi-level concept (based on Geels and Kemp, 2000)

Source: Geels and Kemp, 2000.

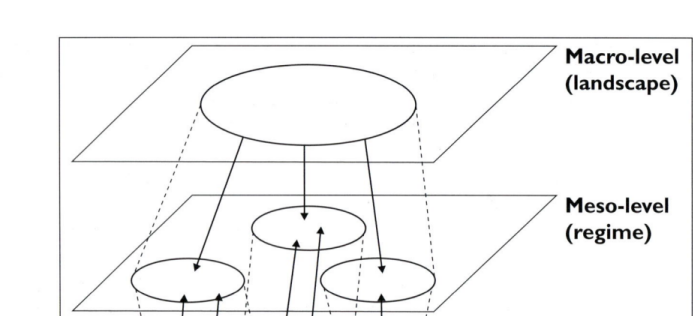

take-off stage, different ideas and perspectives converge into one consistent paradigm that causes an increasingly changing regime that modulates with innovative experiments at the micro-level. This is a highly uncertain period that can possibly result in a lock-in to particular technologies or practices if the regime is not being pushed over the 'edge'. If the result of the take-off stage is a changing regime, the regime enables acceleration through the application of large amounts of capital, technology and knowledge. Dominant practices change rapidly and irreversibly through reinforcements by the three different levels. In the stabilisation stage, a new regime slows acceleration down while resisting new developments. Processes and techniques that were innovative at the beginning of the transition process have become mainstream by now in the cycle. Thus, a new societal dynamic equilibrium is reached that could be the beginning of a new transition cycle.

In the flood-risk management sphere, there has historically been shifts in regime in relation to the way in which rural agricultural land and management practices have been modified to better manage flood risks in downstream urban conurbations.

Practices in many countries have alternated between policies that promote the need to maintain food production by protecting agricultural land from flooding and policies that use flooding of agricultural land as a means of flood relief for downstream areas. These shifts in priority occur largely in response to ambient economic conditions and the national priority given to the security of food supplies.

11.3.5 Transitioning from entrapment to resilient approaches

Traditionally, flood-risk management (FRM) has been dealt with by 'silo thinking' methodologies with the various constituent organisations responsible for different elements of the flooding system being interested only in their defined part of the system and even where there has been some integration, any wider linkages with other urban systems, such as transport, have been largely lacking. Even within organisations, communication and cooperation amongst colleagues of differing disciplines is often lacking. For example, engineers within a municipality dealing with land drainage often report a lack of cooperation with the engineers responsible for roads within the same organisation. Advances in technology now mean that communication between and within organisations is no longer the barrier that it was—although institutional barriers and rivalries to more effective inter-organisational communication still remain. Furthermore, drivers such as climate change, believed to be responsible for increasingly uncertain future global and local impacts, now necessitate timely involvement and effective communication among and within organisations in order to provide solutions that are most able to cope with global (and local) changes—the magnitude of which, as yet, can only largely be guessed at. Such responses need to be resilient (i.e. maintain sustainable urban areas and systems) and usually comprise approaches that are adaptable and flexible enough to cope with and respond to advancing knowledge about global (and local) change impacts and about the effectiveness of responses.

As outlined above, the majority of barriers to such change are rooted in traditional thinking approaches and some commentators call this the 'entrapment effect'. Entrapment concepts define how large and technological systems may become embedded in decision-making pathways which, although perhaps potentially reversible (a key element of adaptability), are not simple to modify. Often such decision pathways run counterproductive to the needs of the system. The framework in Figure 11-4 shows the stages in which large socio-technical systems, such as that used traditionally for FRM, have to pass through in order to reach the stage of sustainable flood-risk management. This framework has emerged through the study of the effectiveness and efficiency of Non-Structural Responses (NSR) to flood management, a case study funded by the Scottish Government and prompted by severe flooding in east Glasgow in 2002.

The vertical boxes depict apparent stages in the process of transition. The boxes above these represent the state of the drivers to which the stages are a response.

11.3.6 Entrapment characteristics

Box 1 in Figure 11-4 depicts the entrapment stage. The drivers which dictate the responses have been relatively fixed (in the context of climate change), and in response have led to responses comprising mainly traditional solutions, which have a mainly structural character. Mechanisms of and benefits from public involvement may not be understood and barriers to any innovation come from the 'we know it works' school of thought where tried and tested experience is relied upon. The actors in this phase tend to be the professionals; solutions may be imposed rather than being part of a consultation process to identify the most appropriate response. Even in this stage, drivers emerge which appear to indicate that alternatives to the traditional approach are appropriate. For example, it is now clear that the supposed fixed drivers are actually non-stationary and system dynamics due to climate change and other factors such as urbanisation and demography render this static approach extremely risky.

11.3.7 Transition characteristics

Box 2 in Figure 11-4 depicts the intermediary stage. The key to initialising this transition stage is capacity building; removal of barriers associated with traditional thinking and a climate of cultivation of ideas. So-called champions are often the driving forces behind transitions that raise awareness and build capacity in their own organisations and amongst the wider community. However, champions rely heavily on the support

FIGURE 11-4

From entrapment to resilience

FLAGS: Flood Liaison Advisory GroupS—comprising only professional actors

Source: Richard Ashley.

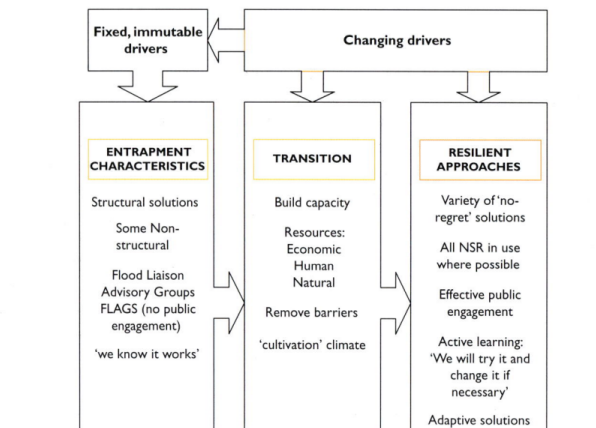

and trust within their organisations and the wider community. Pilot projects are often another stimulant for transitions, because they act as showcases for newly developed technologies. Because of the local character of most pilot projects, transitions benefit by a 'bottom-up' approach to governance. Unfortunately, economic constraints by governments rarely allow sufficient freedom from 'top-down' direction to facilitate the generation of innovative ideas by bottom-up initiatives.

11.3.8 Characteristics of resilient approaches

Box 3 in Figure 11-4 depicts the final stage in the transition process. In addition to the existing traditional approaches, non-structural response alternatives are a fundamental element. An attitude of 'no regret' and 'active learning' must be adopted where solutions are selected according to the best available information to suit current conditions. As with Box 2 in Figure 11-4, the drivers themselves are constantly changing as a result of (largely) unknowns such as climate change. The 'actors' in this phase are more likely to include everyone in the process.

Key Questions

Assessing resilience in flood-risk management

1. What is the difference between restorative and adaptive resilience?
2. Flood risk can be assessed in terms of expected annual damages. Why is an assessment of graduality, next to expected annual damages, required for characterising restorative resilience in flood-risk management?
3. What is the advantage of assessing external and internal conditions within the characterisation of adaptive resilience?

Transitioning from entrapment

1. What is a transition in the context of regimes for flood-risk management?
2. What is meant by the 'entrapment effect' and what are the key novel aspects of adopting a resilient approach?
3. What are the critical stages through which large socio-technical systems have to pass in order to reach the stage of sustainable flood-risk management?

FURTHER READING

- Bruijn, K. de (2005). *Resilience and flood risk management. A systems approach applied to lowland rivers.* IOS Press. ISBN: 978-90-407-2600-2.
- Burby, R.J., Deyle, R.E., Godschalk, D.R. & Olshansky, R.B. (2000). Creating hazard resilient communities through land-use planning, *Natural Hazards Review, 1(2)*, 99–106.
- Cashman, A. & Ashley, R.M. (2008). Costing the long-term demand for water sector infrastructure, *Foresight, 10(3)*, 9–26. ISSN 1463-6689. Emerald Group publishing.
- Edelenbos, J. (2005). Institutional implication of interactive governance: Insights from Dutch practice. *Governance, 18(1)*, 111–134.
- Folke, C. *et al.* (2002). *Resilience and Sustainable Development: Building Adaptive Capacity in a World of Transformations.* Scientific background paper on resilience for the process of the World Summit on Sustainable Development on behalf of The Environmental Advisory Council to the Swedish Government. Sweden: Ministry of the Environment. Accessed Advisory Council to the Swedish. Available from: http://www.unisdr.org/eng/risk-reduction/wssd/resilience-sd.pdf.
- Folke, C., Hahn, T., Olsson, P. & Norberg, J. (2005). Adaptive governance of social-ecological knowledge, *Annual Review of Environment and Resources, 30*, 441–473.
- Gallopin, G.C., Funtowicz, S., O'Connor, M. & Ravetz, J. (2001). Science for the twenty-first century: from social contract to the scientific core, *International Journal of Social Science, 168(2)*, 219–229.
- Mostert, E. (2006). Integrated water resources management in the Netherlands: How concepts function. *Journal of Contemporary Water Research and Education, 135*, 19–27.
- Rees W.E. (1996). Revisiting carrying capacity: Area-based indicators of sustainability. *Population and Environment, 17*, 195–215.
- Thorne, C.R., Evans, E.P. & Penning-Rowsell, E. (eds) (2007). *Future Flooding and Coastal Erosion Risks.* London, UK: Thomas Telford. ISBN 978-0-7277-3449-5.
- WRc plc. (2010). *Sewerage Rehabilitation Manual* 4th Edition. Available from: http://srm.wrcplc.co.uk.

Capacity building and governance

Learning outcomes

In this chapter, you will learn about the following:

- Perceptions of risk play a dominant role in the decisions people make.
- The perception of risk is a function of awareness, worry, and preparedness.
- People are willing to accept flood risk when they perceive themselves to possess sufficient control over the risk. However, the perception of 'sufficient control' is inherently subjective and has to be based on sound information, assessment, evaluation and decision-making.

12.1 RISK PERCEPTION, ACCEPTANCE AND COMMUNICATION

12.1.1 General

To understand more about why people accept certain risks and reject others it is necessary first to define what risk acceptance is. Risk acceptance generally depends on how people perceive the risks to which they assume they are exposed. In other words, risk perception is characterised as the intuitive judgement of individuals and groups of risks in the context of limited and uncertain information. People tend to make quantitative judgements about the current and the desired level of regulation for each type of hazard. Flood risk is typically assumed to be objectively quantifiable by flood risk assessment. In Chapter 6 it was explained how an objective determination is in fact not possible due to what will always be incomplete knowledge and other uncertainties arising from the intrinsic randomness characteristic of the flooding system.

Multiple mechanisms contribute to risk perception. A workable approach to study risk perception is based upon the relationship between awareness, worry, and preparedness as shown in Figure 12-1.

Flood-risk awareness can be defined as knowledge or consciousness of the flood risk (see Chapters 7 and 12) to which an individual or a group of individuals is exposed. Awareness is a precondition for worry and may in turn result in higher preparedness. Worry also results from the lack of preparedness for this risk. In turn, worry can lead to actions taken by individuals or groups to reduce the risk. A high level of worry

FIGURE 12-1
Four types of risk characteristics
(after Raaijmakers et al., 2008)

Source: Ruud Raaijmakers, 2008.

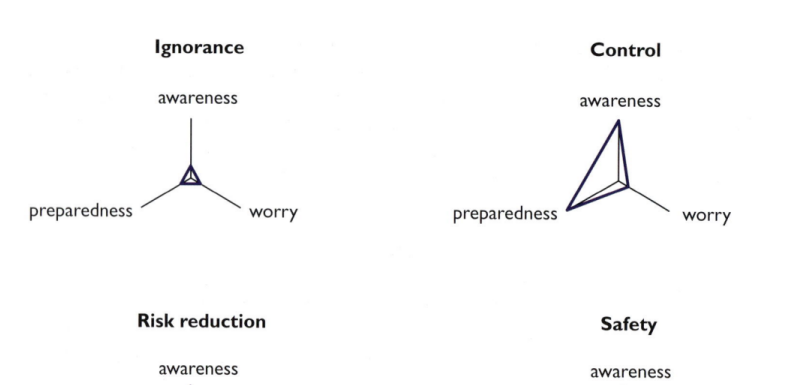

concerning a flood risk is often a prerequisite for having the societal commitment to invest in risk-reducing measures. Awareness and preparedness can be directly influenced by public policy. Preparedness is both the capability of coping with a flood during the flooding (anticipation or pre-flood) and the capability to recover from the impacts of the flooding (post-flood).

Building on the combination of awareness, worry and preparedness, four types of risk characteristics can be distinguished:

- Ignorance—an ignorant individual will not worry about and will not be prepared for the risk.
- Safety—an individual who imagines him or herself to be safe will not worry and is thus not prepared for a risk, because the risk is acceptable, small or believed to be small.
- Risk reduction—an individual who is highly aware, worries and badly prepared will demand risk reduction. When an individual considers exposure to a hazard as involuntary, he or she will assume responsibility for preparing.
- Control—when an individual feels prepared, then he or she has a sense of control over the risk and is, as a consequence, less worried.

12.1.1.1 Different meanings of risk

The term 'risk' has different connotations which stem from the multiple conceptions of risk. It is used to refer to a hazard (which risk should we rank?), as a probability (what is the risk of getting swine flu spread through air?) and as a consequence (what is the risk of letting your parking meter ticket expire?). This notion illustrates the complexity of the risk concept itself and explains in part why the acceptance of flood risk is perceived quite differently by the public and even by experts. Perceptions of risk play an important role in the decisions people make which highlights the relevance of this issue.

12.1.1.2 Voluntary and involuntary flood risk

An important distinction is often made between voluntary and involuntary risks due to the characteristics of risk. This is defined as the freedom of choice a person has to expose him or herself to a particular risk. The freedom of choice is highest if there are a large number of alternatives with similar perceived benefits. When an individual voluntarily decides to expose him or herself to a risk the individual is assumed to have the option of avoiding the risk. Often natural risks are characterised as involuntary. Depending on the situation, some flood risks are considered/supposed to be involuntary and others voluntary. Flood risks associated with low-probability, high consequence events are generally supposed to be involuntary. The same holds true for pluvial flood events which may happen anywhere. Typical examples of voluntary flood-risk situations are developments in floodplains and other evidently risky areas such as low-lying coastal areas and areas designated for water retention. Here, knowledge and awareness are important, as often the occupiers of properties in floodplains are unaware of the risks unless flooding has occurred there in recent history. For example, strenuous efforts are being made to ensure that occupiers of properties in the floodplains behind the dykes in the Loire river basin in France are aware that were the dykes to fail or the river overtop them, flooding would be extensive and catastrophic. As no major flooding has occurred in the Loire basin for more than a century, inhabitants have forgotten that they might be at risk during extreme events. So for this example, the definition of 'voluntary' or 'involuntary' as an assumption about attitudes to risk are unclear.

TEXTBOX 12-1

Concerns about flood management in the Red River Valley

One of the most important problems in flood protection, and one that is a common occurrence around the world, is the short memory of government officials that are responsible for allocating funds for maintenance and upgrading flood protection measures. It has been the typical experience in Manitoba (Canada) that drought years dim the enthusiasm of officials and lead to erosion of funding to maintain vigilance against the inevitable recurrence of flood years.

This problem has been offset to some extent by establishment of legislation that mandates adequate funding for critical protection infrastructure. However, there is still a problem with local governments that are reluctant to devote adequate funding and initiatives to flood emergency, preparedness, and planning.

Source: R.W. Carson (2006)

FIGURE 12-2
The Red River Flood of 1997 occurred in April and May 1997, along the Red River of the North in North Dakota, Minnesota, and Southern Manitoba. Flooding in Manitoba resulted in over $500 million in damages, although the Red River Floodway, an artificial waterway, saved Winnipeg from flooding

Source: Bill Kamphuis, 1997.

TEXTBOX 12-2
Flood preparedness in the Netherlands

Risk communication could be a suitable means of achieving certain public policy goals related to flood preparedness in the Netherlands, as the ultimate purpose of risk communication is to inform, persuade and consult in order to enhance knowledge, change attitudes and behaviour and provide effective conditions for dialogue and conflict resolution. However, in order to be effective, risk communication should not merely explain the characteristics of flood risk in people's local environments; risk communication should also explain what measures people can take in order to cope with the risk (i.e. flood preparedness actions).

A theoretical framework that can be used to develop risk communication is the Protective Action Decision Model (PADM). In particular, the PADM provides a social–psychological perspective on how people decide whether or not to prepare for disasters. The core of the model consists of five successive questions that people typically ask themselves when preparing for disasters:

1. Is there a real threat that I need to pay attention to?
2. Do I need to take protective action?
3. What can be done to achieve protection?
4. What is the best method of protection?
5. Does protective action need to be taken now?

It is possible to gain insight into this decision-making process by investigating the psychological variables related to each of these five questions. In a recent study, the PADM was applied to flood preparedness in the Netherlands using survey questionnaires in a number of risk areas along the Dutch coast and in the pluvial risk areas. The questions attempted to assess the psychological variables used to gain insight into people's protective action in relation to decision making. From the responses of 3,559 Dutch citizens, it is clear that more than 90 per cent has done nothing to prepare for potential floods. Moreover, few citizens intend to prepare for floods in the near future. Thus, the large majority failed to give an affirmative answer to the question 'Does protective action need to be taken now?' The first question is: why? The answer to this question is threefold. First and foremost, many Dutch citizens fail to identify flood risk as a threat to their safety. They have great confidence in the flood defences. As a result, they perceive a low flood probability and are unconcerned about flood risk. Second, although Dutch citizens generally believe that they will suffer damage if there is a flood, they regard the government as primarily responsible for taking care of flood damage. Third, citizens have little faith in protective actions such as sand bags to protect their property against flood damage. However, emergency information such as evacuation routes and safe shelter locations in their neighbourhood are regarded as more effective flood preparedness actions. The second question is: how can Dutch flood-risk authorities use communication to stimulate flood preparedness among citizens? A first prerequisite for developing effective risk communication is therefore to emphasise that the flood defences can never completely prevent floods. Second, risk communication should explain to people that the government has only a limited ability to help people in case of flood disasters. People, therefore, are personally responsible to prepare themselves for floods. Third, communication should explain that the risk authorities are there to help them in preparing for floods. It is of utmost importance that local authorities explain the local consequences of floods, and advise people about the flood preparedness actions that they can obtain in order to prepare for these consequences.

FIGURE 12-3
Risk communication should also explain what measures people can take in order to cope with the risk

Source: Ellen Brandenburg, 2005.

12.2 ADAPTIVE CAPACITY

Adaptive capacity should be seen as the ability of a system to adapt to change, an essential component of developing resilient systems. Managing urban flood risk requires thinking about whole systems. With regard to flood-risk management, the system can be considered as three interlinked systems: ecological (broadly, the characteristics of hydrology, land and ecosystems within the urban area and the wider territory in which it is situated); technical (the infrastructure developed by people to manage flood risk); and social (the people who make decisions, use and manage land and infrastructure, and who are affected by flooding).

Flood-risk management has traditionally tended to focus on the technical solutions to flood risk. More recently, the importance of integrating ecological aspects into flood-risk management, from the local to whole catchment scale, has been recognised. The need to consider ecological approaches along with 'hard' technical solutions has not only been recognised but also encouraged by shifts in policy, such as the European Union Water Framework Directive. This section focuses on the social aspects of flood resilience, often the most neglected aspect in traditional flood-risk management, but increasingly seen as essential for developing effective approaches.

12.2.1 Social component of adaptive capacity

The social aspects of the system need to be considered in planning and design (before flooding events), response to flooding (what happens during the event) and in redevelopment and reconstruction (what happens after the event). An adaptive approach implies that the aim is to maintain resilience of the overall system not necessarily of every component and part of the system. It may also imply changes to the system to better cope with stresses in the future. Aspects of the system may well be designed slightly differently or function differently in the reconstruction after a flood, implying learning from the flood. Such a response was not evident in the media furore following the flooding and damage from Hurricane Katrina. The clarion calls to rebuild New Orleans exactly as it was before the disaster represented a missed opportunity to rebuild

FIGURE 12-4
Effective implementation of UFM approaches is encouraged when people are involved in planning; there is more buy-in, this leads to greater acceptance and uptake

Source: Richard Ashley, 2007.

the urban area to make it more resilient to flooding in the future. In other words, adaptive capacity may well require the system to be adapted to new circumstances, with the aim of enhancing long-term resilience.

There are several reasons why it is important to consider the social aspects of adaptive capacity in urban flood-risk management. These are summarised below:

- The need for appropriate solutions, that are more likely to work in a particular place, culture and context; appropriate solutions require local knowledge to be incorporated into flood resilience measures.
- The need for integrated approaches and innovative solutions that take into account different perspectives; more professions and disciplines need to be involved in developing flood resilient solutions.
- Effective implementation is encouraged when people are involved in planning; there is more buy-in, this leads to greater acceptance and uptake—in particular by homeowners, Local Authorities and the insurance industry.
- New systems are more likely to be used and maintained—building resilience to flood-risk management requires new practices and behaviours, in other words increased adaptive capacity amongst a wide range of social actors; this requires careful consideration of how to encourage different attitudes to risk and responsibility, so that more people are actively involved in flood resilient measures.
- Synergistic benefits—considering the wider social systems within which flood management sits provides an opportunity to consider what other infrastructure changes can enhance flood resilience? This helps maximise the value for flood resilience of investments and changes to infrastructure, which may not be carried out with the express purpose of flood-risk management, and reduces missed opportunities and unintended consequences from changes in other parts of the system, such as in transport infrastructure.

Involving community members and other stakeholders in managing and adapting to flood risk helps to develop effective solutions. Often there are legislative requirements to involve stakeholders in planning decisions. A challenge for flood-risk practitioners is to make the most of such engagement to enhance awareness of ways to build flood resilience.

Communication is essential to build resilient systems and enhance adaptive capacity, but ways of encouraging effective communication are often poorly understood. Handing out brochures and making announcements may provide information but such one-way flows by no means ensure that effective communication has taken place. Professionals involved in flood-risk management need to develop new skills and abilities, so they can interact with community members and stakeholders and develop more integrated systems. There is a need to learn to learn, and to think beyond the boundaries of disciplines.

12.3 CHARACTERISTICS OF EFFECTIVE LEARNING INITIATIVES

Building adaptive capacity requires effective learning approaches. Such learning needs to promote the taking of action, and to take cognitive and emotional dimensions into account. This is clearly an important issue for flood-risk management, which can be a highly stressful and emotive topic for those affected or threatened by flooding.

Four main characteristics of effective learning initiatives for system-wide learning have been identified as outlined below.

12.3.1 Raising awareness

Building resilience will require a change in mental attitudes and behaviour. Good plans and new practices are of limited value if they are not actually taken up and applied. This requires a shift in the locus of responsibility from 'others' being responsible for managing flood risk to awareness that a flood resilient city will require very many actors to be responsible for coping with the risk.

In the change management literature, a sense of 'urgency' is considered an important pre-requisite for change. Learning theories suggest that there needs to be an element of perturbation or discomfort to jog people out of their established thought patterns. A common difficulty encountered in implementing new approaches to flood-risk management is that if people have not recently experienced a flood in their area, they are often reluctant to make investments or put time and energy into change.

Technical knowledge that flood-risk managers take as widely understood can be hard for community members or members of other professions to understand. To communicate such knowledge effectively, there is a need for graphic examples, practical demonstrations and innovative ways of making technical knowledge available in layperson's terms. Flood-risk professionals need to think how to make information more available and accessible.

Raising awareness is not just concerned with understanding risk. Many actors who are not directly involved in flood-risk management, including planners, architects and civil engineers, are not necessarily aware of the options available to enhance flood resilience. Ideas which seem simple and may be taken-for-granted by the flood-risk professional may be unknown to laypeople or practitioners in other professions. It is important to think about how to raise awareness of the options for enhancing resilience. This can include making resilience methods more visible in the urban fabric, especially in demonstration projects. Signs and information boards can help increase the value of such projects as educational tools.

Knowledge of the facts alone is seldom enough to encourage people to make changes. If communication is focused mainly on the negative aspects of flooding, it can even inhibit people from feeling that they can make a difference. This is exemplified by the recent psychological overload of populations by constant bombardment from all kinds of media exhorting lifestyle changes to cope with climate change. Many people hold the view that the problem is so big that they are in fact powerless to do anything about it and hence ignore these messages. A further characteristic to enable change is taking an appreciative, positive approach to learning.

12.3.2 Appreciative

For community members, flooding is only one aspect of what happens in their lives and neighbourhoods, they don't see the world in the same way as people whose job it is to manage flood risk. It is important to start any communication process with questions aiming to understand people's own perceptions of their areas, risks and their own coping strategies.

12.3 CHARACTERISTICS OF EFFECTIVE LEARNING INITIATIVES

Exploring and reinforcing the positive aspects of a system helps generate greater energy and enthusiasm for change. Such a focus on the positive is emphasised in asset-based development in community planning, where it is seen that starting with the positive aspects of a community enables people to build confidence and commitment to future change. An appreciative approach implies that the skills, knowledge and possible contribution of all people have a potential role to play and should be valued.

Flood-risk professionals need to learn to appreciate the views of others, even when the other actors are not technically competent. In addition to needing to learn from other professionals involved in the built environment, such as planners and architects, about possible ways to build resilience, local people have important ideas and knowledge about the area that can prove important information in designing and maintaining resilient systems. There is a need to listen to people's stories and to build on the successes of what people are already doing and what they know about their area. This helps develop solutions that consider how new measures can best fit into existing systems.

Urban flooding has particularly profound impacts on the vulnerable, especially the elderly, disabled and poorest people in a society, who are the least able to cope with the negative effects of flooding. It could be that flood-risk management professionals have a lot to learn about low-cost solutions and ideas to increase resilience from such people who are living in high-risk areas. It is also important to ensure that the solutions proposed are appropriate to the circumstances of the people they are supposed to be helping; especially with vulnerable populations.

It is important to be aware of people's attitudes and feelings, and to take these into account in planning and communication. For example, important social barriers to building-level flood resilience measures were highlighted in the Heywood case study in Chapter 2, Textbox 2-2 One barrier was the fear of adding flood-risk resilience measures to their houses, as this could lower property values. It was seen

FIGURE 12-5
Gathering local knowledge about Lower Ninth Ward, New Orleans, Spring 2010

Source: Chris Zevenbergen, 2009.

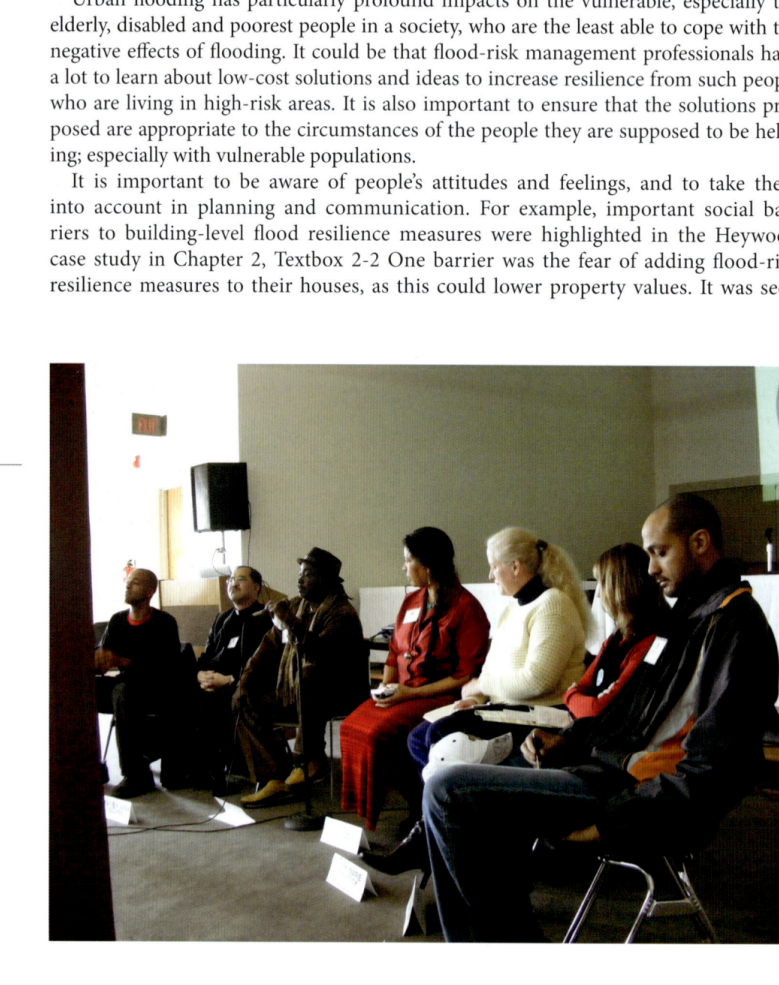

that removable flood barriers could make it obvious when people weren't in their houses (if they needed to be put in place when the occupiers were absent), and therefore make it more likely for the homes to be burgled when people were away. Such local perceptions justifiably impede take-up of flood resilience measures, no matter how technically effective they are. It is important for professionals to learn about and appreciate people's concerns if they are to develop effective strategies for flood-risk resilience. Developing an attitude of listening and taking people's concerns seriously can help build trust between actors, and this can be especially important when dealing with emergency situations.

12.3.3 Action-led

Developing flood resilient practices implies moving towards the unknown; this requires a constant process of learning in attempt to 'manage' such systems. Action-led learning starts with practical action, and invites people to become more reflective, and thus more aware of the intentions and consequences of their day-to-day practice.

The roots of an action-led approach lie in Kolb's (1984) cycle of learning, which starts with experience, followed by reflection; new ideas are assimilated into theory, which is then further tested in experience. Actively engaging people in thinking about the future of their area builds skills in all participants, thus broadening the possibilities for creative solutions and supporting the practice of ongoing adaptive management. An action-led learning approach will be essential if we are going to move to new behaviours, such as active responses to enhancing flood resilience.

12.3.4 Associative

Learning does not only happen at the individual level. Concepts and ideas are shaped by interaction between people and through cultural norms and language. Learning approaches that connect people together, and create opportunities for people to learn together are essential to develop flood resilient systems.

Recent work on developing flood resilience has highlighted the need for local champions to mobilise community action, as well as the need to link into existing community networks if new approaches to flood-risk management are to be successfully implemented.

In order to develop integrated solutions, it is important to ask, 'what may need to be resilient to flooding?' Elements that need to be made resilient include: individual homes; institutional buildings such as schools and hospitals; transport and energy infrastructure and services; hazardous sites (such as chemical works); people's livelihoods and the resources they need to carry them out; communications equipment; and essential infrastructure for e.g. water supply. It is essential that there is learning and communication amongst a wide range of practitioners and the people responsible for managing these systems in order to develop resilience in such systems.

Social learning between different groups is especially important, as flood-resilient systems need to operate across several levels of scale. It is important to think of associative links, creating both horizontal and vertical networks, so that people can learn from others in their team, people in similar roles outside of their usual networks, and from people working at different levels in the urban system. Such vertical and horizontal links have several important outcomes; they enable ideas to be shared more effectively, novel ideas to emerge and different perspectives to be developed, encourage the spread

FIGURE 12-6
Social learning encouraged with hands-on tools (www.ketso.com)

Source: Ketso (UK), 2009.

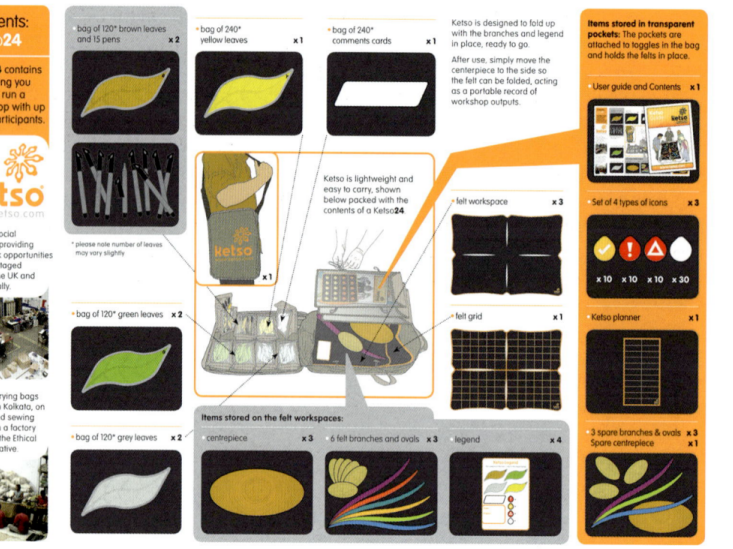

of innovation and can enable people to access the resources needed for effective implementation. These can be seen as 'shadow networks', which enable new forms of organisation and ways of working to be developed and tried before becoming institutionalised in formal structures.

A key lesson is that flood-risk professionals do not have to do this work on their own. There are many other activities going on to engage citizens in planning for their future, for instance in health, land-use planning, landscape management, educational and other emergency services. One question to ask is, how can you link into existing networks and activities in the area you are working in, and use these to engage with people in developing flood resilience? Such engagement may also provide opportunities for integrating flood resilient measures into other programmes and activities, such as infrastructure development or the regeneration of existing neighbourhoods, where flood-risk management is not necessarily the main goal of the project.

12.3.5 New capacities for flood-risk professionals

Participatory approaches build 'ownership' and a collective sense of responsibility. It doesn't really matter how technically brilliant ideas for flood resilience are, if people are not willing to champion them and see that they are implemented. Integrated methods to address flood risk incorporate an array of technologies, systems and tools relevant to both new-build and retrofit. Such integrated approaches require new strategic thinking about multi-functional approaches to flood-risk management. In turn, this needs better communication between many different professional groups and new design and planning processes to encourage innovation. Such communication helps develop intervention mechanisms appropriate to individual urban areas, cognisant of localised restraints and conditions.

As can be seen from the discussion above, flood-risk professionals need to learn ways to interact with different stakeholders, to ask questions and to link people and knowledge together effectively. An interest in engaging with different stakeholders can be pragmatic; it is easier to get things done and to encourage new behaviours and the take up of new practices. On a more fundamental level, however, adaptive capacity implies developing our ability to learn, to communicate and to foster a culture of innovation both in our own practice and in wider groups of people.

Shifts in flood-risk management from rigid, hard engineering approaches to adaptable, flexible approaches require a concomitant development of the social side of adaptive capacity. We will need adaptive institutions, professional networks and informal community networks in addition to adaptive infrastructure in order to build resilient urban areas.

Key Questions

1. What skills beyond technical know-how will be required for flood-risk management professionals to develop adaptive capacity?
2. What other professions and organisations may have important roles to play in developing flood-resilient urban areas?
3. What might be the advantages and challenges for flood-risk management professionals of engaging more closely with other professions involved in the built environment?
4. What might be the advantages and challenges for flood-risk management professionals of engaging more closely with community members?
5. For a particular flooding problem that you have been working on, think of a simple way that you could present this to an audience of laypeople in only five minutes. How would you simplify the technical details so that they make sense, whilst still getting across the important points?

FURTHER READING

Brown, R.R. & Farrelly, M.A. (2007). *Advancing the Adoption of Urban Stormwater Quality Management in Australia: A Survey of Stakeholder Perceptions of Institutional Drivers and Barriers*. Report No. 07/06, National Urban Water Governance Program, Monash University, September 2007, ISBN 978-0-9804298-0-0.

Feldman, M.S. & Khademian, A.M. (2007). The role of the public manager in inclusion: creating communities of participation. *Governance*, April 2007, 305–324.

Jayne, M. (2006). *Cities and Consumption*. New York: Routledge.

Lister, N.E. (1998). A systems approach to biodiversity conservation planning. *Environmental Monitoring and Assessment*, *49*, 123–155.

Mostert, E. (2006). Integrated water resources management in the Netherlands: How concepts function. *Journal of Contemporary Water Research and Education*, *135*, 19–27.

Pahl-Wostl, C. (2008). Requirements for adaptive management. In: Pahl-Wostl, C., Kabat, P. & Moltgen, J. (eds). *Adaptive and Integrated Water Management. Coping with Complexity and Uncertainty*. Springer. ISBN 978-3-540-75940-9.

Pahl-Wostl, C., Sendzimir, J., Jeffrey, P., Aerts, J., Berkamp, G. & Cross, K. (2007). Managing change toward adaptive water management through social learning, *Ecology and Society*, *12*(2), 30. Available from: http://www.ecologyandsociety.org/vol12/iss2/art30/.

Raaijmakers, Ruud and Krywkow, Jörg and Veen van der, Anne (2008). Flood risk perceptions and spatial multi-criteria analysis: an exploratory research for hazard mitigation. *Natural Hazards*, *46* (3), 307–322. ISSN 0921-030X.

Rees, W.E. (1996). Revisiting carrying capacity: area-based indicators of sustainability. *Population and Environment*, *17*, 195–215.

Shelter for all

13

Learning outcomes

In this chapter, you will learn that:

- Planning with a long-term perspective opens the way to develop strategies that are more resilient, adaptable and responsive.
- Redevelopment project are windows of opportunity to adapt the urban fabric to the changing flood risk.
- Despite the significant socio-economic challenges faced by modern cities, there are a number of emerging examples of innovatory initiatives changing the way in which these and other challenges such as climate change are being addressed.
- These initiatives will help to make cities more sustainable even in the face of increasingly uncertain flood risks.

13.1 WHAT DOES THE FUTURE HOLD?

We live in 'yesterday's' cities. Many of the urban patterns that we see today—such as city layouts, buildings, roads and land ownership—are legacies of up to a century and a half of urban policy and decision-making; even longer in some of our cities. Tomorrow's

TEXTBOX 13-1

Topographical shifts at the urban waterfront

Exploring sea level rise as the opportunity for a new diverse edge along San Francisco's industrial eastern waterfront (Rising Tides, International competition for Sea Level Rise, 2009)

How do we build in an area that is dry now, but that may be wet in the future? How do we retrofit existing shoreline infrastructure, such as shipping ports, highways, airports, power plants and wastewater treatment plants? Can we imagine a different shoreline configuration or settlement pattern that allows temporary inundation from extreme storm events? And how do we provide flood protection inland of marshes without drowning the wetland when the water rises?

'Topographical shifts at the urban waterfront' (award recipients: J. Lee Stickles and Wright Huaiche Yang, San Fransisco, California)

Topographical shifts at the urban waterfront explore sea-level rise as the opportunity for a new diverse edge along San Francisco's industrial eastern waterfront. Historically, as

this area was extensively altered from the original coastline, much of it remains on the liquefaction zone and is bay fill. If sea-level rise was to follow the predicted course over the next 100 years, much of this edge would be at the elevation of tidal wetlands. However, sea level rise will constantly be in flux as it adjusts and shifts over time, and any intervention should be able to alter with it.

Currently this industrial land represents the lowest FAR and economic densities in San Francisco, with vast parking lots and large warehouses. With sea-level rise, this underused industrial edge of San Francisco provides the opportunity for the proposed green infrastructure system. The proposed ecological system will allow for migration of wetlands through ecological and urban connectors extending from tidal flats and marshlands to grasslands and upland forest. This pervious landscape will also work to cleanse toxic land before it contaminates the bay and recharge the groundwater system to decrease salinity in the bay.

With extensive new developments planned and constructed along this edge, the eastern waterfront is the area of San Francisco most adaptable to change. All proposed developments along the edge will be built at higher elevations, while the ecological system extends at lower elevations from the waterfront west towards the residential hills and parks of San Francisco. Park corridors with higher densities will be in juxtaposition to the downtown core. New proposed urban development, existing reconfigured port activities, and proposed ecological and cultural habitats will diversify the industrial waterfront edge. The eastern waterfront provides us with the opportunity for coexistence of the natural and built environment.

This complicated edge of the industrial landscape of San Francisco calls for various interventions that recognise the need for change over time and adjustment for future. For our generation, we must decide which areas are crucial to protect, and which areas can take a natural course of change, and shift and adjust as we negotiate with the sea.

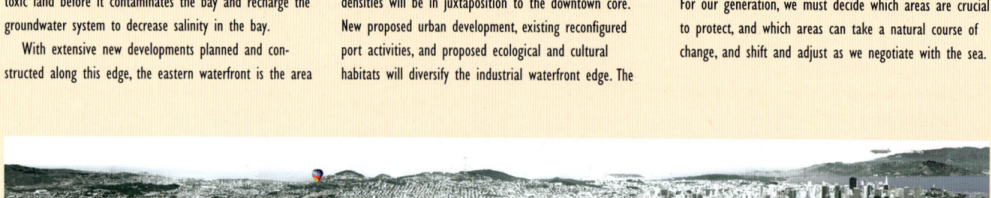

FIGURE 13-1
Shifts at the Eastern waterfront of San Francisco

Source: Lee Stickles, 2009.

cities will also be shaped by the decisions we make today. They must respond to more rapid changes in physical, social, economic and institutional conditions than recent generations have been used to.

Our urban planning systems have changed very little in the past 100 years (see Chapter 3). Linking urban planning with flood management is a relatively new challenge for cities and therefore requires a reappraisal of urban planning. Urban planners, flood managers and a wide range of other related professionals need to find ways of creatively integrating innovations in technology and processes into mainstream urban planning and governance systems in order to better manage urban flooding. However, apart from how best to manage urban floods, cities have to face other often more conspicuous challenges in the near future.

13.2 CHALLENGES AND OPPORTUNITIES

In the previous chapters, areas of significant challenge that have to be taken into account when creating an Urban Flood Management strategy, were identified. A more detailed and comprehensive overview of these challenges and opportunities is summarised below.

13.2.1 Urban areas and living standards

- There is a tendency for increasing urbanisation worldwide with population growth and densification predominantly in the developing world. In highly industrialised countries, smaller families and an ageing population are increasing the demand for smaller sized housing accommodation. These trends increase the economic damage potential of floods. Densification is accompanied by: additional impervious surfaces generating more stormwater runoff; reducing green space and less space for exceedance flow routes.

- Most developments in cities of low and middle income nations are unplanned: much of their physical growth and economic expansion are taking place outside any official plan and regulation. As a consequence, there is a rapidly increasing urban population in the developing world in flood-prone areas that lack provisions for basic infrastructure and services.
- In more affluent societies, increasing expectations about the quality of individual lifestyles are also accompanied by a belief in a 'no risk' living environment that is challenging to those providing services and managing flood risk.
- In the developed cities of the world there is also a belief that protection from disasters can be attained at low cost; reflecting the careless assumption that a risk-free urban environment is achievable in a simple and cheap way. This belief is founded on economies based on low cost carbon fuels and will inevitably change.
- The various sectors of urban services are typically managed individually and separated from each other; thus, energy supplies, transportation, health and other aspects of urban life are managed independently of water and flooding systems. This restricts opportunities that might be available by taking a more holistic and integrated approach to the management of urban living.
- In developed countries, many drainage assets are decaying and require regular maintenance, with condition monitoring to inform difficult and often costly decisions about when to replace them. Whereas in the developing world there are often few or no existing assets and floods and poor sanitation have to be accepted. In each case, there are opportunities to develop new and better ways of managing flood risks through regeneration or new service provision. There are major challenges, however, as to what standards should be provided and maintained. There is now an opportunity to develop a new approach for this based on contemporary ideas and rediscovered thinking on risk-based cost-benefit methodologies.

13.2.2 Governance and institutions

- Changes in the way in which cities are managed mean they are even more important arenas for decision-making than in the past, with many actors involved in the planning of urban communities and management of urban services. This trend can help to construct new links between stakeholders and catalyse new forms of participation and collective action.
- There are many models of city governance especially in relation to urban planning processes and the way in which water is managed. In some cities, all of the key services, including water systems, are managed by the municipality. In other cities all or some of these services, are privately managed by franchises or fully privatised services. No one model is suitable for all cities and circumstances, and it will depend on the administrative arrangements in a particular region or country and the relationship between central government and local action. As flooding occurs at a range of scales from very local to regional and even national, these arrangements are crucial to how we can respond and now also need to be flexible and able to be adapted in the light of new knowledge and circumstances; a need not currently understood by policymakers.
- The governance and way in which cities are managed may restrict their ability to respond autonomously and in the best interests of local citizens. Excessive top-down governance supported by incumbent and self-interested institutional regimes may impose constraints on local communities that prevent the most effective responses

being taken. However, it is essential that a wider perspective is taken due to the potential interactions of local flood-risk management practices across administrative and local catchment boundaries.

- Nonetheless, flexibility and adaptability are also required so that new knowledge gleaned from active learning can be acted upon as rapidly as possible.
- Institutional arrangements have developed as a result of historic events, and as has been noted there is no single best approach. Therefore, the important thing is to integrate the activities of the institutions engaged in flood risk and water management to facilitate flexibility and adaptation.
- The agents within the incumbent regimes and organisations need to be both able and prepared to challenge their own organisation's *common-sense* in order to deliver the significant changes to the way in which all parts of the water cycle are being managed as required for the future. Despite the series of unexpected major extreme flood and drought events that have occurred in the early twenty-first century there is little evidence that this is happening anywhere in the developed world. On the contrary, there is an entrenchment of incumbent regimes with only minor and largely insignificant modifications to custom and practice.
- Where there are distinct divisions of responsibilities for elements of the water and flooding system there is frequently a reluctance to innovate due to lack of incentives and barriers across these man-made boundaries (which floods do not restrict themselves to) and major difficulties in working across organisations to deliver the best response to extreme events. For example, where there is flooding from multiple causes (river, coast and rainfall) the ability of the major institutions to work together and fund the most effective responses across the individual institutional boundaries is difficult and in many cases impossible. This is an example of the often-encountered concept in flood-risk management of 'who-pays' and 'who-benefits' that makes cost optimisation so difficult.

13.2.3 Professional practice

- Urban planning practice is often oblivious to the need to consider water as one of the main priorities; resulting in drainage and flood-risk management being 'added-on' as a material consideration only once the more important planning for layout, roads etc. has been decided. The planning profession needs to become more aware and engaged in the need to place water more centrally in city planning.
- There is a tendency for professionals to rely on tried-and-tested solutions to flood-risk management problems, because 'we know these work', despite the external drivers now changing more rapidly than expected.
- Professional 'silos' of practice prevent a more integrated approach being taken to flood-risk management across disciplines and more cross-disciplinary respect and working together is required to address this.
- Balancing risk management with the maintenance of good environmental water quality in relation to the preservation of ecosystems often causes conflicting requirements and approaches to flood-risk management only at the expense of potential environmental impacts or the other way round. This 'tension' needs to be better recognised and addressed by multi-disciplinary teams. It also needs to be recognised and allowed for by policy and decision-makers.

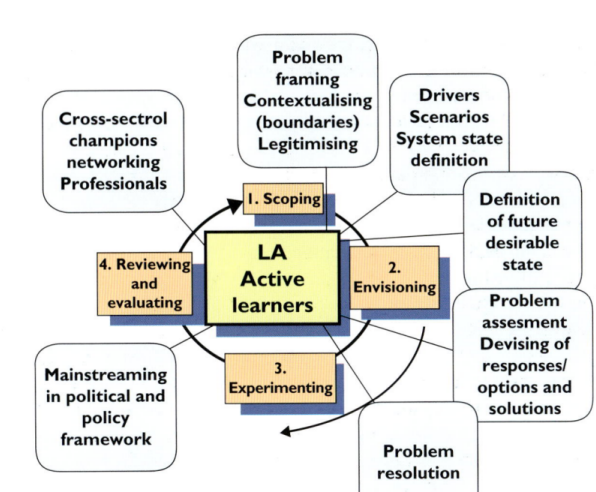

FIGURE 13-2

Delivering more sustainable and integrated water management via a cyclic four-step approach. This approach is based on Integrated Sustainability Assessment (ISA), a cyclical, participatory process of scoping, envisioning, experimenting, and learning through which a shared interpretation of sustainability for a specific context is developed and applied in an integrated manner, in order to explore solutions to persistent problems of unsustainable development

Source: Richard Ashley, 2009.

- Many in the engineering profession still cling to the belief that only large-scale flood defence systems are robust enough to cope with the future challenges. Such systems, whilst being effective up to a specified design standard, risk catastrophic consequences when this is exceeded. They also impose huge cost, energy, resource and carbon footprint burdens during construction and beyond. This 'technological lock-in', defined in social science as 'cognitive lock-in', needs to be challenged if we are to become fit for the future.
- There is reluctance by many in the engineering and physical science professions to engage meaningfully with citizens and policymakers. The need to share information, allow for wide engagement in decision support and making processes has never been so crucial if communities are to appreciate the challenges faced and the difficult choices needing to be made in the light of future uncertainty. This emphasises the need for multi-disciplinary working on flood-risk management projects.

13.3 TURNING IDEAS INTO ACTION

Delivery of the changes needed to make cities more resilient to flooding will depend upon the precise nature of the current regimes for managing water systems within a country or region. A simple model that can be used by those wishing to promote and engage in change is illustrated in Figure 13-1. This has been drawn together from a number of sources defining how to deliver systems that are more sustainable and ideas for sustainability and resilience and the delivery of Integrated Urban Water Management (IUWM). IUWM encompasses simultaneously the appropriate consideration of water as a resource, stormwater management (flooding and quality), sewage treatment and the achievement of good ecological status in urban watercourses.

At the core of the process is the Learning Alliance (LA), comprising as wide a range of stakeholders as needed in order to facilitate change and to engage in the active

¹The cyclic four-step approach has been used from the European Union's MATISSE project, which investigated the delivery of sustainable systems and formulated this simple model.

learning needed to cope with future climate and other change challenges.¹ The core LA can act to both engage in the policy processes and also in the delivery on the ground of the technical and other responses needed to cope with the future challenges. It can also be the main vehicle for providing knowledge, information, techniques and consensual learning.

Some of the key aspects of what and how we may be able to take advantage of the opportunities afforded by the new challenges are considered further in the following sections using the framework in Figure 13-1. It should be appreciated that this is not a static framework, rather one in which the process is continually repeated.

13.3.1 Scoping

Here the understanding of the problem is developed through a shared vision between all the stakeholders. There are issues in this to do with the selection of learning alliance members; their inter-relationships, especially mutual trust; provision of information; engagement; and raising of awareness, especially about the drivers. From this, it is possible to develop a shared vision and assessment of the problems faced both for now and, using scenarios to scope the future, for future conditions that is subscribed to by everyone involved. Importantly, here is a shared vision of where the urban area wishes to get to in the future in terms of flood resilience and robustness across potential scenarios.

TEXTBOX 13-2

The Yorkshire and Humber Learning and Action Alliance (YHLAA)

This was formally established following an initial meeting in January 2009. Flooding in the summer of 2007 had forced the British government to take the threat seriously and to institute a number of changes to the regime and institutional arrangements used to manage flood risk. These placed significantly changed burdens and expectations on municipalities in England in particular. Hence, there was a perceived need by the various players in the Yorkshire and Humber Region of England for support to meet the new challenges of increasing flooding risk, the need to provide 1,000s of additional homes and to maintain the quality of life, health and welfare of their citizens. A group of EU ERDF funded INTERREG projects: FloodResilientCity, MARE and Skint were each tackling aspects of flood and water management and included municipal partners based in Yorkshire. These projects provided resources, a focus and exposure to flood-risk management practices elsewhere in Europe that have helped inform the Yorkshire-based municipalities. Of particular value has been the experience from another EU project, SWITCH, which has Learning Alliances (LAs) at the heart of its activities. In the YHLAA case, however, the main actors from the municipalities wished to include the word 'action'—so they established nested 'Learning and Action Alliances' (LAAs); covering the whole region and separate LAAs for individual river basins within the region. The purpose of these LAAs is to champion the

required transition to resilience and managed adaptive approaches for flood-risk management at three levels:

1. European—through the set-up of a virtual knowledge centre for flood resilience, providing input to relevant policy documents and, through the creation of a nested international LAA, for the mutual review and learning between City LAAs.
2. National level—using demonstration projects to identify and address bridges and barriers for a transition to resilience and managed adaptive approaches for FRM within the present planning, administrative and regulatory and policy framework.
3. Local level—assisting City LAAs to gain deep knowledge of proposed strategies and to comprehensively adopt MARE tools via workshops and trans-national scientific missions between City LAAs.

The LAAs also assist with:

- engaging with and building capacity and involving policymakers, practitioners, key peak groups and the public;
- developing a shared understanding of the flood problems in the case study context (transnationally across case studies) and to identify response options;
- diffusing the 'research' rationale and working methodologies transnationally by reports, a web based portal and workshops;
- engagement with partners in ongoing and former projects, to broaden and strengthen the membership of the LAAs and to link to the outputs of those projects so as to provide a sustainable and transnational legacy beyond the lifespan of the projects;
- actively seeking out other similar networks within each LAAs' country to identify and promote the overall national and transnational potential of the alliances;
- setting up and monitoring leadership (Champion) development programmes for players in the LAAs
- influencing developing policy.

Each partner in the LAA recognises the need to engage in active learning in order to develop the capacity by different stakeholder groups to both accept a different view on risk and performance and also to be able to use different types of response and at different times of implementation. At the same time, it can save costs on adapting to future changes.

The YHLAA now has more than 100 members drawn from each of the institutional players involved in flood-risk management and has begun by developing a list of topics that need to be included in a good practice guide for flood and water management within the Yorkshire and Humberside region.

Within modern cities in developed countries, there are invariably opportunities for engagement by a wide range of stakeholders. However, meaningful engagement can only take place where information is made available and the time is taken to help understanding. This requires the building of capacity in those involved *including the policy and decision-makers as well as citizens*. The degree of commitment to this on the part of the information providers and also the recipients is very variable and requires considerable time and resource allocation on the part of the professionals involved. Hence, experience suggests that the scoping process more often than not fails to include as wide a range of stakeholders as it should. The cities of the future will require as effective as possible citizen and policymaker engagement if the challenges are to be faced and responded to in robust ways. The emergence of learning alliances in a number of areas as a means to address this and as the focus and driver of the process of responding (Figure 13-2) are a promising step forward as illustrated in Textbox 13-2.

13.3.2 Envisioning

This process stage is started in the scoping phase once the common vision of the future desirable state has been defined—this may include flood protection from the rivers in the city with a recurrence interval of several hundred years and several decades for the urban drainage (minor) system. Ideally the standards should be based on a more cost-beneficial approach that seeks to minimise damage costs; resulting in variable recurrence intervals across the urban area. In any case, at this stage the problems are defined and assessed in detail and potential responses to them outlined and evaluated. Here, the idea of 'solutions' is not a good one. With changing external drivers, a 'solution' today may be a problem tomorrow. Therefore, the term 'response measure' is better, as this indicates a rather more temporary approach, recognising that the problem will alter in the future and the solution will need to be re-assessed and potentially added

TEXTBOX 13-3
Sustainable urbanisation versus resilience

The term 'resilience' is often used in discussions about sustainability. For some, resilience is a more useful concept than sustainability, for instance when it is used within the context of sustainable urbanisation. This is partly because resilience embraces explicitly the dynamic nature of (complex) systems such as cities, whereas sustainability is often conceived as a goal to which we should collectively aspire. For others, however, sustainability is an attribute of dynamic, adaptive systems that are able to flourish and grow in the face of change. According to Hester (2006), resilience in cities depends both on its *physical form and characteristics* as well as on the *people's capacity*, and *social behaviour*. Community resilience requires self-reliant, skilled and capable citizens who have 'developed iterative learning with mature face-to-face social networks'. There

is no blueprint for urban sustainability, but there is a growing recognition that innovative planning approaches and processes based on these resilience principles will guide citizens and other stakeholders the way to become co-producers of a *sustainable* community that can respond to change and disruption, and pro-actively reduce vulnerabilities. These approaches (and processes) should not be viewed as models that can be applied in all contexts since they are shaped by the social and cultural norms of particular places.

According to UN-Habitat (2007) environmentally sustainable urbanisation requires that:

- greenhouse gas emissions are reduced and serious climate change mitigation and adaptation actions are implemented;
- urban sprawl is minimised and more compact towns and cities served by public transport are developed;
- non-renewable resources are sensibly used and conserved;
- renewable resources are not depleted;
- the energy used and the waste produced per unit of output or consumption is reduced;
- the waste produced is recycled or disposed of in ways that do not damage the wider environment;
- the ecological footprint of towns and cities is reduced.

TEXTBOX 13-4
Urban Flood Management (UFM) in Dordrecht, The Netherlands

The city of Dordrecht, which is situated on an island, is confronted with the challenges of how to manage changing flood risk in its redevelopments and expansions in at-risk areas outside the primary flood defence system. One of these development areas is De Stadswerven, an area of approximately 30 hectares, located on the edge of the historical Dordrecht city centre, at the confluence of the rivers Beneden Merwede and Wantij. The redevelopment programme foresees the development of approximately

1,600 residential buildings, commercial, cultural and public developments. De Stadswerven adopted UFM results, as they provided new solutions that could improve the spatial quality without increasing costs (short-term gain), addressed the 'new' challenges of climate change, flood risk and urban growth (long-term vision), and because they provided metaphorical fresh air and positive energy without any political burdens in a frustrated process (short-term gain). The UFM project involves a study of a pilot area in the

floodplains of Dordrecht (combination of flood accommodation with urban development). A redevelopment project of a former shipyard area, adopted various UFM design concepts in its new masterplan and will incorporate a pilot project of approximately 100 flood proof buildings. This conforms strongly to the notion that local scale pioneering and experimentation are essential and encourage the belief that these bottom-up initiatives can shape strategy and policy development to cultivate urban flood resilience.

FIGURE 13-3
Redevelopment and expansion area (former shipyard area) outside the primary flood defence system of the city of Dordrecht, The Netherlands

Source: Municipality of Dordrecht, 2008.

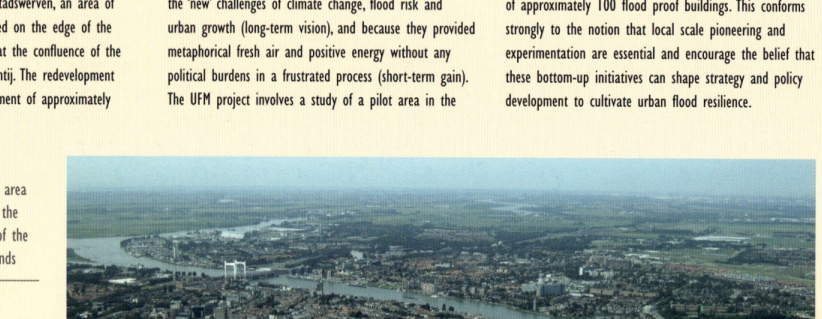

FIGURE 13-4
Envisaged masterplan of the redevelopment and expansion area of Dordrecht. New, flood-proofing developments provide an attractive alternative to a physical barrier in the form of an earthen embankment or concrete wall surrounding these developments to further reduce the flood risk

Source: Baca Architects, 2008.

to incrementally in an adaptive way following the third and fourth steps. The envisioning step should consider all of the options and, importantly, their robustness, flexibility and adaptability. Again, this stage requires that those engaged have the capacity to meaningfully participate in the processes and requires professional support to keep up with the latest opportunities for responding as illustrated below.

13.3.3 Experimenting

In the past, our approach to flood and other risks was a combination of over-design of infrastructure, coupled with a trial and error implementation. We learnt from failure how to build safely. Over-design provided headroom or a 'factor of safety'—the excess capacity added on to the 'design capacity' to allow for present and future uncertainties that could not be resolved at the time. This has more recently been replaced by 'optimisation', where our infrastructure is designed and built to specified performance standards that are now achieved. Unfortunately, the introduction of optimisation has left a legacy of infrastructure that performs just to a standard that copes with what were the perceived external drivers and loadings. With climate and other changes, these external loadings are no longer valid and are increasing; hence, many of our optimised systems will not be able to provide the standard of performance now needed. We need to re-learn how to introduce the excess headroom capacity into the physical/technical system, which will help ensure that the acceptable level of flood risk can be achieved.

Optimisation and the advent of the electronic computer has to some extent, de-professionalised many practitioners. The concept of 'learning by doing' and 'reflection in practice', will need to be re-awakened if professionals, who are competent and confident as a result of real experience, are to become more comfortable with experimenting with options and to be able to justify this to decision-makers.

This will inevitably lead to the greater use of smaller scale, dispersed and non-structural responses to flood risk. Large-scale infrastructure will be too costly and disruptive to experiment with and will need to be used only where the measures required are precautionary and are based on a fairly good understanding of how the future drivers will develop. Therefore, coastal protection and erosion management may be amenable to large-scale infrastructure solutions where sea level rises are expected to be within predictable bounds. On the other hand, urban drainage systems should move away from a reliance on buried piped systems that are less flexible than alternatives such as holding runoff back at source (see Chapter 8).

13.3.4 Reviewing and evaluating

Performance monitoring and post-development evaluation of the success or otherwise of a response is normally a default activity; i.e. if no one is flooded for a while (within the lifetime up to retirement of the designer); the project is deemed to be successful. This approach is no longer tenable. More effort will be needed to assess whether or not a response measure is functioning as expected over a range of drivers, not only if it does not fail. Adaptable responses require semi-constant review by professionals and decision-makers as part of the active learning process. Is it time for example, for a further intervention? When is it best to take the next step? This approach differs from the past when decision-makers could agree to implement a response and then forget about it and move to another area of interest. The latter fits well into a quasi-stationary

TEXTBOX 13-5
Responding to increasing flood risk in the East Riding of Yorkshire

There was severe flooding in England in the summer of 2007 from river and direct rainfall and this occurred at all scales from local to regional. Traditional response measures comprising structural interventions could not be implemented in many instances due to cost and the difficulties of retrofitting in urbanised areas. In the East Riding of Yorkshire, the Municipality, assisted by an innovatory consulting practice, decided to use local adaptations to urban layouts to provide small-scale local surface storage or amendments to flow

pathways to avoid damage. These measures have enabled widespread changes to be made across the region, as they have been low cost. Performance monitoring is also allowing their effectiveness to be assessed for further adaptation and application.

This road junction was completely regraded to ensure that any flows that come out of the drainage ditches will be conveyed to areas where there will be no damage to property.

Source: Yorkshire Council, 2010.

FIGURE 13-5
East Riding of Yorkshire Council Website on Floods. (http://www.eastriding.gov.uk/cr/support-and-procurement-services/humber-emergency-planning/floods/) (Figure 13-5a) and Hull, June 26 2007 after the flood (Figure 13-5b)

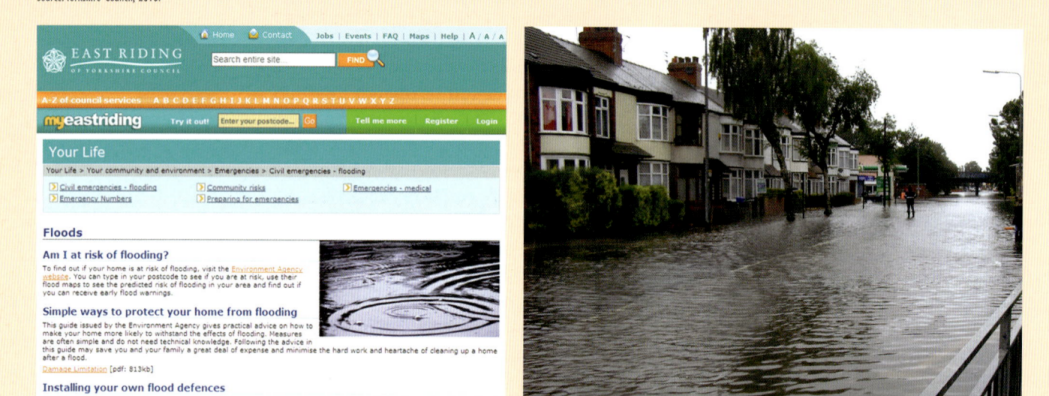

world but not into one that is adversely changing. Therefore, decisions will need to be reviewed and evaluated more frequently than hitherto. This will also require new ideas about performance monitoring and auditing procedures.

13.4 SUCCESS STORIES: SEIZING WINDOWS OF OPPORTUNITY

Many cities around the world are facing the challenges of sustainable living and development and are exploring ways to enhance their ability to manage an uncertain future. Drivers and pressures include relative wealth; population growth; the provision of food; lifestyle expectations; energy and resource use and climate change. These pose new challenges for the way in which we manage urban floods. In the previous chapters we have learned that there is no clear cut, 'best' solution for the avoidance of catastrophic flood events or even how to 'live with (all) floods'. The way forward is thus far from clear although what we can be sure about is that we are rapidly entering a phase of fundamental change and our willingness and ability to adapt to and mitigate the worst effects of this will be critical.

In general, cities are becoming larger and denser (see Chapter 2). Urban expansion is an issue of serious concern and is often placed as a justification for densification. The fundamental question of whether urban expansion should be resisted, accepted or welcomed is still largely unresolved. From the perspective of flooding, concerns for indiscriminate urban expansion or 'sprawl' have captured the attention of both policymakers and academics during the last decade. This is because, alongside climate change, it is considered as the major driver for increased flood risk. Sprawl will occur where unplanned, decentralised development dominates, as is common in developing countries. Where growth around the periphery of the city is coordinated by a strong urban policy, more compact and less vulnerable forms of urban development can be secured. It is evident that these approaches to development have direct consequences for the way floods are managed both in terms of the vulnerability of the urban area and its inhabitants and also in terms of the often indiscriminate effect that urban growth has on the generation of floods in terms of runoff and flood probabilities. At first glance there seem to be conflicting interests between the flood-risk managers who advocate open, green spaces in their cities and those who adhere to the compact cities concept as the sustainable urban form for controlling transport-related greenhouse gas emissions.

Built on the analysis and synthesis of many different planning systems and principles used by cities in the PLUS Network,^{2} a set of planning principles referred to as the PLUS planning cycle (see Textbox 13-6) has been developed which underlie the successful execution of the sustainable cities. The cities are from 14 different countries and represent a broad range of sizes, geography and cultures. These principles may be of value when considering the strategic, long-term risks of cities towards flooding.

^{2} 38 cities participate in network activities of which 34 signed the Memoranda of Agreement.

TEXTBOX 13-6
The Sustainable Cities PLUS Planning Cycle

PLUS Planning Cycle is a generic summary of the principles and stages of planning developed by various cities that are members of a peer-learning network of 38 cities and regions (The Plus Network) engaged in long-term integrated planning and community demonstration projects.

Principle 1: Adopting a long-term lens

The difference between traditional planning and the PLUS Network's planning is the difference between planning from current trends (forecasting) to planning from a vision of a desired future and reflecting back to determine what actions would lead to that future (back-casting). Recognising that it is unrealistic to try to predict the future 100 years from now, the PLUS Network's planning cycle breaks this down into manageable pieces. The 100-year vision does not require highly specific actions. It simply sets out a desired future over three generations or beyond.

(Vancouver is a city of 550,000 residents set within a region—the Greater Vancouver Regional District—that hosts a total population of approximately 2 million. In 2001, Canada's last census year, half of the City of Vancouver's residents were 'visible minorities' (non-Caucasian people who are not Aboriginal in origin), and 46 per cent were immigrants. The largest proportion of the immigrant population comes from Asia—particularly China, India, the Philippines and South Korea. Vancouver is home to 14 per cent of the population of British Columbia but 24.5 per cent of its total immigrants. Less than half the city's population speaks English as a first language.

The City of Vancouver carried out its part of the regional process between 1992 and 1995, when approximately 20,000 residents participated in the CityPlan initiative. As in the overall regional vision, the city's residents expressed a strong commitment to connecting social and environmental principles for achieving sustainability. They requested more efficient public transport, more and safer bicycle routes through the city and more pedestrian-friendly streets. Residents also prioritised the following: development of distinctive neighbourhoods featuring diverse public spaces, affordable housing and access to services; a strong sense of community and increased public safety; and economic development that keeps jobs in the city. The city is also investing millions of dollars to reduce homelessness and drug addiction within the Aboriginal population, which for many years was left out of Canada's development agenda. In October 2005, the Economist Intelligence Unit (EIU) voted Vancouver, British Columbia, the world's most liveable city.

Then we work out a medium-term strategy for the next 30 years, and a comprehensive shorter-term implementation/action plan with 5–10 year timelines. This implementation plan includes a budget and workplan.

Principle 2: Viewing the city as one complex system

Only a holistic view enables planning for long-term sustainability. This means recognising the inter-relations and interactions of the four elements of sustainability—economic, environmental, social and cultural, with each other and with governance—and treating the whole as one system.

Principle 3: Using an integrated and comprehensive approach

Cities tend to manage various systems such as water, transportation, waste and energy on separate tracks. Long-term sustainability requires aligning all these plans and managing them in an integrated fashion. Different cities find their own ways of doing this.

Principle 4: Adaptive management and collective learning

In order to work with a very long-term lens and a systems approach, management systems must operate in a mode of constant learning and experimentation. This is known as adaptive management. It monitors and creates feedback loops, recognising that unexpected outcomes are inevitable and that they represent an opportunity for innovation and learning. This approach enables mid-course corrections, learning from experience and failures and adjusting behaviour to ensure survival. It characterises more resilient cities and communities. Beyond the learning by individual cities, the nature of the Sustainable Cities: PLUS Network is to share these experiences among its members and beyond, thus generating new knowledge and maximising the transfer of learning.

Principle 5: Focusing on a city's bioregion, ecological footprint and neighbours

Humans are an integral part of the natural world and cities must be viewed within their larger ecosystem, bioregion, watersheds and air-sheds. This is often easiest to do when considering the long-term vision and strategy. The city extends its impact beyond its geographic region through its interactions with others in trade, political and cultural processes. The concept of the ecological footprint refers to the area of land and water the city needs to produce the resources it consumes, and to absorb the waste it generates. If everyone in the world consumed at the level of North Americans, we would need more than four planets.

This concept helps people understand the consequences of their own consumption and production decisions. It has been applied as both a research tool and as an educational tool with calculators for individuals, schools and cities.

Principle 6: Participatory engagement

Integrated long-term planning is ongoing, iterative and participatory. Multi-sectoral and multi-stakeholder groups usually lead the process, with the support of multi-disciplinary staff teams. Multi-sectoral groups include leaders from the public, private, academic and not-for profit sectors. Multi-stakeholder groups (such as Roundtables or working representatives of community groups) that have an interest in the planning. Multi-stakeholder approaches improve data gathering, decision-making, priorities and strategies, while ensuring that the process survives leadership changes and elections. General public participation involves different engagement strategies at various stages. Broad public participation is most common in visioning, setting priorities and monitoring progress. Agreement on principles of community engagement, such as acceptance, active listening, collaboration and inclusiveness, is essential to success.

Source: Nola-Kate Seymoar, *The Sustainable Cities Plus Network*, September 2008, Durban Biennial Conference.

FIGURE 13-6
Aerial view of Greater Vancouver

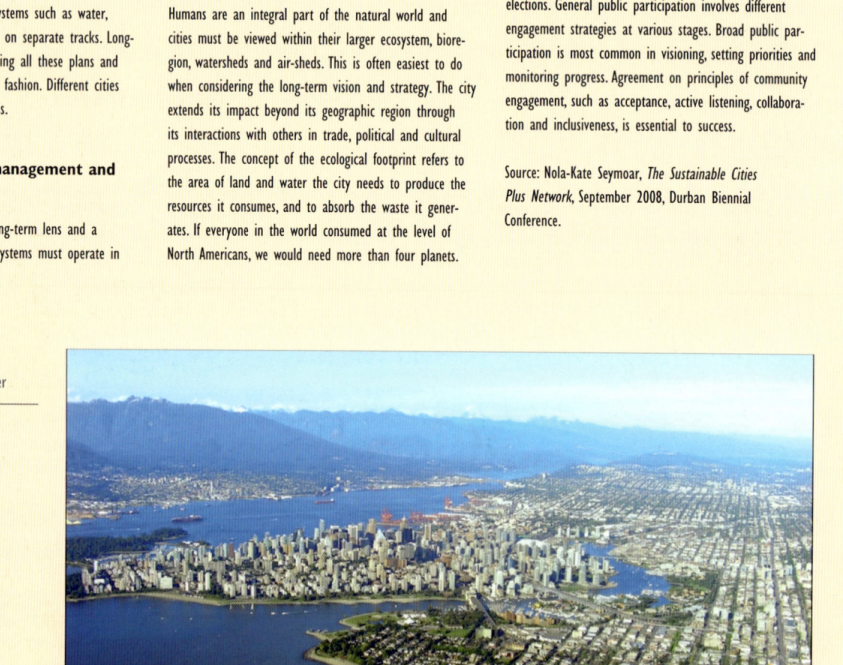

TEXTBOX 13-7
Cities as water catchments

Many of the developed parts of Australia have suffered severe drought conditions for more than a decade and water is precious. Yet Australian cities also suffer from flooding. The recognition there that the appropriate management of the water cycle overall is the best way forward has led to the development of the concept of using the runoff from city areas as a potential source of water for supply. It also has the added benefits of reducing urban temperatures and improving the landscape and liveability of Australian cities. The plan is to enable a hybrid of decentralised and centralised water management solutions that ensure resilience to both water-poor and water-abundant futures and deliver multiple benefits to people and the environment. Much of this will need to be retrofitted into existing urban landscapes. As many of the required approaches are still very much in their infancy, there is a need to find better

ways of combining existing centralised water infrastructure with new, decentralised systems at a range of scales; for households, streetscapes and neighbourhoods. This is what this programme sets out to do. This will also require strong integration into city planning processes. The approach uses 'back-casting' from an envisioned future water-sensitive city, enabling identification of strategic initiatives and implementation pathways that will form the basis of strategic plans for cities; plans covering the immediate and short terms. These initiatives will include development and implementation of planning policies, staging of infrastructure and demonstration projects, community engagement and the development of new and/or adapted urban design principles and processes. The project team comprises a number of Australian Universities led by Monash in Melbourne and is part of the Australian National Urban Water Governance Programme.

FIGURE 13-7
Melbourne Docklands Park WSUD strategy

Source: Jeroen Rijke, 2009.

There is no single 'magic' recipe for successful planning of a city in response to the challenges of sustainability, climate change and flood risks. This is partly because every city has a unique context. What we have learned is that urban design, master planning and the management of buildings, infrastructure, public utilities and green spaces must be included in any urban flood-risk management strategy (see Chapter 9). We also learned of the need for long-term planning. A long-term perspective allows us to identify opportunities for synergy and to overcome barriers for implementation, such as investments that both enhance resilience and provide short-term additional economic, social or environmental benefits. A long-term perspective is also fundamental for incorporating sustainability indicators, such as life cycle cost. Planning with a long-term perspective thus opens the way to develop strategies that are more resilient, adaptable and responsive (see Chapter 11). It also requires skilled and capable stakeholders who are knowledgeable about the systems they live in and are capable of mainstreaming flood-risk management in the process of (re)development (see Section 7.5).

In most industrialised countries, the building stock is mainly ageing and there is much heritage. In the coming decades, the redevelopment (c.f. renovation and modernisation) of the existing stock is a high priority and certainly of higher priority than the provision of new housing. European cities are composed of mixtures of buildings of different ages and life spans, but within 30 years, around one-third of its building stock will probably be renewed. The same holds true for many other cities of the Western world, where continuous restructuring will be common practice. Redevelopment projects may thus provide *windows of opportunity* to make adjustments in the process of urban renewal in order to restore old mistakes and to build in more resilience by adapting and restructuring the urban fabric to new conditions of increased flood risk. The developing world, however, is not constrained by past investments, and much of their 'urbanisation' is to come in the next few decades. There is a huge challenge to exploit this momentum. If we are able to seize these windows of opportunity and share good practices via city-to-city networks stretching across country boundaries and other social networks, than we can create the groundswell for real practical change towards flood-resilient cities on a more global scale. There are a growing number of emerging examples of innovatory initiatives changing the way in which these challenges are being addressed and of which we can learn!

Key Questions

1. What are the key drivers of change and threats to cities at global, regional and local scales?
2. How could further urbanisation and urban transformation be re-directed so that cities become more resilient?
3. Which factors contribute to the adaptive capacity of cities and explain how the drivers and threats can be turned into a powerful trigger for innovations and transitions needed to make cities more sustainable?

FURTHER READING

Bai, X. (2003). The process and mechanism of urban environmental change: an evolutionary view. *International Journal of Environment and Pollution, 19(5)*, 528–541.

Burby, R.J., Deyle, R.E., Godschalk, D.R. & Olshansky, R.B. (2000). Creating hazard resilient communities through land-use planning. *Natural Hazards Review, 1(2)*, 99–106.

Cashman, A. & Ashley, R.M. (2008). Costing the long-term demand for water sector infrastructure, *Foresight*, *10(3)*, 9–26. ISSN 1463-6689. Emerald Group publishing.

FlOODSite (2008). *Long-term Strategies for Flood Risk Management. Scenario Definition and Strategic Alternative Design*. Report No T14-08-01. Available from: http://www.floodsite.net.

Folke, C. *et al.* (2002). *Resilience and Sustainable Development: Building Adaptive Capacity in a World of Transformations*. Scientific background paper on resilience for the process of the World Summit on Sustainable Development on behalf of The Environmental Advisory Council to the Swedish Government. Sweden: Ministry of the Environment. Accessed Advisory Council to the Swedish. Available from: http://www.unisdr.org/eng/risk-reduction/wssd/resilience-sd.pdf.

Godschalk, D.R. (2003). Urban hazard mitigation: creating resilient cities. *Natural Hazards Review*, *4(3)*, 136–143.

Hester, R.T. (2006). *Design for Ecological Democracy*. Cambridge: MIT Press. ISBN: 0262083515.

Jayne, M. (2006). *Cities and Consumption*. New York: Routledge.

Lister, N.E. (1998). A systems approach to biodiversity conservation planning. *Environmental Monitoring and Assessment*, *49*, 123–155.

OECD (2006). *Infrastructure to 2030: Telecom, Land Transport, Water and Electricity*. Paris: Organisation for Economic Cooperation and Development. ISBN 92-64-02398-4.

Thorne, C.R., Evans, E.P. & Penning-Rowsell, E. (eds) (2007). *Future Flooding and Coastal Erosion Risks*. London, UK: Thomas Telford. ISBN 978-0-7277-3449-5.

UN-Habitat (2009). *Planning Sustainable Cities: Global Report on Human Settlements 2009*. United Nations Human Settlements Programme (UN-Habitat). ISBN: 978-1-84407-899-8. Available from: http://www.unhabitat.org.

White, I. (2008). The absorbent city: urban form and flood risk management. *Proc. ICE Urban Design and Planning*, *161(DP4)*, pp. 151–161, December 2008.

References

AC Neilsen, (2003). Disaster Public Awareness Research, for Counter Disaster and Rescue Services, Department of Emergency Services, Queensland.

Advances in Urban Flood Management, [R. Ashley *et al.*, 2007], ISBN: 978-0-415-43662-5.

After Iribarren, The Multivariable Systemic Method. Suarez Bores, P. Revista de Obras Públicas, 2001.

Allenby, B. and Fink, J. (2005). Viewpoint: Toward inherently secure and resilient societies. Science Vol. 309, no. 5737, pp. 1034–1036.

Alley, W.A. and Veenhuis, J.E. (1983). Effective impervious area in urban runoff modeling. *Journal of Hydrological Engineering, ASCE*, 109(2): 313–319.

Allitt, R., Foreman, H., Djordjevié, S. and Chen, A.S. (2009). Integrated Urban Drainage Demonstration Projects, Site 1 (Phase 3) – Cowes, Isle of Wight, UKWIR Report, London.

Andjelkovic, I. (2001). Guidelines on non-structural measures in urban flood management. *IHP-V, Technical Documents in Hydrology*, No. 50, UNESCO, Paris.

Angel, S., Sheppard, S.C. and Civco, D.L. (2005). The dynamics of urban expansion. Transport and Urban Development department. The World Bank, Washington D.C., September.

Arnaud, P. and Lavabre, J. (1999). Using a stochastic model for generating hourly hyetographs to study extreme rainfalls, *Hydrological Sciences Journal*, 44(3): 433–446.

Arnaud, P., Fine, J.A. and Lavabre, J. (2007). An hourly rainfall generation model adapted to all types of climate, *Atmospheric Research*, 85: 230–242.

Arrhenius, S. (1896). On the Influence of Carbonic Acid in the Air upon the temperature of the Ground. Philosophical Magazine and Journal of Science. Ser. 5, Vol. 41, Nº 251.

Ashley, R.M., Newman, R.N., Molyneux-Hodgson, S. and Blanksby, J.R. (2008). Active learning: building the capacity to adapt urban drainage to climate change. Proc. 11th ICUD, Edinburgh.

Ashley, R., Garvin, S., Pasche, E., Vassilopoulos, A., Zevenbergen, C., (Edited) (2007a). Advances in Urban Flood Management. Taylor & Francis Group. Lomdon / Leiden / New York / Philadelphia / Singapore.

Ashley, R., Garvin, S., Pasche, E., Vassilopoulos, A., Zevenbergen, C., (Edited) (2007a). Advances in Urban Flood Management. Taylor & Francis Group. Lomdon / Leiden / New York / Philadelphia / Singapore.

Ashworth, G. (1987). *The lost rivers of Manchester.* Willow Publishing, Altrincham, Cheshire.

Australian greenhouse office (AGO) (2006). Climate change impacts & risk management – A guide for business and government. Department of Environment and Heritage, Australian greenhouse office.

Balmforth, D., Digman, C., Kellagher, R. and Butler, D. (2006). Designing for Exceedance in Urban Drainage – Good Practice, CIRIA Report C635, London.

Barroca, B., Bernardara, P., Mouchel, J.M. and Hubert, G. (2006). Indicators for identification of urban flooding vulnerability. Nat. Hazards Earth Syst. Sci., 6, 553–561.

Batty, M. and Hutchingson, B. (eds) (1980). Systems Analysis in Urban Policy-Making and Planning. New York: Nato Conference Series v. 12, Plenium Press.

Batty, M. (2003). The emergence of cities: complexity and urban dynamics. CASA Working Papers, no.64. ISSN 14671298 Working paper. Centre for Advanced Spatial Analysis, London, UK.

Bedford, T. and Cooke, R. (2001). Probabilistic Risk Analysis. Foundations and methods. Cambridge University Press, Cambridge, UK.

Bernardara, P., Schertzer, D., Sauquet, E., Tchiguirinskaia, I. and Lang, M. (2008). The flood probability distribution tail: how heavy is it? Stoch. Environ. Resear. and Risk Analysis, 22(1): 95–106.

Beven, K.J. (2001). Rainfall-runoff modelling: the primer. John Willey and Sons, Chichester.

Blowes, D.W. and Gillham, R.W. (1988). The generation and quality of streamflow on inactive uranium tailings near Elliot lake, Ontario. *Journal of Hydrology*, 97, 1–22.

References

Bradbook, K., Waller, S. and Morris, D. (2005). National Floodplain Mapping: Datasets and Methods – 160,000 km in 12 months. *Natural Hazards* 36, 103–123.

Brand, S. (1994). How buildings learn. New York: Viking.

BRE (2006). 'Green roofs and facades', BRE, UK.

BRE (2009). 'Long term initiative for flood-risk environments', BRE, UK.

Brooks, K.N., Ffolliott, P.F., Gregersen, H.M. and DeBano, L.F. (2003). Hydrology and Management of Watersheds. Iowa State University Press, Ames.

Bruen, M. (1999). Some general comments on flood forecasting. *Proceedings of the EuroConference on Global Change and Catastrophe Risk Management: Flood Risk in Europe*, June 1999, IIASA, Laxenburg, Austria.

Bruun, P. (1962). Sea Level Rise as a Cause of Shore Erosion. Proceedings ASCE, Journal. Waterways and Harbor Division, Vol. 88, 117–130.

Building Research Establishment (1991). *Dealing with flood damage*, BRE Press Ltd, Watford.

Building Research Establishment (1997). BRE Good Repair Guide 11, *Repairing Flood Damage* Part 1–4, BRE Press Ltd, Watford.

Building Research Establishment (2004). BRE Report BR466, *Understanding dampness*, BRE Press Ltd, Watford.

Building Research Establishment (2006). *Repairing flooded buildings: an insurance industry guide to investigation and repair*, BRE Report.

BWK (1999). Hydraulische Berechnung von naturnahen Fließgewässern, Merkblatt 1/1999.

Canadian Standards Association (CSA) (1991). Risk Analysis Requirements and Guidelines, CAN/CSA Q634-M91, Cited by Salmon G.M. (1997).

Carson, R.W. (2006). Integrated flood management in the Red River Valley—Canada. In: J. van Alphen, E. van Beek and M. Taal (eds) Floods, from Defence to Management (2006). Taylor & Francis Group, London ISBN: 0 415 39119 9.

Chow, V.T., Maidment, D.R. and Mays, L.W. (1988). Applied Hydrology. McGraw-Hill, New York.

Chow, V.T., Maidment, D.R. and Mays, L.W. (1988). *Applied Hydrology*. McGraw-Hill, Singapore, 572 pp.

Chow, V.T., Maidment, D.R. and Mays, L.W. (1988). Applied Hydrology. McGraw-Hill, Singapore, 572 pp.

Christensen, J.H. and Christensen, O.B. (2003). Severe summertime flooding in Europe. *Nature (Lond.)* 421(6925): 805–806.

CIRIA (2007). Design for exceedence of urban drainage. Construction Industry Research and Information Association, London.

CIRIA 'Green roofs – an introduction and overview of benefits', (2008), CIRIA, London, UK.

CIRIA 'SUDs Manual' (2007). CIRIA, London, UK.

Climate Change 2007, The Physical Science Basis, Contribution of Working Group 1 to the Fourth Assessment Report of the Intergovernmental Panel on Climate Change, Cambridge University Press, IPCC 2007.

'Code for sustainable homes' (2008). UK Government.

Cohen, Michael, D., James, G. March, Johan, P. Olsen A Garbage Can Model of Organizational Choice Administrative Science Quarterly, Vol. 17, No. 1. (Mar., 1972), pp. 1–25.[particularly pp.1–3 & 9–13].

Construction Industry Research and Information Association (2005). C623, Standards for the repair of buildings following flooding, CIRIA publication C623, London (authors Garvin S.L. *et al.*).

Corning, P.A. (2002). The re-emergence of "emergence": a venerable concept in search of a theory. Complexity 7(6): 18–30.

Corotis, R.B. (2007). Risk Communication with Generalized Uncertainty and Linguistics, University of Colorado 428 UCB, Boulder, CO 80309-0428.

Covello, V. (2002). Risk communication slides. Center for Risk Communication, 29 Washington Square West, Suite 2A, New York, New York 10011.

Covello, V.T., McCallum D.B., & Pavlova M.T., 1989. Effective risk communication: the role and responsibility of government and non-government organizations. New York (NY): Plenum; 366 pages.ISBN 0-306-43075-4.

CRED (2010). The International Disaster database EMDAT, Centre for Research on the Epidemiology of Disasters.

Crichton, D. (2005). The role of private insurance companies in managing flood risks in the UK and Europe, in Urban Flood Management (eds C Zevenbergen, A Zllosky), Taylor and Francis Group plc, London.

Cutter, S.L., Boruff, B.J. *et al.* (2003). Social Vulnerability to Environmental Hazards. In: Social Sciences Quarterly. Vol. 84(2). pp. 242–261.

Damage to building [W. Roos, June 2003], TNO Bouw, DC1-233-9, "Intermediary report or study", the Netherlands.

Davenport, A., Gurnell, A.M. and Armitage, P.D. (2004). Habitat survey and classification of urban rivers, *River Research and Applications*, 20, 687–704.

Dawson, R.J., Speight, L., Hall, J.W., Djordjević, S., Savić, D. and Leandro, J. (2008). Attribution of flood risk in urban areas, *Journal of Hydroinformatics*, Vol. 10, 4, pp. 275–288.

De Bruijn, K.M. (2004). Resilience indicators for flood risk management systems of lowland rivers. *International Journal of River Basin Management* 2(3), 199–210.

De Bruijn, K.M. (2005). Resilience and flood risk management. A systems approach applied to lowland rivers. PhD thesis. Delft University of Technology, Delft, The Netherlands.

Dean, R.G. (1977). Equilibrium Beach Profiles, U.S. Atlantic and Gulf Coast, Tech Rep No. 12, U. Delaware, Newark.

Decroix, L. and Mathys, N. (2003). Processes, spatio-temporal factors and measurements of current erosion in the French Southern Alps: a review. *Earth Surface Processes and Landforms*, 28, 993–1011.

Defra, (2004). The Appraisal of Human-Related Intangible Impacts of Flooding. R&D Technical Report FD2005/TR.

DETR (2000). *Guidelines for Environmental Risk Assessment and Management, 2nd edition*, The Stationary Office, London, Institute of Environmental Health.

Diamantidis, D. and Bazzurro, P. (2007). Safety acceptance criteria for existing structures.

Diez, J.J. (2005). About El Niño and other concomitant phenomena. Journal of Coastal res. Vol. 21. N. 6. 2005 pp. Xiii–XViii West Palm Beach. Florida.

Diez, J.J. (1982). Littoral Processes in Denia Coasts. Proc. S.E.M.E. pp. 1.81–1.95. Brugge ().

Diez, J.J. (2001). Climatic versus Geomorphologic Changes: influence on Landing Processes in Eastern Coasts of North America. *Journal of Coastal Research*. Vol. 17(3). pp. 553–562. Florida.

Diez, J.J., Esteban, M.D. and Paz, R. (2009). Cancun-Nizuc Barrier. Journal of Coastal Research.

Diez, J.J., Esteban, M.D., Lopez, J.S. and Negro, V. (2008). Natural vs. Anthropogenic in Cancun Barrier Erosion. ICCE'08 (International Conference on Coastal Engineering). Hamburg.

Dinicola, R.S. (1989). Characterization and simulation of rainfall-runoff relations for headwater basins in western King and Snohomish Counties, Washington state. *U.S. Geological Survey Water-Resources Investigation Report* 89- 4052, 52 pp.

Direktorat for sivilt beredskap (DSB) (1994). A guide to risk and vulnerability assessment in the municipality (in Norwegian). Oslo, Norway. Norwegian Ministry for Civil Defence.

Djordjević, S., Prodanović, D., Maksimović, Č., Ivetić, M. and Savić, D. (2005). SIPSON – Simulation of Interaction between Pipe flow and Surface Overland flow in Networks, Water Science and Technology, Vol. 52, 5, pp. 275–283.

Djordjević, S., Vojinović, Z., Dawson, R. and Savić, D.A. (2009). Flood modelling in urban areas, Chapter 10 In: Applied Uncertainty Analysis for Flood Risk Management, K. Beven and J. Hall (Eds.), Imperial College Press and World Scientific, London.

Doob, J.L. (1990) *Stochastic Processes*. John Wiley and Sons, New York.

Duffy, F. (1990). Measuring Building Performance. Facilities, May, p. 17?

Dunne, T. (1978). Field studies of hillslope flow processes, *Hillslope Hydrology*, Ed.: M.J. Kirkby, John Wiley, Chichester.

Dunne, T., Moore, T.R. and Taylor, C.H. (1975). Recognition and prediction of runoff producing zones in humid regions. *Hydrological Sciences Bulletin*, 20, S. 305–327.

EEA, (2001). *Sustainable water use in Europe, Part 3: Extreme hydrological events: floods and droughts. Environmental issue report No. 21*, European Environmental Agency, Copenhagen.

Ellis, P.A., Rivett, M.O. and Mackay, R. (2000). Assessing the impacts of a groundwater pollutant plume on the River Tame, West Midlands.

Emery, K.O. & Aubrey, D.G. (1991). Sea Levels, Land Levels, and Tide Gauges. Springer-Verlag, New York, 237 pp.

EN (2007a). Directives on the assessment and management of flood risks. DIRECTIVE 2007/60/EC OF THE EUROPEAN PARLIAMENT AND OF THE COUNCIL of 23 October 2007.

EN (2007b). Adapting to climate change in Europe – options for EU action. GREEN PAPER FROM THE COMMISSION TO THE COUNCIL, THE EUROPEAN PARLIAMENT, THE EUROPEAN ECONOMIC AND SOCIAL COMMITTEE AND THE COMMITTEE OF THE REGIONS. EC 2007:849.

English Heritage (2004). Flooding and historic buildings, Technical Advice Note, English Heritage, London.

Environment Agency (2001), *Lessons learned: Autumn 2000 floods*, EA, London.

Environment Agency (2005). *River Irwell Catchment Flood Management Plan Pilot Study-Consultation Draft April 2005.* The Environment Agency, Warrington.

Environment Agency (2006). *Climate change in the North West.* www.environment-agency.gov.uk accessed 24/9/2006.

Esteban, V., Diez, J.J. and Fernandez, P. (2006). Evolution of the Iberian Peninsula Coast and recent Climatic Changes: Port Facilities and Coastal Defence in the Muslim Domain. Journal of Coastal Research. (JCR) SI 39 pp. 1839–42. ISSN, 0749-0208. Brazil.

Evans, E., Ashley, R., Hall, J., Penning Rowsell, E., Saul, A., Sayers, P., Thorne, C. and Watkinson, A. (2004). *FORESIGHT. Future Flooding. Scientific Summary Volume 1 and 2*, OST, London.

Fairbridge, R. (1961). Eustatic changes in sea level (in Ahrens, L.H. et al. Eds.) Physics and Chemistry of the Earth, 4, 99–187, London. Pergamon Press.

Fairbridge, R. (1983). Isostasy and Eustasy. Ed: D.E. Smith *et al.* Shoreline and isostasy. Institute of British Geographer. Academic Press. London.

Feller, W. (1971). *An Introduction to Probability Theory and its Applications*, vol. 2. Wiley, New York.

Fessenden-Raden, J., Fitchen, J.M. and Heath, J.S. (1987). Providing risk information in communities: Factors influencing what is heard and accepted. Science, Technology, and Human Values, 12, 94–101.

FHRC, Flood Hazard Research Centre (2003). Multicoloured Manual. Middlesex, UK.

Fischer, R.A. and Tippet, L.H.C. (1928). Limiting forms of the frequency distribution of the largest or smalles number of a sample, *Proc. Cambridge Philos. Soc.*, 24(180–190).

Fisher, A. and Chen, Y.C. (1996). Customer perceptions of agency risk communication; 16(2): 177–84.

Fisher, A., Chitose, A. and Gipson, P.S. (1994). One agency's use of risk assessment and risk communication; 14(2): 207–12.

Flood Hazard Research Centre (FHRC) (2008). Multicoloured Manual. Middlesex, UK.

Forrester, J.W. (1973). Churches at the Transition Between Growth and World equilibrium. In Toward Global Equilibrium. Cambridge, Wright-Allen Press.

Frechet, M. (1927). Sur la loi de probabilité de l'ecart maximum, *Ann. Soc. Math. Polon.*, 6(93–116).

Freeze, R.A. and Cherry, J.A. (1979). Groundwater, Prentice Hall, New York.

Fukuzono, T., Sato, T., Takeuchi, Y., Takao, K., Shimokawa, S., Suzuki, I., Zhai, G., Terumoto, K., Nagasaga, T., Seo K. and Ikeda, S. (2006). A better integrated management of disaster risks: Toward resilient society to emerging disaster risks in mega-cities, Eds., S. Ikeda, T. Fukuzono, and T. Sato, pp. 121–134. _c TERRAPUB and NIED, Participatory Flood Risk Communication Support System (Pafrics).

Füssel, H.-M. (2006). Vulnerability: A generally applicable conceptual framework for climate change research, *Global Environmental Change*, 17(2), pp. 155–167.

Gardiner, C.W. (1990). *Handbook of Stochastic Methods for Physics, Chemistry and the Natural Sciences*. Springer, Berlin, 442 pp.

Garvin, S.L. (2005). *Flood damage to buildings*, in Urban Flood Management (eds C Zevenbergen, A Zllosky), Taylor and Francis Group plc, London.

Garvin, S.L. *Flood damage to buildings*, World Water Forum III, Japan, 16–23 March 2003.

Geels, F. and Kemp, R. (2000). Transities vanuit sociotechnisch perspectief (*Transitions from a sociotechnical perspective*), Maastricht, The Netherlands.

Geert, R. Teisman, Models for research into decision-making processes: on phases, streams and decision-making rounds (2000), in: Public Administration, 78:4(937–9556).

Ghosh, S.N. (1999). Flood control and drainage engineering. A.A. Balkema Publishers, Brookfield.

Giammarco, P.D., Todini, E. and Lamberti P. (1995). A Conservative Finite Element Approach to Overland Flows. The Control Volume Finite Element Formulation, Institute for Hydraulic Construction University Bologna, Italy.

Gill, S., Handley, J. and Ennos, R. (2006). *Greenspace to adapt cities to climate change.* Unpublished paper for the Engineering and Physical Science Research Council's Adaptation Strategies to Climate Change in the Urban Environment (ASCCUE) project.

Gnedenko, B.V. (1943). Sur la distribution limite du terme maximum d'une série aléatoire, *Ann. of Math.*, 44(2): 423–453.

González, M.A. and Galán, P.C. (1990). Probable Future Coastal Behaviour and ClimaticTrends in the Argentine Republic. *Journal of Coastal Research, Special Issue*, 9. 1, 121–126. Skagen, Dinamarca.

Graaf, R.E. de, F.H.M. van de Ven en N.C. van de Giesen (2007). Alternative water management options to reduce vulnerability for climate change in the Netherlands. Natural Hazards. Available online: http://www.springerlink.com/content/0921-030X . DOI 10.1007/s11069-007-9184-4.

Graaf, R.E. de, and F.L. Hooimeijer (eds) (2008). Urban Water in Japan. Urban Water Series volume 11. ISBN 9780415453608. Taylor and Francis, London, UK.

Griffen, C. (2006a). *Flash floods bring sewage.* Sale and Altrincham Messenger, July 13th 2006.

Griffen, C. (2006b). *Call for action on sewage threat.* Sale and Altrincham Messenger, September 13th 2006.

Grigg, N.S. (2008). Total water management: Practices for a sustainable future. American Water Works Association. ISBN-10: 1-58321-550-6.

Gumbel, E.J. (1958). *Statistics of the Extremes*. Colombia Univ. Press, New York, 371 pp.

Haimes, Y.Y. (1998). Risk Modeling, Assessment, and Management. New York: John Wiley & Sons, Inc.

Hall, J.W., Sayers, P.B. and Dawson, R.J. (2005). National-scale Assessment of Current and Future Flood Risk in England and Wales. *Natural Hazards* 36, 147–164.

Hallegatte, S. (2008). An Adaptive Regional Input-Output Model and its Application to the Assessment of the Economic Costs of Katrina, *Risk Analysis* 28(3), pp. 779–799.

Hance, B.J. (1988). Improving dialogue with communities: a risk communication manual for government. Trenton (NJ): New Jersey Department of Environmental Protection.

Hicks, S. (1973). Trends and variability of Yearly Mean Sea Level 1893–1971. NOAA Technical Memorandum n. 12; U.S. Department of Commerce, Washington.

Hoffman, J.S. (1984). Estimates of Future sea level rise. (In Barth & Titus, Eds.) Greenhouse effect and sea level rise. Cap.3 pp. 79–104. Van Nostrand Reinhold. N.York.

Hollis, G.E. (1986). Water management. In: Clout, H.; Wood, P. (eds.), *London: Problems of Change*. 101–110. Longman, Harlow.

Holman, L.P., Loveland, P.J., Nicholls, R.J., Shackley, S., Berry, P.M., Rounsevell, M.D.A., Audsley, E., Harrison, P.A. and Wood, R. (2002). *REGIS- Regional Climate Change Impact Response Studies in East Anglia and North West England.* Defra, London.

Hooker, C. (2007). Discussion 7: Performance Indicators for Adaptiveness, QCAT Technology Transfer Centre. University of Newcastle, Australia.

Horner, M.W. and Walsh, P.D. (2000). Easter 1998 floods. *Water and Environmental Management*, 14, 415–418.

Horton, R.E. (1933). The role of infiltration in the hydrological cycle. *EOS Transactions American Geophysical Union*, 14, 460–466.

Ingham, A., Ma, J. and Ulph, A. (2006). Theory and Practice of Economic Analysis of Adaptation. Tyndall Centre for Climate Change Research, university of East Anglia.

Institution of Civil Engineers (2001). *Learning to live with rivers*, ICE, London.

IPCC (2001). Impacts, adaptation, and vulnerability for climate change, third assessment report of the IPCC. Cambridge University Press.

IPCC, (2000). Special Report on Emissions Scenarios (SRES), IV Evaluation Report. Eds, N. Nakicenovic & R. Swart. Cambridge University Press, Cambridge. UK.

IPPC (2001). *Climate Change 2001: The Scientific Basis*, Cambridge University Press, Cambridge.

J. Krywkow, (2009). *A methodological framework for participatory processes in water resources management*, Universiteit Twente.

Jaffee, D. (2001). Organization Theory: Tension and Change, McGraw-Hill, Boston.

Kaplan, S. (1997). The words of risk analysis. *Journal of Risk Analysis*, 17(4), 407–417.

Kaplan, S. and Garrick, B.J. (1981). On the quantitative definition of risk. *Journal of Risk Analysis*, 1(1), 11–27.

Kieffer Weisse, A. and Bois, P. (2002). A comparison of methods for mapping statistical characteristics of heavy rainfall in the French Alps: the use of daily information, *J. Hydrological Sciences*, 47(5): 739–752.

Kirkby, M.J. and Chorley, R.J. (1967). Throughflow, overland flow and erosion. *Bulletin International Association of Hydrological Sciences*, 12 (3), 5–21.

Kiureghian, A. and Ditlevsen, O. 2007. Aleatory or epistemic? Does it matter?

Kohane, R. (1991). Berechnungsmethoden für Hochwasserabfluß in Fließgewässern mit überströmten Vorländern, Mitteilg. Institut für Wasserbau, Universität Stuttgart, Heft 73.

Kolb, D.A. and Fry, R. (1975). Toward an applied theory of experiential learning, in C. Cooper (ed.) Theories of Group Process, London: John Wiley.

Kolbezen, M. (1991). Flooding in Slovenia on November 1, 1990. *Ujma* 5, Ljubljana, 16–18.

Kolen, B. Thonus, B., Zuilekom, K.M., De Romph, E. (2009). SPOEL: an instrument for training, simulation and testing of emergency planning for mass evacuation—user experiences. Paper presented at the Evacuation Symposium, Delft, 2009.

Laenen, A. (1983). Storm runoff as related to urbanization based on data collected in Salem and Portland, and generalized for the Willamette Valley, Oregon. *U. S. Geological Survey Water-Resources Investigations report* 83- 4238, 9 pp.

Leadbetter, M.R. and Rootzen, H. (1988). Extremal theory for stochastic processes, *Ann. Prob.*, 16(2): 431–478.

Leandro, J., Chen, A.S., Djordjević, S. and Savić, D.A. (2009). A comparison of $1D/1D$ and $1D/2D$ coupled (sewer/surface) hydraulic models for urban flood simulation, Journal of Hydraulic Engineering, Vol. 135, 6, pp. 495–504.

Lekuthai, A. and Vongvisessomjai, A. (2001). Intangible Flood Damage Quantification, *Water Resources Management*, 15(5), pp. 343–362.

Lippert, K. (2005). Analyse von Turbulenzmechanismen in naturnahen Fließgewässern und ihre mathematische Formulierung für hydrodynamische Modelle, Hamburger Wasserbau-Schriften, Heft 4, Hamburg.

Lisitzin, E. (1974). Sea Level Changes. Ocean series 8. Elsevier. Amsterdam.

Lovejoy, S., Schertzer, D. and Allaire, V. (2008). The remarkable wide range spatial scaling of TRMM precipitation, *J. Atmos. Research*, 90: 10–32.

Loynes, R.M. (1965). Extreme value in uniformly mixing stationary stochastic processes, *Ann. Math. Statist.*, 36: 993–999.

Luers, Amy L., Lobell, David B., Sklar, Leonard S., Addams, C. Lee and Matson, Pamela A. A method for quantifying vulnerability, applied to the agricultural system of the Yaqui Valley, Mexico (2003), in: Global Environmental Change, 13(255–267).

Macor, J., Schertzer, D. and Lovejoy, S. (2007). Multifractal methods applied to rain forecast using radar data, *La Houille Blanche*, 4: 92–98.

Majewski. W. (2008). 2001 Urban Flash flood in Gdansk 2001. Annals of Warsaw University of Life Sciences, Warsaw, 2008.

Maksimović, Č., Prodanović, D., Boonya-aroonnet, S., Leitão, J., Djordjević, S. and Allitt, R. (2009). Overland flow and pathway analysis for modelling of urban pluvial flooding, Journal of Hydraulic Research, Vol. 47 (in press).

MATISSE (2006). Integrated Sustainability Assessment of water. An Integrated Project that has been supported by the 6th Framework Program of the European Union.

Mark, O., Weesakul, S., Apirumanekul, C., Boonya-aroonnet, S. and Djordjević, S. (2004). Potential and limitations of 1D modelling of urban flooding, Journal of Hydrology, Vol. 299, 3–4, pp. 284–299.

McGranahan, G. and Marcotulio, P. 2005. Urban Systems. In: *Ecosystem and Human Well-Being Current State and Trends* Volume 1. Island Press.

Milankovitch, M. (1930). Mathematische Klimalehre und Astronomische Theorie der Klimaschwankungen, Handbuch der Klimalogie Band 1. Teil A Borntrager Berlin.

Mille, J.B. (1997). Floods: People at Risk, Strategies for Prevention, United Nations Publication and published in New York and Geneva, ISBN: 92-1-132021-6.

Milly, P.C.D., Dunne, K.A. and Vecchia A.V. (2005). Global pattern of trends in streamflow and water availability in a changing climate. *Nature* 438, 347–350.

Milly, P.C.D., Betancourt, J., Falkenmark, M., Hirsch, R.M., Kundzewicz, Z.W., Lettenmaier, D.P. and Stouffer, R.J. (2008). CLIMATE CHANGE: Stationarity Is Dead: Whither Water Management? Science 319(5863), 573.

Mina Samangooei 'Green spaces in the sky – What role do green roofs play in a 21st century city?', (2006), department of Architecture, Oxford-Brookes University, UK.

Moore, R., Hitchkiss, D. and Black, K. 1993 *Rainfall patterns over London*. Institute of Hydrology, Wallingford, Oxon.

Mörner, N. (1976). Eustasy and Geoid Changes. Journal of Geology, 84 (2), pp. 123–151.

Mörner, N.A. (1995). Sea Level and Climate- The Decadal-to-Century Signals. JCR. SI Nº 17: "Holocene Cycles" 2: Climate, Sea Levels, and Sedimentation. pp. 261–268. Florida.

Mörner, Niels, A. (1976). Eustasy and Geoid Changes. *Journal of Geology*, 84 (2), pp. 123–151.

Morris, D.G. and Flavin, R.W. (1996). *IH Report No. 130: Flood Risk Map for England and Wales*, Institute of Hydrology, Wallingford, UK.

Motoyoshi, T. (2006). A better integrated management of disaster risks: Toward resilient society to emerging disaster risks in mega-cities, Eds., S. Ikeda, T. Fukuzono, and T. Sato, pp. 121–134. _c TERRAPUB and NIED, Public Perception of Flood Risk and Community-based Disaster Preparedness.

National Research Council (NRC). *Urban Stormwater Management in the United States*. The National Academies Press, Washington DC, 2009.

Newman *et al.* (2008). Using non-structural responses (NSR) to better manage flood risk in Glasgow. 11ICUD conference proceedings, Edinburgh, Scotland.

Nicholls, R.J., Hanson, S., Herweijer, C., Patmore, N., Hallegate, S., Corfee-Morlot, J., Chateau, J. and Muir-Wood, R. Rankong port cities with high exposure and vulnerability to climate extremes. OECD ENV/WKP (2007)1.

Nie, L.M. (2004). Risk Analysis of Urban Drainage Systems (Ph.D. thesis). Norwegian University of Science and Technology. Dr.ing thesis 2004:19. ISBN 82-471-6240-7.

Norwegian Technology Standards Institution (NTS) (1998). Risk and Emergency Preparedness Analysis, published by NTS, Oslo, Norway.

Obras Maritimas Exteriores, Servicio de Publicaciones de la ETS Ingenieros de Caminos, Canales y Puertos, Universidad Politécnica de Madrid, 1980.

Okada, N. (2004). Urban diagnosis and integrated disaster risk management. Journal of Natural Disaster Science, Vol. 26, 2, pp. 49–54.

Pandey, G., Lovejoy, S. and Schertzer, D. (1998). Multifractal analysis including extremes of daily river flow series for basins one to a million square kilometers. *J. Hydrology*, 208(1–2): 62–81.

Pappenberger, F., beven, K., Frodsham, K., Romanovicz, R. and Matgen, P. Grasping the unavoidable subjectivity in calibration of flood inundation models: A vulnerability weighted approach. *Journal of Hydrology*, 333, 275–287.

Parker, D.J. (1995). Floodplain development policy in England and Wales, *Applied Geography*, 15, 341–363.

Parker, D. and Tapsell, S. (1995). Hazard transformation and hazard management issues in the London megacity. *GeoJournal*, 37, 313–328.

Parker, D.J., Green, C.H. and Thompson, P.M. (1987). Urban Flood Protection Benefits, a project appraisal guide "The Red Book", Gower Publishing Company, Brookfield, USA.

Pasche, E. and Geisler, T.R. (2005). *New strategies of damage reduction in urban areas proned to flood*, in Urban Flood Management (eds C Zevenbergen, A Zllosky), Taylor and Francis Group plc, London.

Pasche, E., Kraus, D. and Manojlovic, N. (2006). "Kalypso Inform—A Web-Based Strategy for Integrated Flood Management", 7th HIC Nice, September 2006.

Pasche, E., Manojlovic, N., Behzadnia, N. (2008). "Floods in SUCAs: Hydrological Characterization, Risk Assessment and Efficient Integrative Strategies of Mitigation", proc. 11th ICUD, Edinburgh.

PASCHE, E. (1984). Turbulenzmechanismen in naturnahen Fließgewässern und die Möglichkeiten ihrer mathematischen Erfassung, Mitteilungen Lehrstuhl und Institut für Wasserbau und Wasserwirtschaft, Nr. 52, RWTH Aachen.

Pasche, E., Brüning, C., Plöger, W. and Teschke, U. (2004). Möglichkeiten der Wirkungsanalyse anthropogener Veränderungen in naturnahen Fließgewässern, *Erschienen in Proceedings zum Jubiläumskolloquium 5 Jahre Wasserbau an der TUHH Amphibische Räume an Ästuaren und Flachlandgewässern*", Hamburger Wasserbau-Schriften, Heft 4, Hrsg. Erik Pasche, Hamburg.

Pasche, E., Brüning, C., Plöger, W. and Teschke, U. (2005). Möglichkeiten der Wirkungsanalyse anthropogener Veränderungen in naturnahen Fließgewässern, Hamburger Wasserbau-Schriften, Heft 4, Hrsg. Erik Pasche, Technische Universität Hamburg-Harburg.

Paskoff, R. (1985). Les littoraux. Impact des amenagements sur leur evolution. Masson. Paris.

Peltier, W.R. (1999). Global sea level rise and glacial isostasic adjustment. Global Planet Change, 20, 93–123.

Perry, T. and Nawaz, R. (2008). An investigation into the extent and impacts of hard surfacing of domestic gardens in an area of Leeds, United Kingdom. Landscape and Urban Planning, 86:1(1–13).

Peschke, G., Etzberg, C., Müller, G., Töpfer, J. and Zimmermann, S. (1999). Das wissensbasierte System FLAB—ein Instrument zur rechnergestützten Bestimmung von Landschaftseinheiten mit gleicher Abflussbildung.- *IHI- Schriften, Internationales Hochschulinstitut Zittau*, Heft 10, 122 Seiten.

Piégay, H. and Bravard, J.P. (1997). Response of a mediterranean riparian forest to a 1 in 400 year flood, Ouvèze river, Drôme-Vaucluse, France. *Earth Surface Processes and Landforms* 22(1): 31–43.

Pitt (2008). Pitt, M.: Learning lessons from the 2007 floods. Cabinet Office, London.

Planning Policy Statement (PPS) 1 'Delivering Sustainable Development' (2005), UK Government.

Pokrajac, D. and Howard, K. (in press), UGROW—An advanced modelling tool for the transient simulation and management of urban groundwater systems, UNESCO, Paris.

Polajnar, J. (2000). High waters in Slovenia in 1998. *Ujma* 13, Ljubljana, 143–150.

Prysch, E.A. and Ebbert, J.C. (1986). Quantity and quality of storm runoff from three urban catchments in Bellevue, Washington. *U. S. Geological Survey Water-Resources Investigations Report* 86–4000, 85 pp.

Raaijmakers, Ruud and Krywkow, Jörg and Veen van der, Anne (2008). *Flood risk perceptions and spatial multi-criteria analysis: an exploratory research for hazard mitigation*. Natural Hazards, 46 (3). pp. 307–322. ISSN 0921-030X.

Raban, A. (1983). Sumerged prehistoric sites off the Mediterranean coast of Israel: in P.M. Masters and N.C. Flemming, eds., Quaternary Coastlines and marine Archaeology-Towards the Prehistory at land Bridges and Continental Shelves. Academic Press, 215–232. New York.

Read, Geoffrey. F. (1986). *The Development, Renovation and Reconstruction of Manchester's Sewerage System*. Proceedings of the Manchester Literary and Philosophical Society 1895-86,14–30.

Recommendations of Maritime Works, Actions in the design, ROM 0.2/90, Ministry of Public Works, 1990.

Recommendations of Maritime Works, General Criteria, ROM 0.0, Ports of the State, 2001.

Richards, J., White, I. and Carter, J. (2008). "Local planning practice and flood risk management in England: is there a collective implementation deficit?" Urban Environment, vol. 2, no. 1, p. 11–20.

Ridder, D., E. Mostert and H.A. Wolters (red.) (2005). Samen leren om samen te beheren—effectievere participatie in het waterbeheer. Harmonising Collaborative Planning (HarmoniCOP). Osnabrück: University of Osnabrück, Institute of Environmental Systems Research. URL: http://www.harmonicop.uos.de/HCOPmanualdutch.pdf.

Rijke, J., Zevenbergen, C. and Veerbeek, W. (2009). State of the art Klimaat in de Stad. Kennis voor Klimaat report KvK007/09. ISBN 978-94-90070-07-6.

RODI, W. (1993). Turbulence Models and their Application in Hydraulics, Int. Assoc. Hydraul. Res. (IAHR), Delft, 3rd edition.

Rodriguez-Iturbe, I., Cox, D.R. and Isham, V. (1987). Some models for rainfall based on stochastic point processes, *Proceedings of the Royal Society of London*, A 410: 269–288.

Rose, A. (2004). Modelling the Spatial Economic Impacts of Natural Disasters, Springer, Advances in Spatial Sciences, New York, USA.

Rosso, R. and Burlando, P. (1990). Scale invariance in temporal and spatial rainfall, *Annales Geophysicae*, special issue: 145.

Roth, C. (1992). Die Bedeutung der Oberflächenverschlämmung für die Auslösung von Abfluss und Abtrag. *Bodenökologie und Bodengenese*, Heft 6, Inst. f. Ökologie, TU Berlin.

Santos S.L. and McCallum, D.B. (1997). Communicating to the public: using risk comparisons. Hum Ecol Risk Assess 3(6): 1197–214.

Scherer, C. (1991). Strategies for communicating risks to the public. *Food Technology, 45*, 110–116.

Schertzer, D. and Lovejoy, S. (1987). Physical modeling and analysis of rain and clouds by anisotropic scaling of multiplicative processes, *Journal of Geophysical Research*, D 8(8): 9693–9714.

Schertzer, D. *et al.* (2006). Extremes and multifractals in hydrology: results, validation and prospects, *La Houille Blanche*, 5: 112–119.

Schertzer, D. *et al.* (2009b). Hydrological extremes and multifractals: from GEV to MEV? *Stochastic Environmental Research and Risk Assessment*, (in press).

Schertzer, D., Tchiguirinskaia, I., Lovejoy, S. and Hubert, P. (2009a). No monsters, no miracles: in nonlinear sciences hydrology is not an outlier! *J. Hydrological Sciences*, (submitted).

Schueler, T. (1995). The importance of imperviousness. *Watershed Protection Techniques* 1(3): 100–111.

Selboe, O.K. (1997). Risk and Vulnerability analysis in the Municipality, draft report from the working group on flooding (In Norwegian). Orkanger, Norway, Orkdal municipality, Plan and Environment works. Report No. 000815/97.

Shackley, S., Wood, R., Hornung, M., Hulme, M., Handley, J., Darier, E. and Walsh, M. (1998). *Changing by Degrees: The Impacts of Climate Change in the North West of England.* A Report Prepared for the "Climate Change in the North West" Group, Supported by the North West Regional Association, Environment Agency, The National Trust and Sustainability North West, Manchester.

Sheppard, S.C. (2007). Infill and the microstructures of urban expansion. Homer Hoyt Advanced Studies Institute, January 12.

Shoven, J.B. and Whalley, J. (1992). Applying General Equilibrium. Cambridge, UK.

Sieker, F., Bandermann, S., Holz, E., Lilienthal, A., Sieker, H. Stauss, M. and Zimmermann, U. (1999). *Innovative Hochwasserreduzierung durch dezentrale Maßnahmen am Beispiel der Saar* – Zwischenbericht.- Deutsche Bundesstiftung Umwelt, DBU, Projekt AZ 07147, Osnabrück.

Sly, T. (2000). The perception and communication of risk: a guide for the local health agency. Can J Public Health;91(2):153–6.

Smith, K. and Ward, R. (1998). *Floods, Physical Processes and Human* Impact, John Wiley & Sons Ltd, Chichester, England.

Stalenberg, B. and Redeker, C. (2007). Urban flood protection strategies. p. 70–71 In: Proceedings NCR-days 2006; (ISSN 1568-234X).

Stanbridge, H.H. History of Sewage Treatment in Britain. The Institute of Water Pollution Control, Kent, 1976.

STARDEX (Statistical and Regional dynamic Downscaling of Extremes for European regions) (2005). *STARDEX scenarios information sheet: How will the occurrence of extreme rainfall events in the UK change by the end of the 21st century?*. www.cru.uea.ac.uk/projects/stardex/ deliverables/ accessed 24/9/2006.

STEIN, C.J. Experimentelle Untersuchung und numerische Simulation -, Mitteilungen Lehrstuhl und Institut für Wasserbau und Wasserwirtschaft, RWTH Aachen, Heft 76, 1990.

Stirling, A. (2003). Risk, uncertainty and precaution: some instrumental implications from the social sciences. In: Negotiating Environmental Change, F. Berkhout, M. Leach and I. Scoones (eds), 33–76. Cheltenham, UK: Edward Elgar.

Szöllösi-Nagy, A. and Zevenbergen, C. (Eds.)(2005). Urban Flood Management. A.A. Balkema Publishers. ISBN 04 1535 9988.

Tanaka, A., Yasuhara, M., Sakai, H. and Marui, A. (1988). The Hachioji experimental basin study – Storm runoff processes and the mechanism of its generation. *Journal of Hydrology*, 102, 139–164.

Tapsell, S.M. and Tunstall, S.M. (2003). An examination of the health effects of flooding in the United Kingdom. Journal of Meteorology. 28, nr 283.

Tapsell, S.M., Penning-Rowsell, E.C., Tunstall, S.M. and Wilson, T.L. (2002). Vulnerability to flooding: health and social dimensions. Philosophical Transactions of the Royal Society of London, 360, pp. 1511–1525.

Ten Raa, T. (2005). The Economics of Input-Output Analysis. Cambridge, UK.

Tessier, Y., Lovejoy, S., Hubert, P., Schertzer, D. and Pecknold, S., 1996. Multifractal analysis and modeling of rainfall and river flows and scaling, causal transfer functions. *J. Geophy. Res.*, 101(D21): 26427–26440.

Tessier, Y., Lovejoy, S., Hubert, P., Schertzer, D. and Pecknold, S. (1996) Multifractal analysis and modeling of rainfall and river flows and scaling, causal transfer functions, *J. Geophy. Res.*, 101(D21): 26427–26440.

The multivariate method, Suarez Bores, P. Revista de Obras Públicas. Abril 1979.

Thevenot D. (ed.). DayWater: An adaptive decision support system for urban stormwater management. IWA publishing 2008. ISBN 1843391600.

Tippett, J. and Griffiths E.J. (2007). New approaches to flood risk management - implications for capacity-building. *In*: Advances in Urban Flood Management. R. Ashley, S. Garvin, E. Pasche, A. Vassilopoulos, and C. Zevenbergen (Eds.). Taylor and Francis. ISBN: 978 0 415 43662 5.

Tricart, J. (1975). Phénomènes démesurés et régime permanent dans des bassins montagnards (Queyras et Ubaye, Alpes Françaises). *Revue de Géographie Alpine* **23**: 99–114.

Tucci, (2007). Urban Flooding WMO.

TUCCI, C.E. (2008). Inundações Urbanas ABRH 450p. (in portuguese).

Turner, B.L., Kasperson, R.E., Matsone, P.A., McCarthy, J.J., Corell, R.W., Christensene, L., Eckley, N., Kasperson, J.X., Luers, A., Martellog, M.L., Polsky, C., Pulsipher, A. and Schiller A. (2003). A framework for vulnerability analysis in sustainability science. PNAS 100, 8074–8079.

UN, (2009). Urban and Rural http://www.un.org/esa/population/publications/wup2007/2007urban_ rural.htm accessed in 01/16/2009.

UN-Habitat (2007). Sustainable Urbanization: local actions for urban poverty reduction, emphasis on finance and planning. 21st Session of the Governance Council, 16–20 April, Nairobi, Kenya.

UNISDR (2009). 2008 disasters in numbers. UN International Strategy for Disaster Reduction.

United Nations (2006). World Population Prospects: The 2005 Revision, United Nation Population Division, Department of Economic and Social Affairs, United Nations, New York

Van Kampen, N.G. (1981) *Stochastic Processes in Physics and Chemistry*. North-Holland Physics Publishing, Amsterdam.

Vatn, J. (2007). Description for Risk and vulnerability analysis—InfraRisk, a computation tool for risks. Dept. of production and quality Control, Norwegian University of Science and Technology.

Veerman, C. *et al.* (2008). Samen *werken* met water—Een land dat leeft, bouwt aan zijn toekomst. Bevindingen van de Deltacommissie 2008.

Vrijling, J.K. (2001). Probabilistic design of flood defence systems in the Netherlands. *Reliability Engineering and System Safety*, Vol. 74 Issue 3 pp. 337–344.

Wainwright, J. (1996). Infiltration, runoff and erosion characteristics of agricultural land in extreme storm events, SE France. *Catena* 26(1–2): 27–47.

Walker, W. (2000). Entrapment in large technology systems: Institutional commitments and power relations. Research Policy, 29, 883–846.

Warner, K., Kuhlicke, C., de Vries, D.H., Sakdapolrak, P., Wutich, A., Real, B., Briones Gamboa, F., Verjee, F., Sosa Rodriguez, F.S. and Olson, L. (2007). Perspectives on Social Vulnerability. SOURCE Publication No. 6. UNU Institute for Environment and Human Security (UNU-EHS).

Washington, D.C.: Center for Mental Health Services, Substance Abuse and Mental Health Services Administration, 2002. Communicating in a Crisis: Risk Communication Guidelines for Public Officials.

Waymire, E. and Gupta, V.K. (1981). The mathematical structure of rainfall representations, 1, A review of stochastic rainfall models, *Water Resources Research*, 17(5): 1261–1272.

Weng, Q. (2008). *Remote sensing of impervious surfaces*. CRC Press, ISBN: 1420043749.

WENKA, T. (1992). Numerische Berechnung von Strömungsvorgängen in naturnahen Fließgewässern mit einem tiefengemittelten Modell, Dissertation Universität Karlsruhe, 1992.

White, G.F. (ed) (1974). Natural Hazards. Oxford University Press, New York.

White, I. and Howe, J. (2002). Flooding and the role of planning in England and Wales: A critical review, *Journal of Environmental Management and Planning* 45 (5), 735–745.

White, I. and Howe, J. (2004a). Like a fish out of water: The relationship between planning and flood risk management, *Planning Practice and Research* 19 (4), 415–425.

White, I. and Howe, J. (2004b). The Mismanagement of Surface Water, *Applied Geography* 24 (4), 261–280.

Wigmosta, M.S., Burges, S.J. and Meena, J.M. (1994). Modeling and monitoring to predict spatial and temporal hydrologic characteristics in small catchments. *Report to U.S. Geological Survey, University of Washington Water Resources Series Technical Report No.* 137, 223 pp.

WMO, 2004. *Water and disasters. WMO-No. 971*, Geneva, Switzerland.

Wong, T.H.F. Introduction. In: Wong THF (ed.) Australian runoff quality—a guide to water sensitive urban design. Engineers Australia, Sydney, Australia, 2006.

Yu, D. and Lane, S.N. (2005). Urban fluvial flood modelling using a two-dimensional diffusion-wave treatment, part 1: mesh resolution effects. *Hydrological Processes*, 20, 1541–1565.

Zevenbergen, C. (2007). Adapting to change: towards flood resilient cities. Inaugural address, UNESCO-IHE, Institute for Water Education. Delft, the Netherlands.

Zevenbergen, C., Veerbeek, W., Gersonius, B. and Van Herk, S. (2008). Challenges in Urban Flood Management: travelling across spatial and temporal scales, Journal of Flood Risk Management 1:2(81–88).

Zuidema, P.K. (1985). Hydraulik der Abflussbildung während Starkniederschlägen.- *Mitteilungen der Versuchsanstalt für Wasserbau, Hydrologie und Glaziologie*, Nr. 79, ETH Zürich.

Abbreviations and acronyms

Abbreviation	Full Form
ABI	Association of British Insurers
BFE	Base Flood Elevation
BMP	Best Management Practice
BPP	Best Planning Practices
CAS	Complex Adaptive System
CGE	Computation General Equilibrium
CIKR	Critical infrastructure and key resources
CIRIA	Construction Industry Research and Information Association
CLG	Communities and Local Government (Department of UK Government)
CSO	Combined Sewer Overflows
Defra	Department for Environment Food and Rural Affairs
DEM	Digital Elevation Model
DFE	Design Flood Elevation
DTM	Digital Terrain Model
EAD	Expected Annual Damage
EIA	Environmental Impact Assessment
EVT	Extreme Value Theory
FEMA	Federal Emergency Management Agency
FFWRS	Flood Forecasting, Warning and Response System
FLAGS	Flood Liaison Advisory GroupS
FRM	Flood-risk management
GIS	Geographic Information Systems
HS	Hydraulic System
IDF	Intensity Duration Frequency
IFRM	Integrated flood-risk management
IPCC	Intergovernment Panel of Climate Change (UN)
IUH	Instantaneous Unit Hydrograph
IUWM	Integrated Urban Water Management
IWRM	Integrated Water Resources Management
KOSTRA	Koordinierte Starkniederschlags Regionalisiering—Auswertungen
LA	Learning Alliance
LAA	Learning and Action Alliance
LIA	Little Ice Age
LIDS	Low Impact Developments
MMC	Modern Methods of Construction
NFIP	National Flood Insurance Program
NIPP	National Infrastructure Protection Plan
NSR	Non-Structural Responses
OECD	Organisation for Economic Cooperation and Development
PADM	Protective Action Decision Model
PES	Physico-environmental Subsystem
POT	Peak Over Thresholds
PPS25	Planning Policy Statement 25
PRA	probabilistic risk analysis
PRUDENCE	Prediction of Regional scenarios and Uncertainties for Defining EuropeaN Climate change risks and Effects

Abbreviations and acronyms

Abbreviation	Full Form
QRA	Quantitative Risk Assessment
RAMCAP	Risk Assessment Methodology for Critical Asset Protection
SAR	synthetic aperture radar
SCS-CN	Soil Conversation Service—Curve Number
SES	Social-economic Subsystem
SHIRA	Strategic Homeland Infrastructure Risk Assessment
SSOs	storm sewer overflows
SUDACS	Sustainable Drainage and Conveyance Systems
SUDS	Sustainable Urban Drainage Systems
TIN	Triangular Irregular Network
UBA	Umweltbundesamt
UFM	Urban Flood Management
UHI	Urban heat islands
UKCIP	UK Climate Impacts Programme
URBEM	Urban River Basin Enhancement Methods
USGS	United States Geological Survey
WRC	Water Resources Council
WSUD	Water Sensitive Urban Design
WTP	Willingness-to-pay

Glossary

Accuracy Closeness of a measurement, calculation and estimate to its true value.

Aims The objectives of groups/individuals/organisations involved with a project. The aims are taken to include ethical and aesthetic considerations.

Asset "An asset is a resource controlled by the enterprise as a result of past events and from which future economic benefits are expected to flow to the enterprise." (International Accounting Standards Board, 2005)

Attenuation (flood peak) Lowering a flood peak (and lengthening its base).

Basin (river) (see catchment area) The area from which water runs off to a given river.

Bias The disposition to distort the significance of the various pieces of information that have to be used.

Capacity "A combination of all the strengths and resources available within a community, society or organization that can reduce the level of risk, or the effects of a disaster. Capacity may include physical, institutional, social or economic means as well as skilled personal or collective attributes such as leadership and management. Capacity may also be described as capability." (UN/ISDR, 2004)

Capacity, adaptive Is the ability to plan, prepare for, facilitate, and implement adaptation options. Factors that determine a adaptive capacity include economic wealth, availiable technology and infrastructure, the information, knowledge and skills that it possesses, the nature of its institutions, its commitment to equity, and its social capital. Adaptive capacity refers to a longer timeframe than coping capacity.

Capacity, coping "Refers to the manner in which people and organisations use existing resources to achieve various beneficial ends during unusual, abnormal, and adverse conditions of a disaster event or process. The strengthening of coping capacities usually builds resilience to withstand the effects of natural and other hazards." (ESPON, 2003)

Catastrophe A disruption of society that may cause a total breakdown in day-to-day functioning. One aspect of catastrophes, is that most community functions disappear; there is no immediate leadership, hospitals may be damaged or destroyed, and the damage may be so great and so extensive that survivors have nowhere to turn for help (Quarantelli, 1994). In disaster situations, it is not unusual for survivors to seek help from friends and neighbors, but this cannot happen in catastrophes. In a disaster, society continues to operate and it is common to see scheduled events continue. . ." (Tobin and Montz, 1997; quoted in Blanchard, 2005)

Catchment area The area from which water runs off to a river, sewer or other draining entity. Instead of catchment area, the term watershed is often used in the United States.

Characterisation The process of expressing the observed/predicted behaviour of a system and it's components for optimal use in decision making.

Confidence interval A measure of the degree of (un)certainty of an estimate. Usually presented as a percentage. For example, a confidence level of 95% applied to an upper and lower bound of an estimate indicates there is a 95% chance the estimate lies between the specified bounds. Confidence limits can be calculated for some forms of uncertainty (see knowledge uncertainty), or estimated by an expert (see judgement).

Consequence An impact such as economic, social or environmental damage/ improvement that may result from a flood. May be expressed quantitatively (e.g. monetary value), by category (e.g. High, Medium, Low) or descriptively.

Correlation Between two random variables, the correlation is a measure of the extent to which a change in one tends to correspond to a change in the other.

Damage potential A description of the value of social, economic and ecological impacts (harm) that would be caused in the event of a flood.

Decision uncertainty The rational inability to choose between alternative options.

Defence system Two or more defences acting to achieve common goals (e.g. maintaining flood protection to a floodplain area/ community).

Dependence The extent to which one variable depends on another variable. Dependence affects the likelihood of two or more thresholds being exceeded simultaneously. When it is not known whether dependence exists between two variables or parameters, guidance on the importance of any assumption can be provided by assessing the fully dependent and independent cases (see also correlation).

Design discharge See Design standard and Design flood.

Design objective The objective (put forward by a stakeholder), describing the desired performance of an intervention, once implemented.

Design standard A performance indicator that is specific to the engineering of a particular defence to meet a particular objective under a given loading condition. Note: the design standard will vary with load, for example there may be different performance requirements under different loading conditions.

Deterministic process/method A method or process that adopts precise, single-values for all variables and input values, giving a single value output.

Disaster For a disaster to be entered into the database of the UN's International Strategy for Disaster Reduction (ISDR), at least one of the following criteria must be met:

- a report of 10 or more people killed
- a report of 100 people affected
- a declaration of a state of emergency by the relevant government
- a request by the national government for international assistance" (IRIN/OCHA, 2005).

Discharge (stream, river) As measured by volume per unit of time.

Efficiency In everyday language, the ratio of outputs to inputs; in economics, optimality.

Emergency management The ensemble of the activities covering emergency planning, emergency control and post-event assessment.

Epistemology A theory of what we can know and why or how we can know it.

Ergonomics The study of human performance as a function of the difficulty of the task and environmental conditions.

Error Mistaken calculations or measurements with quantifiable and predictable differences.

Evacuation scheme Plan for the combination of actions needed for evacuation (warning, communication, transport etc.).

Event (in context) Events are the conditions which may lead to flooding. An event is, for example, the occurrence in Source terms of one or more variables such as a particular wave height threshold being exceeded at the same time as a specific sea level, or in Receptor terms a particular flood depth. When defining an event it can be important to define the spatial extent and the associated duration.

Expectation Expectation, or expected value of a variable, refers to the mean value the variable takes.

Expected annual frequency Expected number of occurrences per year (reciprocal of the return period of a given event).

Expected value See Expectation.

Exposure "People, property, systems, or functions at risk of loss exposed to hazards." (Multihazard Mitigation Council, 2002)

Extrapolation The inference of unknown data from known data, for instance future data from past data, by analysing trends and making assumptions.

Failure Inability to achieve a defined performance threshold (response given loading). "Catastrophic" failure describes the situation where the consequences are immediate and severe, whereas "prognostic" failure describes the situation where the consequences only grow to a significant level when additional loading has been applied and/or time has elapsed.

Failure mode Description of one of any number of ways in which a defence or system may fail to meet a particular performance indicator.

Flood A temporary covering of land by water outside its normal confines.

Flood control (measure) A structural intervention to limit flooding and so an example of a risk management measure.

Flood damage Damage to receptors (buildings, infrastructure, goods), production and intangibles (life, cultural and ecological assets) caused by a flood.

Flood forecasting system A system designed to forecast flood levels before they occur.

Flood hazard map Map with the predicted or documented extent of flooding, with or without an indication of the flood probability.

Flood Insurance Specific type of insurance that offers coverage against property loss from flooding, often based on susceptibility of topographical areas to flood risk.

Flood level Water level during a flood.

Flood management measures Actions that are taken to reduce either the probability or the consequences of flooding or some combination of the two.

Flood peak Highest water level recorded in the river during a flood.

Flood plain Part of alluvial plain that would be naturally flooded in the absence of engineered interventions.

Flood protection (measure) To protect a certain area from inundation (using dikes etc).

Flood risk zoning Delineation of areas with different possibilities and limitations for investments, based on flood hazard maps.

Flood risk management Continuous and holistic societal analysis, assessment and mitigation of flood risk.

Flooding System (in context) In the broadest terms, a system may be described as the social and physical domain within which risks arise and are managed. An understanding of the way a system behaves and, in particular, the mechanisms by which it may fail, is an essential aspect of understanding risk. This is true for an organisational system like flood warning, as well as for a more physical system, such as a series of flood defences protecting a flood plain.

Flood warning system (FWS) A system designed to warn members of the public of the potential of imminent flooding. Typically linked to a flood forecasting system.

Fragility The propensity of a particular defence or system to fail under a given load condition. Typically expressed as a fragility function curve relating load to probability of failure. Combined with descriptors of decay/deterioration, fragility functions enable future performance to be described.

Functional design The design of an intervention with a clear understanding of the performance required of the intervention.

Hazard "A potentially damaging physical event, phenomenon or human activity that may cause the loss of life or injury, property damage, social and economic disruption or environmental degradation." (UN/ISDR, 2004)

Hazard mapping The process of establishing the spatial extents of hazardous phenomena.

Hazard, natural "Natural processes or phenomena occurring in the biosphere that may constitute a damaging event." (UNDP, 2004)

Human Security "Human Security is about attaining the social, political, environmental and economic conditions conducive to a life in freedom and dignity for the individual." (Hammerstad, 2000)

Ignorance Lack of knowledge.

Integrated risk management An approach to risk management that embraces all sources, pathways and receptors of risk and considers combinations of structural and non-structural solutions.

Integrated Water Resource Management IWRM is a process which promotes the co-ordinated management and development of water, land and related resources, in order to maximise the resultant economic and social welfare in an equitable manner without compromising the sustainability of vital ecosystems. (Global Water Partnership—Technical Advisory Committee, 2000)

Intervention A planned activity designed to effect an improvement in an existing natural or engineered system (including social, organisation/defence systems).

Inundation Flooding of land with water. (NB: In certain European languages this can refer to deliberate flooding, to reduce the consequences of flooding on nearby areas, for example. The general definition is preferred here.)

Likelihood A general concept relating to the chance of an event occurring. Likelihood is generally expressed as a probability or a frequency.

Natural variability Uncertainties that stem from the assumed inherent randomness and basic unpredictability in the natural world and are characterised by the variability in known or observable populations.

Pathway Route that a hazard takes to reach Receptors. A pathway must exist for a Hazard to be realised.

Performance The degree to which a process or activity succeeds when evaluated against some stated aim or desired objective.

Performance indicator The well-articulated and measurable objectives of a particular project or policy. These may be detailed engineering performance indicators, such as acceptable wave overtopping rates, rock stability, or conveyance capacity or more generic indicators such as public satisfaction.

Post-flood mitigation Measures and instruments after flood events to remedy flood damages and to avoid further damages.

Precautionary Principle Where there are threats of serious or irreversible damage, lack of full scientific certainty shall not be used as a reason for postponing cost-effective measures to prevent environmental degradation.

Precision Degree of exactness regardless of accuracy.

Pre-flood mitigation Measures and instruments in advance to a flood event to provide prevention (reducing flood hazards and flood risks by e.g. planning) and preparedness (enhancing organisational coping capacities).

Preparedness The ability to ensure effective response to the impact of hazards, including the issuance of timely and effective early warnings and the temporary evacuation of people and property from threatened locations.

Probability A measure of our strength of belief that an event will occur. For events that occur repeatedly the probability of an event is estimated from the relative frequency of occurrence of that event, out of all possible events. In all cases the event in question has to be precisely defined, so, for example, for events that occur through time reference has to be made to the time period, for example, annual exceedance probability. Probability can be expressed as a fraction, % or decimal. For example the probability of obtaining a six with a shake of four dice is 1/6, 16.7% or 0.167.

Probabilistic method Method in which the variability of input values and the sensitivity of the results are taken into account to give results in the form of a range of probabilities for different outcomes.

Probability density function (distribution) Function which describes the probability of different values across the whole range of a variable (for example flood damage, extreme loads, particular storm conditions etc.).

Probabilistic reliability methods These methods attempt to define the proximity of a structure to fail through assessment of a response function. They are categorised as Level III, II or I, based on the degree of complexity and the simplifying assumptions made (Level III being the most complex).

Project Appraisal The comparison of the identified courses of action in terms of their performance against some desired ends.

Receptor Receptor refers to the entity that may be harmed (a person, property, habitat etc.). For example, in the event of heavy rainfall (the source) flood water may propagate across the flood plain (the pathway) and inundate housing (the receptor) that may suffer material damage (the harm or consequence). The vulnerability of a receptor can be modified by increasing its resilience to flooding.

Recovery time The time taken for an element or system to return to its prior state after a perturbation or applied stress.

Residual life The residual life of a defence is the time to when the defence is no longer able to achieve minimum acceptable values of defined performance indicators (see below) in terms of its serviceability function or structural strength.

Residual risk The risk that remains after risk management and mitigation measures have been implemented. This may include, for example, damage predicted to continue to occur during flood events of greater severity that the 100 to 1 annual probability event.

Resilience "The concept [of resilience] has been used to characterize a system's ability to bounce back to a reference state after a disturbance and the capacity of a system to maintain certain structures and functions despite disturbance.[...] resilience of the system is often evaluated in terms of the amount of change a given system can undergo (e.g. how much disturbance or stress it can handle) and still remain within the set of natural or desirable states (i.e. remain within the same 'configuration' of states, rather than maintain a single state)." (Turner et al., 2003)

Resilience, social The capacity of a community or society potentially exposed to hazards to adapt, by resisting or changing in order to reach and maintain an acceptable level of functioning and structure. This is determined by the degree to which the social system is capable of organising itself to increase its capacity for learning from past disasters for better future protection and to improve risk reduction measures.

Resistance The ability of a system to remain unchanged by external events.

Response (in context) The reaction of a defence or system to environmental loading or changed policy.

Response function Equation linking the reaction of a defence or system to the environmental loading conditions (e.g. overtopping formula) or changed policy.

Return period The expected (mean) time (usually in years) between the exceedence of a particular extreme threshold. Return period is traditionally used to express the frequency of occurrence of an event, although it is often misunderstood as being a probability of occurrence.

Risk Risk is a function of probability, exposure and vulnerability. Often, in practice, exposure is incorporated in the assessment of consequences, therefore risk can be considered as having two components – the probability that an event will occur and the impact (or consequence) associated with that event. See Section 4.3 above.

Risk = Probability multiplied by consequence

Risk analysis A methodology to objectively determine risk by analysing and combining probabilities and consequences.

Risk assessment Comprises understanding, evaluating and interpreting the perceptions of risk and societal tolerances of risk to inform decisions and actions in the flood risk management process.

Risk management The complete process of risk analysis, risk assessment, options appraisal and implementation of risk management measures.

Risk mapping The process of establishing the spatial extent of risk (combining information on probability and consequences). Risk mapping requires combining maps of hazards and vulnerabilities. The results of these analyses are usually presented in the form of maps that show the magnitude and nature of the risk.

Risk profile The change in performance, and significance of the resulting consequences, under a range of loading conditions. In particular the sensitivity to extreme loads and degree of uncertainty about future performance.

Risk reduction The reduction of the likelihood of harm, by either reduction in the probability of a flood occurring or a reduction in the exposure or vulnerability of the receptors.

Robustness Capability to cope with external stress. A decision is robust if the choice between the alternatives is unaffected by a wide range of possible future states of nature.

Robust statistics are those whose validity does not depend on close approximation to a particular distribution function and/or the level of measurement achieved.

Scale Difference in spatial extent or over time or in magnitude; critical determinant of vulnerability, resilience etc.

Scenario A plausible description of a situation, based on a coherent and internally consistent set of assumptions. Scenarios are neither predictions nor forecasts. The results of scenarios (unlike forecasts) depend on the boundary conditions of the scenario.

Sensitivity Refers to either: the resilience of a particular receptor to a given hazard. For example, frequent sea water flooding may have considerably greater impact on a fresh water habitat, than a brackish lagoon; or: the change in a result or conclusion arising from a specific perturbation in input values or assumptions.

Sensitivity Analysis The identification at the beginning of the appraisal of those parameters which critically affect the choice between the identified alternative courses of action.

Severity The degree of harm caused by a given flood event.

Social learning Processes through which the stakeholders learn from each other and, as a result, how to better manage the system in question.

Source The origin of a hazard (for example, heavy rainfall, strong winds, surge etc.).

Standard of service The measured performance of a defined performance indicator.

Stakeholders Parties/persons with a direct interest (stake) in an issue.

Stakeholder Engagement Process through which the stakeholders have power to influence the outcome of the decision. Critically, the extent and nature of the power given to the stakeholders varies between different forms of stakeholder engagement.

Statistic A measurement of a variable of interest which is subject to random variation.

Statistical inference uncertainty See Knowledge uncertainty.

Strategy (flood risk management) A strategy is a combination of long-term goals, aims, specific targets, technical measures, policy instruments, and process which are continuously aligned with the societal context.

Strategic spatial planning Process for developing plans explicitly containing strategic intentions referring to spatial development. Strategic plans typically exist at different spatial levels (local, regional etc.).

Sustainable Development Is development that meets the needs of the present without compromising the ability of future generations to meet their own needs.

Susceptibility The propensity of a particular receptor to experience harm.

System An assembly of elements, and the interconnections between them, constituting a whole and generally characterised by its behaviour. Applied also for social and human systems.

System state The condition of a system at a point in time.

Uncertainty A general concept that reflects our lack of sureness about someone or something, ranging from just short of complete sureness to an almost complete lack of conviction about an outcome.

Validation Is the process of comparing model output with observations of the 'real world'.

Variability The change over time of the value or state of some parameter or system or element where this change may be systemic, cyclical or exhibit no apparent pattern.

Variable A quantity which can be measured, predicted or forecast which is relevant to describing the state of the flooding system e.g. water level, discharge, velocity, wave height, distance, or time. A prediction or forecast of a variable will often rely on a simulation model which incorporates a set of parameters.

Voluntariness The degree to which an individual understands and knowingly accepts the risk to which they are exposed in return for experiencing a perceived benefit. For an individual may preferentially choose to live in the flood plain to experience its beauty and tranquillity.

Vulnerability Characteristic of a system that describes its potential to be harmed. This can be considered as a combination of susceptibility and value.

Watershed See catchment area.

Subject index

adaptation 15, 26, 30, 32, 39, 104, 157, 161, 168
adaptability 26, 39, 159, 265
adaptive management 19, 39, 260, 277, 279, 292
adaptive responses 39, 167, 182
adaptive strategy 39, 167
adaptive systems 9, 14, 15, 34, 41, 228, 287
alleviation 111, 151, 165, 206, 233
asset management 235–261
assistance 165, 238–248
avoidance 147, 158, 165, 210, 290
awareness 19, 147, 150, 165, 206, 241, 269–275

Best Management Practice (BMP) 98, 148, 171, 184, 201, 309
Building Regulations 65, 66, 161–165
building types 68, 163, 177, 215–218
built environment 71, 115, 171, 177, 188, 218, 276, 279, 282

calibration 113–117
capacity building 171, 217, 218, 269, 277
climate change 79–92, 199, 244, 290, 291, 294
consequences of climate change 80, 88–92
El Niño 83
extreme climatic events 55, 56, 79, 85, 86
global warming 31, 51, 54, 80–85, 92
greenhouse effect 80–87, 184
greenhouse effect on temperature 80–87
hockey stick 80
impact of climate change 85–92
IPCC 24, 87
signs of climate change 85, 86
Combined Sewer Overflows (CSO) 43, 44, 51, 98, 183–200
communication 10, 19, 21, 139, 170, 228, 233, 270, 275–279
compensation 39, 81, 146, 240–243, 248–250
Complex Adaptive System (CAS) 9, 15, 20, 27
contingency planning 171, 235
Coriolis effect 83, 119

decision making 14, 25, 27, 64, 147, 235, 269, 274–300
definition of design storm events 17, 98, 113, 189, 197, 222
delta 73, 74, 100
delta cities 73, 74, 100
Delta Works 17
Design performance standards 159–164, 189
deterministic modelling 109, 246
Digital Elevation Model (DEM) 114, 116, 248
discharge coefficient 108, 113, 190, 195
drainage systems 39, 49, 50, 62, 91, 98, 102, 105, 113, 124, 162, 183–204

early warning 31, 227–233
emergence 10, 15, 16, 19, 33, 34, 40, 48, 139

emergency planning 235–240
emergency planning principles 237
emergency response 170, 236, 237
emergent behaviour 15, 34
entrapment 17, 262–266
entrapment characteristics 262–266
ergodicity principle 54–57
European Union Water Framework Directive 273
evacuation 170, 236–239
evacuation planning 239
evaporation 39, 40, 42, 98–100, 103, 111, 203
external drivers 27–29, 167, 266, 284
extremes 86, 91, 106, 132, 135, 138, 244
extreme rainfall 109, 126
Extreme Value Theory (EVT) 135
Gumbel 56

finite difference method 119
flexibility 26, 27
floating and amphibious structures 211, 212
flood and water resources management 66, 285
flood defence systems 17, 43, 125, 126, 219–223
Flood Directive 164, 202
flood forecasting 159, 229, 231–233
flood forecasting systems 229
Flood Forecasting, Warning and Response System 229, 233
flood hazard mapping 147–151, 248
flood insurance 240–252
bundle system 243
compensation and flood insurance 240–252
Crichton Risk Triangle 247
government-led flood insurance schemes 242
insurance 240–252
insurance premiums 240, 241
market-led flood insurance schemes 242
micro-insurance 250
mixed flood insurance schemes 242
option system 242, 243
reinsurance 240, 249
types of flood insurance 240–249
flood preparedness 41, 171, 236, 269, 272
flood probability reduction measures 171
flood proofing 205–225
active flood proofing 209
passive flood proofing 209
flood proofing methods 205–225
dry flood proofing 209
dry proofing 209

flood openings 214, 215
flood resilient repair 164, 215–217
flood-resistant materials 208, 213–215
floors 215–218
foundations 215
openings 215
pro-active retrofitting 218
retrofitting 215–219, 290
walls 213, 215
wet flood proofing 209, 214
wet proofing 172, 209, 210
flood warning systems 158, 159, 172, 228, 230, 232–234
flooding, types of 42–52
causes of flooding 43, 49, 50
coastal flooding 77, 118–120
conveyance flooding 117
flash floods 50, 106, 107
fluvial flooding 40, 42, 43
inland flooding 42, 43
pluvial flooding 113–117
river flooding 48, 89–91, 113
the great flood 4
flooding system 170, 260, 261, 314
flooding system processes 27, 28, 170
flood-risk assessment 123, 141–146, 151, 235, 236, 245–247
assessment of direct damage 139–141
cross-scale factors 142–144
direct damages 138–140
indirect flood damages 140, 143–146
intangible damages 139, 140, 151
intangible flood damage 139, 140
loss of human life 141, 142
loss of life estimation 141, 142, 151
measuring indirect damages 145
method for loss of life estimation 142
substitution 16, 144–146
flood-risk mapping 146–150
flow velocity 47, 48, 104, 105
Foresight 2004 165, 166
Fourier Transforms 128, 133

governance 24, 25, 147, 269–280
groundwater flow 46, 47, 53, 98, 103, 111

headroom 30, 289
Hurricane Katrina 42, 95, 141, 146, 239, 242, 244, 273
Hydraulic design 189–198
hydraulic conductivity 104
hydraulic design for design storm event 191–197
hydraulic models 23, 108, 115
1D modelling 112, 116, 117
2D modelling 114, 116, 117
dual drainage models 115

hydrology of cities 97–121
complex hydrologic models 108–111
hydrological cycle 97–102
hydrograph 46, 102–110
Instantaneous Unit Hydrograph (IUH) 13, 14, 106, 107
runoff coefficients 187, 191, 197

ignorance 22, 24, 270
infrastructure 15, 29, 40, 73, 88, 89, 177–189
blue infrastructure 177, 178
critical infrastructure 180
green infrastructure 177–179, 201
Integrated Urban Water Management (IUWM) 285
Intensity-Duration-Frequency (IDF) 55, 56, 131, 134, 136
involuntary flood risk 271

Japan 31

Kaplan 123

land subsidence 100, 223
land-use planning 63–67, 173–182
land-use change 67, 86, 106
land use planning framework 66
land-use planning: lessons learnt 20
lead time 231
life cycle cost 218, 258
life cycle cost analysis 258
LISFLOOD model 113
long-term planning 33, 67, 281, 282, 291, 292

managed adaptive approach 26, 261, 287
Markoffian processes 132
metropolitans 70, 74
Montana law 134, 136
Monte Carlo simulations 24
multi-level theory 263–266

NIPP framework 180
non-stationarity 55, 56, 262
non-structural responses 31, 34, 146, 165, 168, 172
non-structural measures 31, 34, 146, 165, 168, 172
non-structural solutions 31, 34, 146, 165, 168, 172
no-regret responses 161, 262

passive and active flood warning systems 228
Peak Over Thresholds 112
performance monitoring 255, 256
performance standards 160, 161, 168
polder 100, 101, 125
pollutants 102
POT 132
PPS25 163
precautionary approach 26, 32, 167–172
precautionary measures 32
preparedness 165, 171, 240, 241, 269, 270, 272

prevention 19, 165
probabilistic modelling 246
probability 17, 21, 26, 124, 126–138
probability density function 54

rainfall intensity 55, 109, 131, 189
raising awareness 275
recovery and lessons learned 165, 170, 237, 260
recovery rate 260
resilience 29, 32, 164–167, 260, 316
adaptive resilience 29, 260, 261
amplitude 260
assessing resilience 259–261
characteristics of resilient approaches 29, 30
flood resilience measures 30, 261–266
graduality 259, 260
materials resilience 217
resilience indicators 260
restorative resilience 259
responses 157–182
risk 24–26, 123–126
fault tree 124, 126
probabilistic risk analysis 125
risk analysis 123–125
risk perception 269–273
risk reduction 19, 32, 148
robustness 18, 29, 30, 32, 164–167
robustness-focused approach 164–167
runoff 49, 92, 103–122
runoff modelling methods 107–113

safety 15, 29, 219, 222, 239, 270
Scenario analysis 23, 24, 56
sea level 39, 73, 84, 85, 92, 245
sea level rise 17, 38, 44, 80, 84, 91, 100, 119, 233, 261
seepage 101, 216
sensitivity analysis 23
sewer system 183–185, 257
site design 187, 205–210
social learning 24, 278
Learning Alliance (LA) 202–204, 285–287
socio-economic scenarios 24, 164
Source-Pathway-Receptor 159, 170
stakeholder groups 66, 287, 292
stakeholder involvement 21, 66, 67
stakeholder participation 21, 66, 67
standards of protection 18
state involvement in compensation for flood survivors 243, 244
stationarity 54, 128, 138
storage 75, 100, 103, 108, 112
storm-surge barriers 17
stormwater management 187, 199, 285
Floodways 185, 186, 200

green roof 189–195, 207
hydraulic design of a green roof 189
hydraulic system of a swale 194
LIDS 186–198
Low Impact Development 186–198
SUDACS 200
SUDS 186–198
Sustainable Drainage and Conveyance Systems 200
Sustainable Urban Drainage Systems 186–198
swales 194
urban drainage systems 103–113, 124, 183–204
structural factors 215
structural responses 169, 265, 288
subsidence 18, 74, 84, 87, 100, 223
subsurface flow 46, 47, 104
surface runoff 104, 107–117
sustainability 25, 67, 159, 164–167, 287
SWITCH project 201
system approach 15, 26–34
PES 26, 30, 33, 219–233
physico-environmental subsystem 26, 30, 33, 219–233
SES 26, 30, 33, 219–233
socio-economic subsystem 26, 30, 33, 219–233

transition characteristics 265, 266
transition theory 266
trauma 139, 141

uncertainty 9, 20–32, 47
dimensions of uncertainty 22
epistemic uncertainty 22
level of uncertainty 23–26, 30
methods for dealing with uncertainty 23–26, 30
types of uncertainty 22, 23
urbanisation 63–78
urban catchment 47, 102, 106–108, 121
urban core 68, 69
urban design 98, 184, 187, 198–204
urban development 173, 178, 201, 288, 291
urban fabric 201, 205–223, 281, 294
urban flood defences 219–224
urban fringe 3, 68, 69
urban growth 11, 63, 67, 71, 73
urban heat islands 57
urban patterns 14
urban population 9–11, 13, 71, 76
urban sprawl 13, 57, 73
urban typologies 67–70
urban vulnerability 9–14, 247

variability in climate 85
vertical evacuation 238, 239
voluntary flood risk 270, 271

vulnerability 18–20, 123–154
    adaptive capacity 18, 19, 21, 27, 28, 30, 273, 274
    coping capacity 19
    recovery capacity 19
    threshold capacity 19

warning 227–234
water 4
    drinking water 97–99
    groundwater 99, 100, 109, 111, 113

stormwater 21, 48, 50, 66, 88, 89, 97, 98
surface water 97–112
wastewater 97, 98, 102, 199, 229
water balance 99, 100, 104, 108
water cycle 38, 97, 98
water quality 102, 103
Water Sensitive Urban Design 198–204